Contemporary Calculus
through applications

Contemporary Calculus
through applications

The Department of Mathematics and Computer Science
The North Carolina School of Science and Mathematics

Principal Authors

Kevin G. Bartkovich

John A. Goebel

Julie L. Graves

Daniel J. Teague

Contributing Authors

Gloria B. Barrett

Helen L. Compton

Dorothy Doyle

Jo Ann Lutz

Donita Robinson

Karen Whitehead

Janson Publications Inc., Dedham Massachusetts

The Contemporary Calculus through Applications project was supported, in part, by the National Science Foundation (Grant No. ESI-9,252901). Opinions expressed are those of the authors and not necessarily those of the foundation.

Library of Congress Cataloging-in-Publication Data

Contemporary calculus through applications/principal authors,
  Kevin G. Bartkovich...[et al.]; contributing authors, Gloria B.
  Barrett... [et al]..
      p.   cm.
    Includes index.
    ISBN 0-939765-87-X
    1. Calculus. I. Bartkovich, Kevin G. II. Barrett, Gloria B.,
  1948– .
  QA303.C74   1995
  515--dc20   95-36878
                                                CIP

Printed in the United States of America
9 8 7 6   2 3 4 5 6 7 8 9

Design Malcolm Grear Designers

Images ©1995 PhotoDisc, Inc.

Dedication

Those of us who teach at the North Carolina School of Science and Mathematics have two families, our own and our students. We dedicate this book first to our own families who have supported us throughout the long process of writing and revising this text. We also dedicate this book to our students, whose enthusiastic support of this course from its inception and throughout its many revisions has inspired us.

Contents

Preface

In 1985 the Department of Mathematics and Computer Science at the North Carolina School of Science and Mathematics began its precalculus curriculum reform project. For the next five years, the entire Department worked to develop a precalculus course that was applications-oriented and that took full advantage of technology. The textbook, *Contemporary Precalculus through Applications*, was published in 1991.

Even before the precalculus text was published, we saw the need to revise the course that follows precalculus. We wanted to continue with the major themes that had guided us through the development of our precalculus course and incorporate these themes into a calculus course. In some ways this new task was much more difficult than it had been with precalculus. Unlike precalculus, which is very loosely defined, calculus (even called "The Calculus" by some), seems to be much more rigorously defined. Nearly all calculus books contain exactly the same topics, and the Advanced Placement Syllabus defines what will be tested, and therefore, what will be included in most calculus textbooks.

Technology, more than anything else, dictates that calculus as we know it must change. We took this as a serious challenge and attempted to do more than layer technology on top of the existing curriculum. As a result, the use of technology is not an optional part of this text. Instructors cannot demonstrate ideas and students cannot do homework or complete labs or take tests without the use of technology.

The Tulane Conference held in January of 1986 sounded the warning that college calculus was not meeting the needs of many students. The follow-up conference, sponsored by the National Academy of Sciences and the National Academy of Engineering in the Fall of 1987, began to offer some suggestions for reform as noted in the Mathematical Association of America's publication *Calculus for a New Century*. The National Council of Teachers of Mathematics *Curriculum and Evaluation Standards for School Mathematics* called for significant changes to both the mathematics curriculum and existing pedagogy. This textbook takes seriously the recommendations of the NCTM Standards and the MAA Notes. In particular, the text responds to the following three frequently cited concerns.

1. Students should be able to apply mathematics to problems that are stated in new and different contexts. In other words, students should understand mathematics, in contrast with merely substituting numbers into formulas.

2. Mathematics, and calculus in particular, is applied to virtually every field of human endeavor. Students should appreciate and understand a broad range of these applications, and they should be able to use mathematics as a tool for modeling a variety of real-world phenomena.

3. The numeric, symbolic, and graphic capabilities of computers and calculators have given technology a crucial role in the mathematics classroom.

This text defines an applications-oriented, investigative calculus course in which students are provided tools for understanding the world in which they live. The textbook involves students in both the development of problem statements and in their solution. Students learn to use the concepts of calculus to solve problems in a variety of contexts, many of which are discussed over extended periods of time. Concepts are presented in the context of real-world applications; calculators and computers are used to develop concepts and to solve problems; and the interpretation of problem solutions is given strong emphasis.

The fabric of the course is woven with seven themes that are an indispensable part of virtually every segment of the course. These themes address important student needs and are described here.

1. *Developing Understanding*: Calculus offers students a repertoire of new techniques for describing the world around them and should be accessible to as wide an audience as possible. For that reason, this text encourages intuitive understanding of many topics. Although mathematical rigor may be appropriate, it should not be an obstacle to the success of students. Every student may not be able to prove the theorems of calculus, however each should obtain a solid understanding of the concepts of calculus.

    To help students develop their conceptual understanding, graphical, numerical, and algebraic interpretations are used together whenever possible in each section of the text. Students are led to discover concepts for themselves as often as is feasible, primarily through mathematics laboratory experiences.

2. *Data, Applications, and Mathematical Models*: Much of our information about the world comes to us in the form of data. Problems presented in this textbook describe realistic situations that can be modeled using calculus. As much as possible, major concepts are introduced through data, applications, and mathematical modeling. These applications come from a wide variety of areas in which calculus is used as a tool to solve problems. While the physical sciences provide a wealth of applications of calculus, examples are drawn from many other areas of study. As students see a wide variety of applications they will begin to appreciate the power of calculus.

3. *Computer and Calculator as Tools*: Much of traditional calculus involves extensive paper-and-pencil manipulation. While students must be able to do some manipulations, technology makes obsolete the skills required for solving difficult problems with paper and pencil. Computer technology empowers us to solve problems with "messy" data and large numbers of computations, which is exactly how problems appear in the real world.

    Technology also is used to experiment with different values for parameters in problems, to try different strategies for solution, to test out various conjectures, and to ask "What if...?" questions that otherwise would not be feasible to investigate. Students must, however, know how to interpret results, when a result is reasonable, and when technology is not going to provide the best solution to a particular problem. Each of these requirements further emphasizes the need for students to develop a conceptual understanding of calculus.

4. *Numerical Algorithms*: Approximations for the derivative, the definite integral, and solutions to differential equations can be found using numerical algorithms. Since the techniques used in these algorithms reinforce the broad concepts of calculus, the algorithms provide more than just the solution to a problem. Numerical algorithms do have limitations, and the text will help students develop an understanding of the accuracy and the appropriateness of various numerical methods.

5. *Discrete Phenomena*: Calculus involves the study of phenomena that are represented by continuous functions. Because computations can be carried out very rapidly, it is possible to study these phenomena using discrete techniques. Recursive techniques, difference quotients, and other discrete concepts help students move from the discrete to the continuous domain. Links between discrete and continuous phenomena are emphasized frequently in the textbook.

6. *Computer and Calculator Laboratory Experiences*: Computer and calculator lab work is the centerpiece of the entire course. Students use technology as a tool for investigation and discovery. Labs activities allow students to discover concepts, to develop their intuition related to calculus, and to investigate extended problems in groups and individually. These labs are designed to culminate in written reports that summarize and analyze a particular investigation.

7. *Writing about Mathematics*: This textbook requires that students write about mathematics, including both concepts and the interpretations of these concepts. The students will use the language of mathematics and communicate their ideas to other individuals familiar with the subject. Writing about mathematics completes a process that begins with the translation of a problem statement into mathematical notation, then uses mathematics to investigate the problem, and ends with translating the results back into a verbal explanation and summary.

The NCSSM Contemporary Calculus project began in 1988 as a joint curriculum project, Project CALC, with Drs. David Smith and Lawrence Moore at Duke University, and funded by the National Science Foundation. The format for this text and much of the content areas grew out of early work with Project CALC. In 1992 the authors received funding, also from the National Science Foundation, to complete the work on a separate reformed calculus course and text. This text, *Contemporary Calculus through Applications,* is the direct result of that grant.

Acknowledgments

This text was developed under two grants from the National Science Foundation. The first grant was to Duke University; we worked as a subcontractor to write modules and to field-test and modify materials. The second grant to the Department of Mathematics and Computer Science at the North Carolina School of Science and Mathematics (NCSSM) gave us the opportunity to take what we had learned from our work with Duke and write our own course.

Our thanks to Professors David Smith and Lawrence Moore of Duke University for including us in Project CALC. Drs. Smith and Moore were among the first university mathematicians to recognize the need for reforming calculus instruction. They included us in their first planning grant and again in their first major grant. Our initial involvement with Duke University included writing some sections, field-testing their materials in our classes and modifying their materials for our students. After we received funding to continue our project, Drs. Smith and Moore continued to work with us as members of our Advisory Board.

In addition to David Smith and Lawrence Moore, our Advisory Board included Henry Pollak, retired from Bell Labs, Joan Countryman of The Lincoln School, and Tom Tucker of Colgate University. Their careful reading of our materials reassured us that we were making good progress and resulted in changes which have greatly improved our efforts.

All of the members of the Department of Mathematics and Computer Science at the North Carolina School of Science and Mathematics, both current and former, have made significant contributions to the development of this text. John Goebel, Kevin Bartkovich, and Lawrence Gould worked with Duke during the first two years of the first grant. Lawrence attended planning meetings with Duke, observed classes at Duke, wrote some of the early units, and tested materials in our calculus classes. His untimely death in 1990 deprived us of his insight. Mary Malinauskas, Tracey Harting, and Julie Allen wrote solutions for many of the exercises and labs. Maria Hernandez, Peggy Craft, Marilyn Schiermeier, and Joey Sinreich all taught the course during the writing of this text, made suggestions, and helped with the many tasks associated with preparing the text, the *Solutions Manual*, and the *Instructor's Guide*. In addition to members of the Mathematics Department, we would like to thank Dr. John Kolena of the NCSSM Physics Department and Dr. Myra Halpin of the Chemistry Department for their help with the physics and chemistry applications in the text.

Our thanks to our field testers who taught all or part of the course and sent us valuable input. These field testers were: Sharon Adam, Pensacola High School, Pensacola, FL; David Bannard, Collegiate Schools, Richmond, VA; Michael Lutz, Elkhart Central High School, Elkhart, IN; Nannette Dyas, Montgomery Blair High School, Silver Spring, MD; Gale Farmer, Wichita Collegiate School, Wichita, KS; Landy Godbold, Westminster School, Atlanta, GA; Ted Gott, Southern High School, Harwood, MD; Peter Gufstafson, The Pennington School, Pennington, NJ; Cheryl Haley, Norview High School, Norfolk, VA; Roger Jardine, Caribou High School, Caribou, ME; Beverly Johnson, Webb School of Knoxville, Knoxville, TN; John King, Dwight-Englewood School, Englewood, NJ; Beth Krantz, Vallivue Sr. High, Caldwell, ID; Brad Morris, Friends Central School, Wynnewood, PA; Donavon Nagel, John Marshall High School, Rochester, MN; Mel Noble, Olympic High School, Bremerton, WA; Steve Unruhe, Riverside High School, Durham, NC.

Special thanks to Steve Unruhe who has taught the entire course at Riverside High School in Durham, NC, for several years, has given us a wealth of feedback and input, and has helped us with the *Instructor's Guide*. Special thanks also to Ted Gott, who has also taught the course at Southern High School in Harwood, MD, for several years and has demonstrated that the course can be taught effectively using graphing calculators.

Many instructors participated in the Calculus workshops we taught during the summers from 1991 through 1995. We thank these instructors for their reactions to preliminary drafts of the text that helped to shape the final outcome.

We appreciate the efforts of our publisher Barbara Janson and our editor Eric Karnowski in helping us create a book from a manuscript that has been undergoing continuous change.

We would like to thank Sally Berenson and Glenda Carter of the Center for Research in Mathematics and Science Education at North Carolina State University who were the formal evaluators for our project.

We would also like to thank the NCSSM administration for helping us pursue funding for this project, the personnel and business offices for cheerfully accepting the extra work that the project required of them, and Terry Brown, the department secretary, who helped us complete all of our school responsibilities in the midst of our work on this book.

Our students would have been without a photocopied text, our field testers would have been stranded, and the project would have ground to a halt if not for the support provided daily by Pam Smith, the secretary for this project. We thank her for her commitment to the success of our development effort.

Features of this Text

The seven major themes detailed in the Preface permeate this text but the following features of the text deserve special mention.

Laboratory Experiences

Throughout this text students will be expected to complete lab assignments which will either aid them in the discovery of important concepts or give them the opportunity to use the calculus they have learned to solve real-world applications. We strongly encourage our students to work in pairs on the labs. These labs are an integral part of this text and are not optional. We have left the labs somewhat open-ended so that students will have the freedom to explore and use different techniques to discover the concepts or solve the problems in the labs. This open-endedness may be unfamiliar to some students, but one of the major goals of this text is to empower students with the skills and confidence they will need to solve problems on their own. For this reason, suggested solutions to the labs are not included in the *Solutions Manual*, but are available in the *Instructor's Guide*. This will allow instructors to provide hints to students as needed. We recommend that these hints be given very sparingly.

Use of Technology

Just as the labs are essential to the successful completion of this text, so is the use of technology. In the earliest stages of development of this text we depended almost entirely on Mathcad®, an electronic mathematics scratchpad, to do all of the labs and much of the classwork and homework. The course has been successfully taught both with a graphing calculator that handles data well and also with a spreadsheet that has good graphing capabilities.

End of Chapter Exercises

We have provided summary exercises at the end of each of Chapters 2 through 6. These exercises, though not exhaustive, summarize the important concepts in the chapter and give the students an opportunity to review these concepts out of the context of the section in which they were taught. In some cases, the ideas of the chapter are extended in these exercises.

Investigations

The end of Chapter 3 and the entirety of Chapter 8 consist of investigations. These "super labs", like the regular labs, come from a wide variety of fields, but are intended to be more open-ended than the other labs and are not specifically based on material covered in any one section of the text. The intent is for students, working in groups, to have an opportunity to think about all of the calculus they have learned up to the point in the text where the investigations appear and to use the appropriate calculus to develop a reasonable solution to the problem. Many of the investigations can be solved by more than one method, and we encourage multiple solutions. We also encourage students to try the extensions that are often provided, or to look for their own extensions to the problems. In these investigations, students have the opportunity to pull together the calculus they have studied, to demonstrate their problem-solving skills, and to communicate their ideas both orally and in writing.

CHAPTER 1

How Things Change

1.1 Introduction

Why is mathematics, and calculus in particular, important? One of the main reasons we study mathematics is because it helps us to understand the phenomena we experience in the world around us. This understanding often allows us to predict what will happen in the future or explain what happened in the past. Prediction and explanation are two ways that mathematics helps us to solve problems.

Calculus is especially important for solving problems involving how phenomena change because *calculus is a mathematical language of change*. How crucial is change in understanding phenomena in the world around us? Consider the following observation:

At the beginning of 1987, the world population passed 5 billion.

This statement tells us the size of the world population at a particular time, but says nothing about what the population was before 1987 or will be after 1987. The fact that the world population was 5 billion in 1987 may be less significant than the fact that the population doubled between 1950 and 1987. Assessing the significance of the current population level is difficult without knowing the rate at which the population is changing. The mathematics of populations addresses a number of questions of special relevance to our generation. Are we in the midst of a global population explosion? If so, what approaches might be effective in dealing with this problem? Suppose there is no impending population explosion. Will the world population eventually level off? If so, at what level? Will the size of the limiting population be so large that we will have a host of other problems in the future? If the population does not level off, will the population oscillate between different levels? If the population does oscillate, will the oscillations include wild swings between very high and very low levels, or will the swings be relatively small? All of these questions involve the concept of the rate of change of population and can be investigated mathematically with calculus.

Numerical information presented by the media often has the characteristic that the rate of change is more significant than the current value. For instance, scientific reports about the ozone layer usually describe the rate at which ozone is being destroyed. Rarely do reports mention the actual amount of ozone that currently exists. In fact, such a figure would probably be meaningless to the average reader. What is significant is the rate at which ozone is being depleted. A decrease in the rate of ozone depletion is considered good news by most people. Unfortunately, such a decrease does not mean that the ozone layer is being replenished, but only that the ozone layer is being destroyed less quickly than in the past.

Calculus encompasses much mathematics used for answering the questions mentioned above, as well as related questions concerning economics, medicine, and the analysis of data. Furthermore, calculus is essential for understanding the motion of atomic particles, automobiles, satellites, rockets, and galaxies. Calculus is basic to the study of flow in rivers, currents in the oceans, and movement of air over airplane wings. Businesses use calculus in cost and profit analysis. This list can go on and on, yet the characteristic common to all of these phenomena is the central importance of how quantities change.

On a mathematical level, these examples share another common theme. They all involve related variables (such as population and time), but the information given by the rate of change of one variable with respect to the other is more important than the values of the variables. In precalculus courses, the study of functions often emphasizes graphs and ordered pairs. For instance, if $f(x) = x^2 - 3x$, we know that $f(2) = -2$ and that the point $(2, -2)$ is an ordered pair of the function. The calculus view of functions is more dynamic. In calculus courses we want to know more than, "If $x = 2$, then $y = -2$." We will ask, "Around $x = 2$, how are the y values changing? Are they increasing or decreasing? At what rate are the y values changing?" One of the primary purposes of calculus is to quantify rates of change. As the above examples illustrate, rates of change are important because knowing how a phenomenon changes gives valuable information about the phenomenon.

1.2   Phenomena Modeled by Discrete Change Expressions

Figure 1.1 shows a graph of 1995 first class postage costs versus weight. Notice that the value for the weight of a letter can be any positive number. On the other hand, the set of values for the postage costs is discrete, since the possible values for postage costs belong to the set {32, 55, 78, 101,…}. The discrete nature of the set of postage costs gives the graph a stair-step appearance. How does the U.S. Postal Service treat mail that weighs exactly some integer amount? A letter that weighs less than or equal to 1 ounce costs 32 cents to mail first class; however, a letter that weighs over 1 ounce but less than or equal to 2 ounces is charged the next highest rate, which is 55 cents. Graphically this is indicated by the use of open endpoints on the left end of each step and closed endpoints on the right.

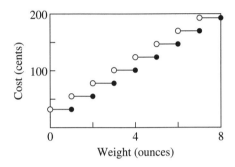

Figure 1.1 Postage cost versus weight

Suppose you have a checking account that earns no interest. The balance in the account changes only when you deposit or withdraw money. Assume that on January 1 the balance is $200. At the end of each month you make a deposit of $1000, but then $850 is automatically withdrawn to pay your bills; therefore, the account has a net increase of $150 each month. On February 1 you will have $350, on March 1 you will have $500, on April 1 you will have $650, and so on. In general, we find the balance in any month by adding the change in the balance to the previous month's balance. Since the change occurs at the constant rate of $150 per month, each month's balance is simply the previous month's balance plus $150. Figure 1.2 shows the balance in the account during the seven months after January 1. As in the postage stamp example, one variable, the number of months, can have any non-negative value; the other variable, balance, has only certain discrete values. The step graph in Figure 1.2 shows that the balance in your account is the same from the beginning of any month to the end of that month.

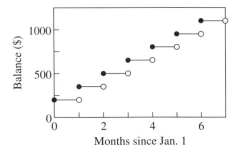

Figure 1.2 Monthly balance of checking account

Since there is no change in the balance within a month, we can modify our representation of this phenomenon by letting the domain consist of only integer values of the number of months. The table in Figure 1.3 lists monthly account balances, and the corresponding graph consists of seven distinct points. Each point represents the balance in the account n months after January 1, for integer values of n from zero to six. Notice that the points graphed in Figure 1.3 correspond to the left endpoints of the steps graphed in Figure 1.2.

n (months)	Balance ($)
0	200
1	350
2	500
3	650
4	800
5	950
6	1100

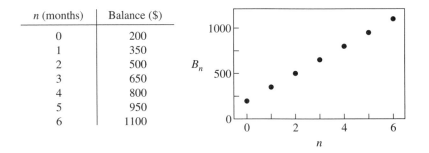

Figure 1.3 Monthly balance for integer values of n

constant rate of change   The relationship between the balance in the checking account and the number of months exhibits a ***constant rate of change***. The balance changes by a constant amount from one month to the next. To calculate the balance in each successive month, we add $150 to the balance from the previous month. We represent this in symbols with the equation

$$B_n = B_{n-1} + 150, \tag{1}$$

in which B_{n-1} represents the account balance for month $n-1$, and B_n represents the account balance for month n, which is the next month after month $n-1$. Equation (1) *recursive equation*   is an example of a ***recursive equation***, which is used to generate a sequence of values, each value defined in terms of the preceding value (or values). For example,

$$B_1 = B_0 + 150,$$
$$B_2 = B_1 + 150,$$

and so on. Each step using a recursive equation to generate a value from the preceding value is called an ***iteration*** of a recursive equation.

iteration

Equation (1) does not provide a complete description of the account balance. The equation tells us that we should add $150 to the balance each month, but we also need to state a starting balance. The system

$$B_0 = 200$$
$$B_n = B_{n-1} + 150, \tag{2}$$

specifies a beginning account balance and thus is a complete recursive description of the checking account balance over time in months.

The graph in Figure 1.4 is a plot of the account balance versus month. The points on the graph are the ordered pairs (n, B_n), where n is the month and B_n is the balance in month n. Notice that the points appear to lie along a line. The form of the recursive equation for B_n shows that the next value B_n equals the previous value B_{n-1} plus $150, so that the balance after 1 month is given by

$$B_1 = B_0 + 150,$$

and the balance after 2 months is

$$B_2 = B_1 + 150.$$

Substituting for B_1 in the last equation yields

$$B_2 = (B_0 + 150) + 150$$
$$= B_0 + (150)(2).$$

Since $B_3 = B_2 + 150$, we can write

$$B_3 = B_0 + (150)(3).$$

This pattern continues so that in general

$$B_n = B_0 + 150n,$$

which shows that B_n is a linear function of n. Though the more familiar way to write such a function is in the form $B(n) = B_0 + 150n$, we will remain consistent and use the subscripted form initially introduced in the recursive equations.

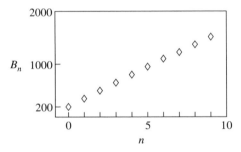

Figure 1.4 Account balance versus month

The line containing the points in the scatter plot in Figure 1.4 has an intercept on the vertical axis that is the starting balance $200. The slope of the line is $150 per month, which is the constant rate of change of the balance. The ordered pairs (n, B_n) therefore fit the equation

$$B_n = 200 + 150n, \tag{3}$$

where n is a non-negative integer. The line containing the points in the scatter plot in Figure 1.4 also contains many other points; however, only the points (n, B_n) with n a non-negative integer make sense in the context of the account balance that we are modeling.

closed form expression   Equation (3) is an example of a ***closed form expression*** for the account balance B_n, in contrast to the recursive equation for the account balance given by the equations in (2). Notice that the rate of change of balance, $150 per month, is embedded in equation (3). The closed form expression is useful because it allows us to find a value for the account balance after any month by simply substituting a whole number value for n; finding an account balance using a recursive equation requires us to generate a table of all prior account balances. However, in some situations a closed form expression is much more difficult to derive than a recursive equation, and with technology, recursive equations are generally sufficient to generate whatever values are desired.

In this chapter we will concentrate on recursive equations involving rate-of-change expressions. These will allow us to study a phenomenon if we know how it changes.

Example 1

As we add weights of equal mass to a spring, the spring's displacement from equilibrium increases. Suppose that for every weight added, the spring stretches by two centimeters. After the addition of four weights, how much would you expect the spring to stretch? Write a recursive system that describes the stretch of the spring for each additional weight. Also write a closed form expression for this phenomenon.

Solution   With a single weight, the spring stretches 2 cm from equilibrium. With two weights, the spring stretches 4 cm. Add another weight (three total) and the spring stretches 2 more cm for a total stretch of 6 cm. Finally, with four weights attached to the spring, the stretch increases by 2 cm from the stretch with three weights, and the total stretch is 8 cm. Let S_n represent the stretch in centimeters caused by n weights attached to the spring. The initial stretch is 0, so $S_0 = 0$. Each additional weight causes an added stretch of 2 cm, which the recursive equation $S_n = S_{n-1} + 2$ represents. The recursive system that represents the stretch of the spring is

$$S_0 = 0$$
$$S_n = S_{n-1} + 2. \tag{4}$$

Figure 1.5 shows a table of values generated by six iterations of equation (4), together with a graph that shows a plot of the ordered pairs (n, S_n). The points on the graph lie along a line through the origin with slope 2; therefore, a closed form expression for S_n is

$$S_n = 2n,$$

where n is a non-negative integer. ∎

n (# of weights)	S_n (stretch in cm)
0	0
1	2
2	4
3	6
4	8
5	10
6	12

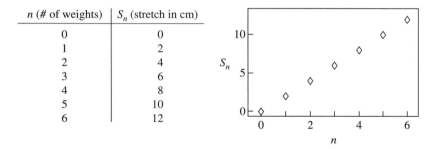

Figure 1.5 Stretch of spring versus number of weights

Both of the recursive systems we have examined thus far describe a phenomenon that changes at a constant rate. With both systems, we determine successive values of the subscripted variable by adding a constant to the previous value. In general, a recursive equation for constant rate of change has the form

$$A_n = A_{n-1} + k,$$

where k is a constant that represents the rate of change. The graphs in Figures 1.4 and 1.5 show that k is also the slope of the line that contains the ordered pairs (n, A_n); therefore, the closed form expression for a function with a constant rate of change is

$$A_n = kn + A_0,$$

where n is a non-negative integer.

What have we accomplished in the examples we have investigated thus far? We have found that we are able to generate values of a variable by using an expression for how the variable changes. In other words, knowing how a phenomenon changes enables us to determine a lot about the phenomenon. Knowing the rate of change, we are able to predict a value for some time in the future, and we can generate intermediate values as well.

Example 2

Suppose you open a savings account that earns 3% annual interest and place $100 in the account. At the end of each year, the bank adds the interest for that year to the account balance, thus producing a new balance that earns 3% interest during the subsequent year. How much money will accumulate in your account from year to year?

Solution   In this case, the change in the balance from year to year is equal to the interest the account earns, and the interest is equal to 3% of the current balance. This implies that the change in the balance is equal to 0.03 times the current balance. If we define the subscripted variable A_n as the amount in dollars in the account after n years, then $A_0 = 100$, since the initial balance is $100. We know that the balance in the account after any year is equal to the balance from the previous year plus 0.03 times the balance from the previous year. This relationship is expressed by the recursive equation

$$A_n = A_{n-1} + 0.03A_{n-1}. \tag{5}$$

Equation (5), coupled with the initial value $A_0 = 100$, enables us to generate the table of values and accompanying graph shown in Figure 1.6. Once again we have used information about how a quantity changes to generate values of the quantity. ■

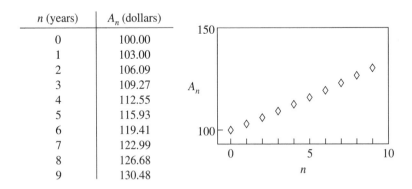

n (years)	A_n (dollars)
0	100.00
1	103.00
2	106.09
3	109.27
4	112.55
5	115.93
6	119.41
7	122.99
8	126.68
9	130.48

Figure 1.6 Savings account balance versus year

When a bank computes savings account balances, it truncates the balance after each computation, dropping any digits after the hundredths place. A bank computes the next year's balance A_n from the truncated balance A_{n-1} of the previous year, and A_n is likewise truncated. Our computations, on the other hand, do not involve truncation of the account balances. The values of A_n shown in the table are rounded to the nearest hundredth; however, we used additional digits of A_{n-1} in the recursive calculations of A_n. We adopt this convention to keep our iterative system simple; we do not include the additional step of truncating A_n after each iteration. In some situations, this will lead to results that are slightly different from the balances computed by a bank.

Can we determine the equation of the function whose graph contains the points shown in Figure 1.6? If we can, we will have found a closed form expression for A_n. The points graphed in Figure 1.6 appear to lie along a line; however, the table of values in Figure 1.6 and equation (5) both reveal that the rate of change is not constant. In the table we see that balances increase by varying amounts from year to year, and in the equation we see that the monthly change is $0.03A_{n-1}$, a variable amount. Thus the ordered pairs (n, A_n) are not linear. The right side of equation (5) is factorable, giving us the equivalent equation $A_n = (1 + 0.03)A_{n-1}$, or simply $A_n = (1.03)A_{n-1}$. This form of the recursive equation shows that the next value A_n equals the previous value A_{n-1} multiplied by 1.03. Specifically,

$$A_1 = 1.03A_0,$$

and

$$A_2 = 1.03A_1.$$

Substituting for A_1 in the last equation yields

$$A_2 = 1.03(1.03A_0) = (1.03)^2 A_0.$$

Since $A_3 = 1.03A_2$, we can write

$$A_3 = (1.03)^3 A_0.$$

This pattern continues so that, in general,

$$A_n = (1.03)^n A_0, \tag{6}$$

which shows that A_n is an exponential function of n. This means that the points (n, A_n) in Figure 1.6 lie along an exponential function with base 1.03 and intercept A_0. The exponential curve that contains the points given by equation (6) also contains many other points; however, only the points on the curve with n a non-negative integer make sense in the context of our situation.

The data plotted in Figure 1.6 appear linear, yet our analysis demonstrates that the account balances follow an exponential curve. We can see the curvature in the plot of the account balances by looking at a graph over a longer interval of time. Figure 1.7 shows a graph of the account balances over a 50-year interval. We perceive curvature when our viewing window is relatively large, whereas the data appear linear in a small viewing window. The concept that curves appear linear when viewed over small intervals is significant, and we will investigate this concept in depth in Chapter 2.

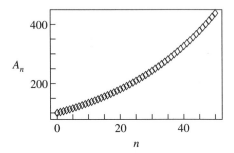

Figure 1.7 Graph of savings account balances for first 50 years

In equation (5), the change from the previous balance A_{n-1} to the subsequent balance A_n is $0.03A_{n-1}$. This change is proportional to the previous balance A_{n-1}, and the constant of proportionality is 0.03, the interest rate. The relationship between the savings account balance and time is an example of ***proportional rate of change***. This term signifies that the rate of change of a quantity is proportional to the current value of the quantity. Recursive equations for proportional rate of change have the general form

proportional rate of change

$$A_n = A_{n-1} + kA_{n-1},$$

where k is a constant that represents the rate of proportional growth. In equation (6), we observed that proportional change leads to an exponential closed form expression, which has the general form

$$A_n = (1+k)^n A_0. \tag{7}$$

Example 3

Suppose a rabbit population grows in such a way that when it is measured at the end of each month, the population doubles in size compared with the previous month. Assume that the population has unlimited space to grow and unlimited resources. What is the recursive equation that describes the behavior of the population on a month-by-month basis? What is the recursive equation if the population grows by 50% of the population every month? by 10% every month? Write a closed form expression for a population growing at a rate of 10% of the population each month.

Solution   A population that doubles in size every month grows according to the equation

$$P_n = 2P_{n-1},$$

where P_n is the size of the rabbit population n months after the population is initially measured. We can also write this equation in the form

$$P_n = P_{n-1} + P_{n-1},$$

which makes clear that the change in the population in a month is equal to the size of the population at the beginning of the month. If the population grows by 50% each month, then the recursive equation is

$$P_n = P_{n-1} + 0.5P_{n-1}.$$

If the population grows by 10% each month, then the recursive equation is

$$P_n = P_{n-1} + 0.1P_{n-1}.$$

Referring to equation (7), a closed form expression for the 10% growth situation is

$$P_n = (1.10)^n P_0.$$

The recursive equations in this example are potentially misleading. In the case of the last equation, the population does not suddenly add 10% to its size at the end of a month. Increases (as well as decreases) in the population occur throughout the month; however, this model assumes that the net effect of the births and deaths in the population is equivalent to adding 10% to the population at the end of the month. Another way to think about this model is that we only measure the population at the end of each month, and at those times we always find that the population has increased by 10% over the population measured at the end of the previous month. ∎

Example 4

A forest service harvests a preserve of 100,000 acres of forest at a rate of 10% of the existing forest each year. In addition, 5000 acres of new forest are planted each year. What is the long-term trend in the number of acres of forest in the preserve? If, at a later date, the forest service increases the rate at which new forest is planted to 7000 acres per year, what effect will that have on the size of the forest?

Solution   Assuming that no trees are gained or lost due to natural causes, there are two components to the rate of change of the size of the forest. Due to harvesting, the acreage decreases each year by 10% of the size of the forest at the beginning of the year. Due to planting, the acreage increases by 5000 acres per year. Thus, the rate of change in the size of the forest during year n is $-0.1F_{n-1} + 5000$ acres per year, where F_{n-1} is the number of acres of forest at the beginning of year $n-1$. The number of acres of forest can be described by the recursive system

$$F_0 = 100,000,$$
$$F_n = F_{n-1} - 0.1F_{n-1} + 5000. \tag{8}$$

We can use a calculator or spreadsheet to find that $F_1 = 95,000$, and $F_2 = 90,500$, and so on. The recursive equation (8) combines change that is proportional to the existing population, represented by $-0.1F_{n-1}$, and change that is constant, represented by 5000. Figure 1.8 shows the first ten iterations of equation (8) and a graph of the ordered pairs (n, F_n). The values in the table are rounded to the nearest acre, although additional digits are used in the calculations. Figure 1.8 shows that the values of F_n are decreasing, but the amount of the decrease is less each year. If we continue iterating recursive equation (8), the value of F_n levels off at 50,000 acres.

n (in years)	F_n (in acres)
0	100,000
1	95,000
2	90,500
3	86,450
4	82,805
5	79,525
6	76,572
7	73,915
8	71,523
9	69,371
10	67,434

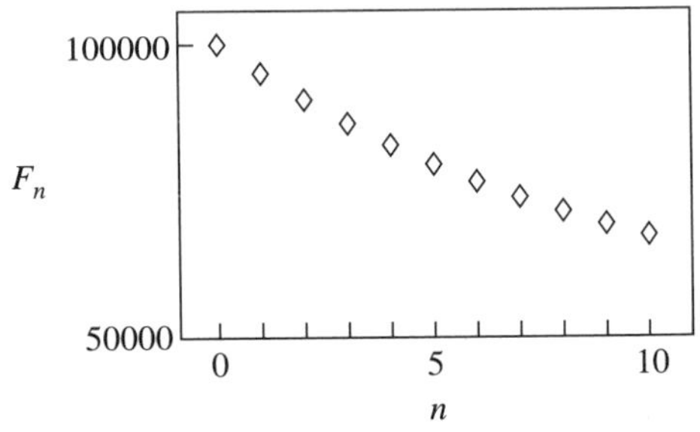

Figure 1.8 Forest size versus year

With a forest of 50,000 acres, if we then change the planting policy to 7000 acres per year, the recursive equation for number of acres in the forest becomes

$$F_n = F_{n-1} - 0.1F_{n-1} + 7000.$$

If we iterate this equation, starting with $F_0 = 50,000$, the values begin to rise, first to 52,000, then to 53,800, and so on, and eventually level off at 70,000 acres.

A closed form expression for the size of the forest in terms of year is much less obvious than in the previous examples. Since our main interest is with recursive equations, we will not attempt to find a closed form expression for the size of the forest. ■

Exercise Set 1.2

1. Find an example in a newspaper, magazine, or news report of information given as a rate of change. Write a brief summary of the information.

2. The fee for parking at the municipal parking deck is 75 cents per hour, with any fraction of an hour charged for a full hour. Write a recursive system for p_n, the fee for parking for between $n-1$ and n hours, where n is a positive integer. Also write a closed form expression for p_n in terms of n. Over what set of values for n do you believe your model is accurate? Support your answer.

3. Record times for running the mile have been decreasing at an approximate rate of $\frac{1}{3}$ second per year. Write a recursive equation that describes how the record time for running the mile changes from one year to the next. Do you think this model will be accurate for a long time into the future? Explain your answer.

4. A new car worth $15,000 loses one-fifth of its original value, or $3000, each year for 5 years. This kind of loss of property value is known as ***linear depreciation***. Write a recursive system that gives the value of the car after each of the first 5 years. What values for the year variable make sense in your recursive system? What is the slope of the line that contains the ordered pairs (year, value)?

linear depreciation

5. In which account would you rather place your money, one that earns 0.5% interest each month or one that earns 6% interest each year? Why? How does the value of each account differ after 1 year? after 5 years?

6. A certain account earns 5% annual interest, with an initial deposit of $100. Write a recursive system for the account balance at the end of each year. Generate a table of values for the account balance at the end of each year and plot a graph of these values. How long is required for the balance to double? What is the equation of the curve that contains the ordered pairs (year, balance)?

7. Suppose that due to inflation your money is losing 0.5% of its purchasing power each month. This means that prices are increasing in such a way that each month your money can buy 0.5% less than it could the previous month. Write a recursive system for the purchasing power of your money after each month. How much will $100 be worth in 1 year? in 2 years? in 5 years?

8. Jen College's parents finance her college education by depositing $1000 into a tax-exempt account each year on her birthday. They do this every year until Jen turns 18, and the account earns 8% annual interest added at the end of each year. They opened the account when she was born by making an initial deposit of $1000. How much money will be in the account after they make the final deposit when Jen turns 18?

An account such as the one in exercise 8, into which regular deposits are made and which earns interest, is known as an **annuity***. Examples of annuities are some types of life insurance policies and individual retirement accounts (IRAs).*

annuity

9. Suppose you place an initial deposit of $1000 in an annuity that earns 10% each year, and you deposit $500 at the end of each year. Write a recursive system for the balance of the annuity at the end of each year. What will the balance of the annuity be in 10 years? in 20 years? in 30 years?

10. A loan can be thought of as a special type of annuity that reduces in size as you make payments to the bank. Suppose you have a loan of $50,000 in which the outstanding balance increases by 8% each year. Payments are made at the end of each year after the 8% interest is charged against the account.

   a. If you make a payment of $4250 at the end of each year, in how many years will you pay off the loan?

   b. What annual payment will keep the unpaid loan balance fixed at $50,000?

   c. What annual payment is necessary to pay off the loan in 30 years? in 20 years? What is the total paid to the loan agency in each case?

11. A person takes 200 mg of ibuprofen every 4 hours. After 4 hours, the body elimi-
     nates 75% of the drug that is in the bloodstream. Write a recursive system that
     describes this process. What is the long-term level of ibuprofen in the blood-
     stream?

12. An annual count of butterflies on an island reveals that each year the population
     declines by 20% of the previous year's population. To keep the population from
     dying out, an environmental group brings 50 butterflies to the island and releases
     them at the beginning of each year.

     a. Assuming that 500 butterflies are on the island initially, what is the long-term
         trend in the population?

     b. Suppose a severe winter reduces the butterfly population to 100. What will
         happen to the population in subsequent years?

     c. What is the relationship of the initial population to the long-term trend in the
         population?

13. Suppose a rabbit population is 500 in January. Imagine two different rates of
     change for the population:

     (1) The population is increasing at the rate of 50 rabbits per month.

     (2) The population is increasing at the rate of 10% per month.

     Compare the sizes of the rabbit populations over time for these two scenarios.

1.3   Phenomena Modeled by Continuous Change Expressions

So far, we have investigated quantities for which change occurs or is measured at
discrete times. We examined a savings account in which the interest is added to the
account only at the end of each year. We also investigated a rabbit population that is
measured only at the end of each month. Likewise, we assumed that acreage in a
forest changes only at the times that seedlings are planted and trees are harvested.
Given a starting value and information about how a variable changes, we have been
able to predict values for the variable at some future time. In each example to this
point, a certain fixed interval between measurements makes sense in the context of the
problem. For many phenomena, however, the interval between measurements is not
fixed. For example, the height of a baseball thrown into the air can be measured at
every instant of time that it is aloft. World population is so large, and births and deaths
occur so frequently, that we often use the assumption that the world population is
changing at virtually every instant of time. The height of a baseball and the size of the
continuous change   world population are examples of phenomena that can be modeled by ***continuous***
expressions   ***change expressions***.

The distinction between discrete change expressions and continuous change expressions is illustrated by two different ways that we can change the water level in a tank. We can use a bucket to scoop water out of the tank, causing the water level to change at discrete times. In contrast, we can use pipes to allow water to drain out of the tank, causing the water level to change continuously. To be specific, we could use a bucket to scoop out 2 gallons of water at one-minute intervals, causing the water level to decrease by 2 gallons after each minute. The interval between measurements is fixed at one minute. The one minute interval cannot be subdivided, since the change occurs only after each minute elapses. In contrast, we could let water drain continuously at the rate of 2 gallons per minute. In this case, 2 gallons would drain in 1 minute, 1 gallon in $\frac{1}{2}$ minute, $\frac{1}{2}$ gallon in $\frac{1}{4}$ minute, $\frac{1}{3}$ gallon in $\frac{1}{6}$ minute, and so forth. In this case, it makes sense to measure the water level as often as we like, and we will observe a change in the water level between any two measurements.

We will begin our study of continuous change expressions with a finance example. Financial institutions sometimes advertise accounts that offer "continuous compounding of interest." Suppose you place $100 in an account that offers "10% annual interest, compounded continuously." How much money will you have in the account at any point in time during the next ten years?

To answer this question, we need to understand what it means to compound continuously. To do this, we will first consider other types of compounding. If an account has annual compounding of interest, then interest for a particular year is added to the account at the end of that year. During subsequent years, the interest from previous years is considered part of the account balance, and this interest earns interest as well. If an account has semi-annual compounding of interest, some of the annual interest is added to the account every six months. In each year, interest added at the end of the first half-year earns interest during the second half-year. In the case of quarterly compounding, interest is credited to the account after the first quarter, and this interest in turn earns interest during the second quarter. The new interest is then added to the account, thus earning more interest during subsequent quarters. All of these accounts are examples of accounts that earn **compound interest**.

compound interest

What effect does increasing the frequency of compounding have on the account balance? We will answer this question by examining the effect of compounding interest at regular intervals over the course of ten years. We use Δt (in years) to represent the length of the interval between successive times that we compound the interest. The subscripted variable t_n represents the time corresponding to the nth addition of interest. Since successive values of t_n differ by Δt, the values of the elapsed time for n compounding periods are generated by the recursive equations

$$t_0 = 0$$
$$t_n = t_{n-1} + \Delta t.$$

For annual compounding, which implies that $\Delta t = 1$, we compound the interest at the end of each year. The balance in a given year is expressed by the recursive equation

$$A_n = A_{n-1} + 0.10 \cdot A_{n-1}.$$

Table 1.9 contains a list of values generated by this equation. All account balances in this table and in successive tables are rounded to the nearest cent.

n (iterations)	t_n (in years)	A_n (in dollars)
0	0	100.00
1	1	110.00
2	2	121.00
3	3	133.10
4	4	146.41
5	5	161.05
6	6	177.16
7	7	194.87
8	8	214.36
9	9	235.79
10	10	259.37

Table 1.9 Account balances with $\Delta t = 1$

If we now let $\Delta t = \frac{1}{2}$, which is semi-annual compounding, interest is added to the account at the end of each half-year. After the first half-year, the interest is one-half of 10% of the initial balance of $100, which is $5. That $5 is added to the balance, so that for the second half of the year, the interest credited is one-half of 10% of $105, which is $5.25. This interest is added to the half-year balance of $105, so the balance at the end of the year is $110.25. Crediting the interest to the account each half-year yields an additional $0.25 in the first year compared with annual compounding.

What happens to the account balance during successive years with semi-annual compounding? The values of the balance after each half-year are given by the recursive equation

$$A_n = A_{n-1} + 0.10 \cdot A_{n-1} \cdot \frac{1}{2},$$

where A_n is the account balance after n half-years, or n compounding periods. A list of values with the balance after each year is shown in Table 1.10. Notice that when one year has passed, n must equal 2, so that $t_2 = 1$.

n (iterations)	t_n (in years)	A_n (in dollars)
0	0	100.00
2	1	110.25
4	2	121.55
6	3	134.01
8	4	147.75
10	5	162.89
12	6	179.59
14	7	197.99
16	8	218.29
18	9	240.66
20	10	265.33

Table 1.10 Account balances with $\Delta t = \frac{1}{2}$

What is the balance after each year if the interest is compounded monthly (that is, interest is added to the account at the end of each month)? We can generate a sequence of values for the balance after each month beginning with $A_0 = 100$, the initial deposit. For the first month, we add one-twelfth of 10% of the initial deposit to A_0 to get A_1, which is the balance after one month. The value of A_1 is expressed by the equation

$$A_1 = A_0 + 0.10 \cdot A_0 \cdot \frac{1}{12}$$
$$= 100 + 0.10 \cdot 100 \cdot \frac{1}{12},$$

so that $A_1 = 100.83$. The interest credited to the account during the second month is one-twelfth of 10% of A_1, and this interest is added to A_1 to get A_2, so A_2 is given by the equation

$$A_2 = A_1 + 0.10 \cdot A_1 \cdot \frac{1}{12}$$
$$= 100.83 + 0.10 \cdot 100.83 \cdot \frac{1}{12},$$

so that $A_2 = 101.67$. A recursive equation for the values of the account balance is

$$A_n = A_{n-1} + 0.10 \cdot A_{n-1} \cdot \frac{1}{12}.$$

A list of values with the balance after each year is shown in Table 1.11.

n (iterations)	t_n (in years)	A_n (in dollars)
0	0	100.00
12	1	110.47
24	2	122.04
36	3	134.82
48	4	148.94
60	5	164.53
72	6	181.76
84	7	200.79
96	8	221.82
108	9	245.04
120	10	270.70

Table 1.11 Account balances with $\Delta t = \frac{1}{12}$

The general form of the equation we have used to generate the sequence of account balances is

$$A_n = A_{n-1} + 0.10 \cdot A_{n-1} \cdot \Delta t \tag{9}$$

where Δt is the length in years of the interval over which interest is compounded. Notice that the amount by which the balance increases during each compounding period is proportional to the balance A_{n-1} at the beginning of the compounding period, and the constant of proportionality is $0.10\Delta t$.

As the length of the compounding period decreases, the interest earned is added *continuous compounding* to the balance in the account more and more frequently. The term ***continuous compounding*** implies that interest earned on the account is added to the account at every instant in time, which is more frequently than once every day, or once every hour, or once every minute, or even once every second. We can simulate continuous compounding by letting the compounding period shrink toward zero; that is, we add interest to the account a large number of times each year. So far, we have seen that the balances after each year have increased as Δt has decreased. Will the account balances after each year continue to increase as Δt decreases and we simulate continuous compounding?

If we let $\Delta t = \frac{1}{100}$ (the compounding period is $\frac{1}{100}$ of a year), then the balance after one year is found by iterating equation (9) one hundred times. We iterate the equation one thousand times to find the balance after 10 years. For $\Delta t = \frac{1}{1000}$, one thousand iterations of equation (9) are needed to find the balance from one year to the next, and so on for smaller values of Δt. Table 1.12 contains a list of the account balances after each year for various choices of Δt.

t_n (in years)	$\Delta t = 1$	$\Delta t = \frac{1}{12}$	$\Delta t = \frac{1}{100}$	$\Delta t = \frac{1}{1000}$	$\Delta t = \frac{1}{5000}$
0	100.00	100.00	100.00	100.00	100.00
1	110.00	110.47	110.51	110.52	110.52
2	121.00	122.04	122.13	122.14	122.14
3	133.10	134.82	134.97	134.98	134.99
4	146.41	148.94	149.15	149.18	149.18
5	161.05	164.53	164.83	164.87	164.87
6	177.16	181.76	182.16	182.21	182.21
7	194.87	200.79	201.30	201.37	201.37
8	214.36	221.82	222.47	222.55	222.55
9	235.79	245.04	245.85	245.95	245.96
10	259.37	270.70	271.69	271.81	271.83

Table 1.12 Account balances A_n, in dollars, for various Δt

The last column in Table 1.12 is produced by using 5000 compounding periods per year, each of length $\frac{1}{5000}$ year. Notice that the yearly balances in the rightmost column have hardly changed in comparison with the balances from the column for $\Delta t = \frac{1}{1000}$. If we make Δt even smaller than $\frac{1}{5000}$, then the yearly account balances do not change more than a fraction of a cent. The values in the $\Delta t = \frac{1}{5000}$ column, therefore, are correct to the nearest cent no matter how small we choose Δt.

Table 1.12 illustrates that over a fixed period of time, increasing the frequency of compounding increases the balances at the end of each year. The table also shows that the balances at the end of each year do not appear to increase without bound as Δt gets closer and closer to zero. As the frequency of compounding increases, the balances at the end of each year appear to level off. The values to which they level are the balances we would have if we invested $100 at 10% annual interest, compounded continuously. A graphical illustration of this leveling is shown in Figure 1.13, which displays a graph of the year-end balances associated with compounding periods of $\Delta t = 1$, $\Delta t = \frac{1}{12}$, and $\Delta t = \frac{1}{5000}$. The points generated with $\Delta t = \frac{1}{5000}$ are connected by line segments to distinguish them from the points generated with $\Delta t = 1$ and $\Delta t = \frac{1}{12}$.

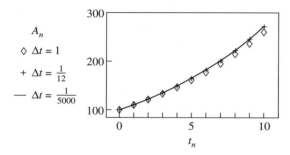

Figure 1.13 Graphs of balances from Table 1.12

When we speak of continuous compounding, we imagine a situation in which the number of compounding periods is infinitely large (or, equivalently, the time Δt between successive compounding is arbitrarily close to 0). With other types of compounding, such as quarterly or monthly, a fixed compounding period is used in our recursive calculations and the resulting balances are exact. In contrast, balances associated with continuous compounding cannot be computed exactly using recursive equations, but can be approximated by using an arbitrarily small compounding period. The smaller the compounding period, the better the approximation for continuous compounding. For example, we have observed that the values in the $\Delta t = \frac{1}{5000}$ column of Table 1.12 are good approximations for the limiting values of the year-end account balances as the frequency of compounding increases without bound. In addition, the graph of account balances versus time associated with continuous compounding is not a discrete set of points, but instead is a continuous curve that represents the account balance at every instant of time. The line segments through the points with $\Delta t = \frac{1}{5000}$ in Figure 1.13 provide a good approximation for the curve that represents continuous compounding.

What function models the account balances that are derived with continuous compounding? Notice that we can rewrite the recursive equation that we have been using to generate the sequence of account balances; that is,

$$A_n = A_{n-1} + 0.10 \cdot A_{n-1} \cdot \Delta t$$
$$= A_{n-1}(1 + 0.10 \cdot \Delta t).$$

If we begin with an initial deposit of $100 and iterate this equation to get successive balances, we obtain

$$A_1 = 100(1 + 0.10 \cdot \Delta t)$$

and

$$A_2 = A_1(1 + 0.10 \cdot \Delta t).$$

Substituting for A_1 in the expression for A_2 yields

$$A_2 = [100(1 + 0.10 \cdot \Delta t)](1 + 0.10 \cdot \Delta t)$$
$$= 100(1 + 0.10 \cdot \Delta t)^2.$$

Since $A_3 = A_2(1 + 0.10 \cdot \Delta t)$, we can write

$$A_3 = 100(1 + 0.10 \cdot \Delta t)^3.$$

This pattern continues so that, in general

$$A_n = 100(1 + 0.10 \cdot \Delta t)^n. \tag{10}$$

We can now use the closed form expression in (10) to verify the balances that were generated recursively in Table 1.12. For example, to get the balance at the end of one year with $\Delta t = \frac{1}{12}$ we need to let $n = 12$ and evaluate

$$A_{12} = 100\left(1 + 0.10 \cdot \frac{1}{12}\right)^{12}$$
$$= 110.47.$$

This result equals the one we obtained by iterating equation (9) twelve times with $\Delta t = \frac{1}{12}$. When $\Delta t = \frac{1}{1000}$, we can compute the balance at the end of the year by letting $n = 1000$. In this case,

$$A_{1000} = 100\left(1 + 0.10 \cdot \frac{1}{1000}\right)^{1000}$$
$$= 110.52,$$

which also agrees with the result in Table 1.12. In fact, every entry in the second row of Table 1.12 can be calculated using the formula

$$Balance = 100(1 + 0.10 \cdot \Delta t)^n,$$

where n is the number of compounding periods in a year, so that $n = \frac{1}{\Delta t}$. For example, when $\Delta t = \frac{1}{5000}$, we need 5000 compounding periods to obtain the end-of-year balance for the first year.

Since the initial value is the same for all computations, increases in the balances in any particular row of Table 1.12 result from differing values of $(1 + 0.10 \cdot \Delta t)^n$. This is equivalent to $(1 + 0.10 \cdot \Delta t)^{1/\Delta t}$ if we limit our attention to the second row. Since these balances appear to have some limiting value, we can conclude that the quantity $(1 + 0.10 \cdot \Delta t)^{1/\Delta t}$ must have some limiting value as Δt becomes very small. Table 1.14 provides values of this quantity for Δt values associated with increasing frequency of compounding. Correct to five decimal places, the limiting value appears to be 1.10517.

	$\Delta t = 1$	$\Delta t = \frac{1}{12}$	$\Delta t = \frac{1}{100}$	$\Delta t = \frac{1}{1000}$	$\Delta t = \frac{1}{5000}$
$(1 + 0.10 \cdot \Delta t)^{1/\Delta t}$	1.10000	1.10471	1.10512	1.10517	1.10517

Table 1.14 Values of $(1 + 0.10 \cdot \Delta t)^{1/\Delta t}$

If we use a calculator to compute the value of $e^{0.10}$ and compare this number with the values in Table 1.14, we observe that the values of $(1 + 0.10 \cdot \Delta t)^{1/\Delta t}$ for small values of Δt approach the value of $e^{0.10}$. Recall that 0.10 reflects the 10% annual interest rate. Suppose now the annual interest rate is 8%, so that $(1 + 0.08 \cdot \Delta t)^{1/\Delta t}$ is used to compute year-end balances. As Δt becomes very small, does this expression have a limiting value that is approximately $e^{0.08}$? Changing the interest rate r and comparing the value of $(1 + r \cdot \Delta t)^{1/\Delta t}$ to the value of e^r (where r represents the annual interest rate expressed as a decimal) should convince you that e^r is the limiting value of $(1 + r \cdot \Delta t)^{1/\Delta t}$ as Δt shrinks toward zero.

We can conclude that $100 \cdot e^{0.10}$ gives the account balance after one year of continuously compounded interest, assuming an initial deposit of $100 and an annual rate of 10%. This means that the initial balance is multiplied by $e^{0.10}$ to obtain the balance one year later. To find the balance after two years, assume that we begin with an initial amount of $100 \cdot e^{0.10}$ instead of 100. We multiply $100 \cdot e^{0.10}$ by $e^{0.10}$ to obtain $(100 \cdot e^{0.10}) \cdot e^{0.10}$, or $100 \cdot e^{0.10(2)}$. Similarly, we can show that the balance after three years is $100 \cdot e^{0.10(3)}$. We suspect that the function $A(t) = 100 \cdot e^{0.10t}$ is a closed form expression for the continuously compounded account balance at any time t years. Graphing the function $A(t) = 100 \cdot e^{0.10t}$ on the same axes as a scatter plot of the final column of balances in Table 1.12 confirms that this is a good model.

Table 1.12 shows how the frequency of compounding affects the rate at which the account balance grows. For instance, if the initial deposit of $100 earns 10% interest compounded annually, then the balance after one year is $110.00. However, if the interest is compounded monthly, then the balance after one year is $110.47. This means that 10% annual interest compounded monthly yields the same balance as 10.47% interest compounded annually. Another way of expressing this information is that the *effective annual yield* of 10% annual interest compounded monthly is 10.47%. If an initial deposit of $100 is earning 10% annual interest compounded continuously, then interest is credited to the account at each instant in time, and the balance after one year is about $110.52. Thus, the effective annual yield for an account with a 10% annual interest rate compounded continuously is approximately 10.52%.

effective annual yield

Example 1

A mosquito population, with an initial size of 1000, grows to 1200 in one week. Assuming that the weekly growth rate is proportional to the population at the beginning of the week, what will the population be if the same weekly growth rate continues for 10 weeks? What growth rate could be continuously compounded and have the same effect on the population?

Solution   The first question assumes that the population can be modeled discretely. The time interval between successive measurements of the population size is fixed at one week. An increase from 1000 to 1200 in one week corresponds to a weekly growth rate of 20% of the population, assuming growth is proportional to the size of the population. We can generate values for the size of the population using a method similar to the one we used to determine bank account balances. To calculate the size of the mosquito population at the end of week n, we increase the population from the end of week $n - 1$ by 0.2 times itself. We need to iterate $P_n = P_{n-1} + 0.2 \cdot P_{n-1}$ ten times, beginning with a population of $P_0 = 1000$ to determine the population after ten weeks. This gives us $P_{10} \approx 6200$, so the population reaches about 6200 mosquitoes in ten weeks.

The values generated by the recursive equation $P_n = P_{n-1} + 0.2 \cdot P_{n-1}$ can also be generated by the closed form expression $P(n) = 1000(1.2)^n$ where $n = 0, 1, 2, \ldots$ represents the number of weeks. Based on our previous work with the data in Table

1.12, if we express $P(n) = 1000(1.2)^n$ as an exponential function with base e, then the coefficient of n in the exponent will give the continuously compounded rate we seek. To find this rate, we need to solve the equation

$$1000(1.2)^n = 1000e^{kn}$$

for k. Dividing both sides of the equation by 1000 and then taking the nth root of both sides gives the equation

$$1.2 = e^k,$$

which can be solved by taking natural logarithms. We find that

$$k \approx 0.1823.$$

Thus, $1000(1.2)^n \approx 1000e^{0.1823n}$, which tells us that an 18.23% rate compounded continuously yields an effective rate of about 20%.

Example 2

The population of Newtonville is currently 50,000 and growing at a 2% annual rate due to births, a growth rate that is compounded continuously. This rate is applied at every instant of time, so the effective annual growth rate of the population exceeds 2% per year. In addition, people are moving to Newtonville from out of town at an average rate of 1000 per year. (This number is actually a net immigration rate calculated from the number of new arrivals per year minus the number of people moving away per year.) How will the population of Newtonville increase over the next ten years?

Solution   As with the bank account balances, we generate a sequence of population values by computing a new value of P_n after each interval of length Δt using the equation

$$P_n = P_{n-1} + (0.02 \cdot P_{n-1} + 1000) \cdot \Delta t. \tag{11}$$

Equation (11) states that in each time interval Δt, the existing population grows by an amount that is given by a proportion Δt of the 2% growth rate. In addition, we also increment the population by a proportion Δt of 1000, the yearly immigration rate. We do this because we assume that the immigrants arrive at a constant rate throughout the year; thus, $1000 \cdot \Delta t$ of them arrive during each time interval of length Δt.

Table 1.15 gives the values of the population generated by equation (11) using various values of Δt. The table includes only the values at the end of each year for the next ten years. All populations are rounded to the nearest whole number. We can make Δt as small as we like because we are using a continuous model. Notice that the values of the population appear to be approaching a set of limiting values as Δt gets closer to zero; however, finding a closed form expression for the limiting values of the population is difficult and will be omitted for the present. ∎

t_n (in years)	$\Delta t = 1$	$\Delta t = 0.5$	$\Delta t = 0.1$	$\Delta t = 0.01$	$\Delta t = 0.001$
0	50,000	50,000	50,000	50,000	50,000
1	52,000	52,010	52,018	52,020	52,020
2	54,040	54,060	54,077	54,081	54,081
3	56,121	56,152	56,177	56,183	56,184
4	58,243	58,286	58,320	58,328	58,329
5	60,408	60,462	60,506	60,516	60,517
6	62,616	62,683	62,736	62,748	62,750
7	64,869	64,947	65,011	65,026	65,027
8	67,166	67,258	67,332	67,349	67,351
9	69,509	69,615	69,700	69,720	69,722
10	71,899	72,019	72,116	72,138	72,140

Table 1.15 Population estimates P_n for various Δt

We have learned in this chapter that if we know the rate of change of a phenomenon and initial values of the independent variable x and the dependent variable y, then we can learn a lot about the phenomenon. We can generate values using the following equations:

$$x_n = x_{n-1} + \Delta x$$
$$y_n = y_{n-1} + (\text{rate of change}) \cdot \Delta x.$$

These values are exact if they are being used to model a phenomenon that changes discretely, whereas these values are approximate if the phenomenon changes continuously.

Given an expression for rate of change, we can generate ordered pairs numerically and visualize these values graphically. Sometimes we can recognize a function that models the ordered pairs we have generated. For example, we have learned that constant rate of change generates a linear set of values and that proportional rate of change generates an exponential set of values. In general, finding a function with a given rate of change is a difficult problem. The inverse problem, finding a rate-of-change expression from an algebraic expression, is not so difficult. That is where we begin the next chapter.

Exercise Set 1.3

1. Write a paragraph explaining what is happening to the account balances in Table 1.12 as Δt decreases. Consider the following question: What values would you expect if you added a column in Table 1.12 for $\Delta t = \frac{1}{10,000}$? Include an explanation of your understanding of continuous compounding and continuous change.

2. What is happening to the estimates of the population of Newtonville given in Table 1.15 as Δt decreases? What values would you expect if you added a column in Table 1.15 for $\Delta t = 0.0001$?

3. A savings account earns 4% annual interest, compounded continuously, with an initial deposit of $200.

    a. What is the effective annual yield of this account?

    b. Write an equation for a curve that fits a scatter plot of the account balances over the next ten years.

4. How large does the annual interest rate r need to be so that the effective annual yield for an account with continuous compounding is 1% more than the yield with annual compounding? 5% more? 10% more?

5. Use techniques of data analysis to fit an exponential function with base e to the continuous growth data from the rightmost column of Table 1.12. How does the coefficient in the exponent relate to the annual interest rate?

6. An account with an initial deposit of $100 earns 12% annual interest, compounded monthly.

    a. What is the effective annual yield of this account?

    b. How long does it take for the account balance to double?

    c. How much will be in the account in 10 years?

    d. What annual rate compounded continuously has the same effective annual yield?

7. Which savings account would you rather have, one that earns 6% annual interest compounded annually or one that earns 5.9% annual interest compounded continuously? Justify your answer.

8. The number of atoms in a radioactive substance decreases at a continuous rate of 0.693% per year. How long will it be before half of the atoms are gone?

9. Suppose that due to inflation your money is losing value at a continuously compounded rate of 0.5% per month. How much will $100 be worth in 1 year? in 2 years? in 5 years?

10. A person is given a drug continuously through an intravenous solution at a rate of 100 mg per hour. The body eliminates the drug from the bloodstream at a continuous rate of 10% of the remaining drug per hour.

    a. Write a recursive equation to generate values for B_n, which is the level of drug in the bloodstream at time t_n.

    b. The person is given an initial dose of 200 mg, which you may assume is immediately absorbed into the bloodstream. Use the recursive equation from part a to generate a table of values for the level of drug in the bloodstream over the first 10 hours with a step size of $\Delta t = 1$ hour. Repeat the calculations with $\Delta t = 0.1$ and $\Delta t = 0.01$.

c. Obtain a graph of the limiting curve for the values in part b, with the level of drug in the bloodstream plotted versus time.

11. The population of Batesville has a continuously compounded growth rate of 3% of the current population per year. Due to its pleasant environment, Batesville averages 25 immigrants each year. The present population is 10,000.

a. Write a recursive system to generate values for P_n, the population at time t_n.

b. Use the recursive system from part a to generate a table of values for the population over the next 10 years with a step size of $\Delta t = 1$ year. Repeat the calculations with $\Delta t = 0.1$ and $\Delta t = 0.01$.

c. Obtain a graph of the limiting curve for the values in part b, with the population plotted versus time.

12. A butterfly population on an island declines at a continuous rate of 20% of the current population per year. To keep the population from dying out, 50 butterflies per year are brought to the island and released throughout the year.

a. Write a recursive equation to generate values for P_n, the population at time t_n.

b. Assuming the initial population is 500, use the recursive equation from part a to generate a table of values for the population over the next 20 years with a step size of $\Delta t = 1$ year. Repeat the calculations with $\Delta t = 0.1$ and $\Delta t = 0.01$.

c. Obtain a graph of the limiting curve for the values in part b, with the population plotted versus time. What is the long-term trend in the population?

d. Suppose a severe winter reduces the butterfly population to 100. What will happen to the population in subsequent years assuming that the 20% decline continues and 50 new butterflies are introduced each year?

e. What appears to be the relationship between the initial population and the long-term trend in the population?

CHAPTER 2

All About the Derivative

2.1 Investigating Average Rates of Change

In Chapter 1 we saw that knowing the rate of change of a function and an initial value enables us to generate values of the function. In some situations it was easy to recognize the function we created—for instance, a constant rate of change created a linear function. In other instances, however, it was not easy to recognize the function whose values we had created. To gain more insight about the relationship between functions and their corresponding rates of change, it is helpful to know how to quantify the rate of change of a known function. For example, suppose we learn that a certain type of function f has a rate-of-change function g. Then whenever we find a phenomenon whose rate of change at a point x is given by $g(x)$, we will be able to conclude that the phenomenon can be modeled by a function similar to f.

One rate of change that we are familiar with is velocity, which is the rate of change of position with respect to time. We find numerical values for velocity by dividing the change in position between two points by the change in time between the same two points. In general, if y is a function of x, we find a numerical value for the rate of change of y with respect to x from the point (x_i, y_i) to the point (x_j, y_j) by dividing the difference between the two y values by the difference between the two x values, as in the expression

$$\frac{y_j - y_i}{x_j - x_i}.$$

This expression is an example of a ***difference quotient***. In Chapter 1 we used the uppercase Greek letter *delta*, specifically Δt, to symbolize the length of a time interval.

difference quotient

We will now expand our use of Δ to symbolize any change. Using *delta* notation, a difference quotient that gives the rate of change of y with respect to x is written $\frac{\Delta y}{\Delta x}$.

average rate of change

A difference quotient represents an ***average rate of change*** over an interval determined by two points. For example, suppose Lee drives the 150 miles from Durham to Charlotte in 2.5 hours. Lee's average speed for this trip is 60 miles per hour. He actually may have been driving at 60 mph for very little of the trip; however, despite variations in speed, he arrived in Charlotte in the amount of time the trip would have taken if he had driven at precisely 60 mph for the entire trip.

The function $d = f(t)$ shown in Figure 2.1 represents Lee's distance from Durham as a function of elapsed time t for part of his trip to Charlotte. The difference quotient between points A and B on the graph is the ratio $\frac{\Delta d}{\Delta t}$, where the quantities Δd and Δt are labeled on the graph. The ratio $\frac{\Delta d}{\Delta t}$ is the change in d values $(d_2 - d_1)$ divided by the change in t values $(t_2 - t_1)$ from A to B. This ratio is the same as the slope of the line

secant line

through A and B. A line that contains two points on a graph is a ***secant line***, meaning that a difference quotient represents the slope of a secant line.

In Figure 2.1, point A has coordinates $(1, 55)$ and point B has coordinates $(2, 120)$. The difference quotient representing the average rate of change between A and B is $\frac{120-55}{2-1}$, or simply 65. This means that Lee's average speed for the hour between A and B is 65 mph. If Lee actually had driven at a constant rate of 65 mph for that entire hour, then the rate of change of his distance with respect to time would be constant. In that case, the graph from A to B would be a straight line segment with slope 65 mph.

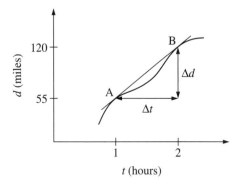

Figure 2.1 Lee's distance from Durham versus time

Example 1

Suppose we shoot a projectile into the air so that the function $h(t) = -10t^2 + 100t$ approximates the relationship between height in meters and time in seconds.

Calculate $\frac{\Delta h(t)}{\Delta t}$ between $t = 1$ and $t = 2$, between $t = 4$ and $t = 6$, and between $t = 7$ and $t = 8$. Explain what these difference quotients represent.

Solution   The graph in Figure 2.2 shows the height of the projectile versus time.

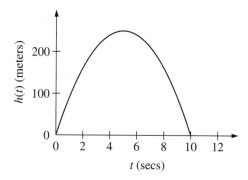

Figure 2.2 Height of projectile versus time

When t changes from 1 to 2, $h(t)$ changes from 90 to 160, so the difference quotient between $t = 1$ and $t = 2$ has the value

$$\frac{\Delta h(t)}{\Delta t} = \frac{160 - 90}{2 - 1} = 70 \text{ m/sec.}$$

It is common to drop the dependent variable from the numerator of a difference quotient, so we will write $\frac{\Delta h}{\Delta t}$ rather than $\frac{\Delta h(t)}{\Delta t}$.

When $t = 4$ and when $t = 6$, the h values are both 240, so the difference quotient between $t = 4$ and $t = 6$ has the value

$$\frac{\Delta h}{\Delta t} = \frac{240 - 240}{6 - 4} = 0 \text{ m/sec.}$$

When t changes from 7 to 8, $h(t)$ changes from 210 to 160, so the difference quotient between $t = 7$ and $t = 8$ has the value

$$\frac{\Delta h}{\Delta t} = \frac{160 - 210}{8 - 7} = -50 \text{ m/sec.}$$

Each value of $\frac{\Delta h}{\Delta t}$ represents the average rate at which the height changes over an interval of time. Each value of $\frac{\Delta h}{\Delta t}$ also represents the slope of a secant line through two points on the graph of $h(t) = -10^2 + 100t$. These three secant lines are shown in Figure 2.3. ∎

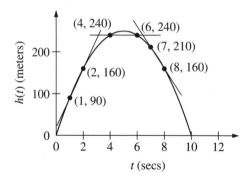

Figure 2.3 Height versus time and three secant lines

For the quadratic function that describes the height of the projectile in the previous example, the average rate of change of the height with respect to time varies from interval to interval. When the height is increasing and the projectile is going up, the average rate of change is positive. When the height is decreasing and the projectile is coming down, the average rate of change is negative. When the height has a net change of zero from the beginning to the end of an interval, then the average rate of change is zero over that interval.

Example 2

Do the difference quotients for the linear function with equation $y = 2.5x - 6.25$ exhibit a particular pattern?

Solution   Table 2.4 lists the average rate of change of y with respect to x over four successive intervals. In this table, Δx is the difference between successive x values, which we can calculate as $\Delta x = x_i - x_{i-1}$. Likewise, we can compute $\Delta y = y_i - y_{i-1}$. In the tables for Examples 2 through 4, we will use equally spaced intervals to make it clear how the average rate of change varies.

x_i	y_i	Δx	Δy	$\frac{\Delta y}{\Delta x}$
−2	−11.25			
		1	2.5	2.5
−1	−8.75			
		1	2.5	2.5
0	−6.25			
		1	2.5	2.5
1	−3.75			
		1	2.5	2.5
2	−1.25			

Table 2.4 Average rates of change of $y = 2.5x - 6.25$

As we expect from our work in Chapter 1, all of the average rates of change in Table 2.4 are equal to 2.5, which is the slope of the line with equation $y = 2.5x - 6.25$. ∎

Example 3

What is the pattern of the difference quotients for the quadratic function with equation $y = x^2$?

Solution   Table 2.5 shows difference quotients for some successive intervals near zero with $\Delta x = 0.5$. In contrast to a linear function, the average rate-of-change values for a quadratic function are not constant, as also observed in Example 1. Holding Δx constant allows us to see a pattern in the difference quotients. The values of $\frac{\Delta y}{\Delta x}$ increase by 1 unit for each change of 0.5 units in the value of x. We will explore this pattern further in Lab 1. ∎

x_i	y_i	Δx	Δy	$\frac{\Delta y}{\Delta x}$
−1.5	2.25			
		0.5	−1.25	−2.5
−1	1			
		0.5	−0.75	−1.5
−0.5	0.25			
		0.5	−0.25	−0.5
0	0			
		0.5	0.25	0.5
0.5	0.25			
		0.5	0.75	1.5
1	1			
		0.5	1.25	2.5
1.5	2.25			

Table 2.5 Average rates of change of $y = x^2$

Example 4

What is the pattern of the difference quotients for the exponential function with equation $y = 2^x$?

Solution   Table 2.6 gives difference quotients for the exponential function with equation $y = 2^x$ for values of x near zero with $\Delta x = 2$. We can observe the following pattern in the average rates of change for this function: the difference quotients are multiplied by 4 for each increase of 2 in the x values. ∎

x_i	y_i	Δx	Δy	$\dfrac{\Delta y}{\Delta x}$
0	1			
		2	3	1.5
2	4			
		2	12	6
4	16			
		2	48	24
6	64			
		2	192	96
8	256			

Table 2.6 Average rates of change of $y = 2^x$

Exercise Set 2.1

1. Construct two tables similar to Table 2.4 to investigate the rate of change of the function $y = 2.5x - 6.25$. In one table use $\Delta x = 0.5$, and in the other use $\Delta x = 2$. How does the size of Δx appear to influence the average rates of change?

2. Construct two tables similar to Table 2.5 to investigate the rate of change of the function $y = x^2$. In one table use $\Delta x = 1$ and in the other table use $\Delta x = 2$. How does the size of Δx appear to influence the average rates of change?

3. Construct two tables similar to Table 2.6 to investigate the rate of change of the function $y = 2^x$. In one table use $\Delta x = 0.5$ and in the other table use $\Delta x = 1$. How does the size of Δx appear to influence the average rates of change?

4. On the same axes, graph the function $y = 2^x$ and some of the secant lines whose slopes are represented by $\dfrac{\Delta y}{\Delta x}$ in Table 2.6. Write a few sentences to describe how the secant slopes appear to be changing.

5. A particle is moving so that its distance d, measured from its original position, depends on time t according to the equation $d = t(10 - t)$. Find the average rate of change of distance with respect to time (that is, the average velocity) for each time interval.

   a. $t = 1$ to $t = 3$ b. $t = 1$ to $t = 2$ c. $t = 1$ to $t = 1.5$

   d. $t = 1$ to $t = 1.05$ e. $t = 1$ to $t = 1.001$

6. In Chapter 1 we used the recursive system

$$x_n = x_{n-1} + \Delta x$$
$$y_n = y_{n-1} + (\text{rate of change}) \cdot \Delta x$$

   to generate x and y values from a known expression for rate of change. Write a paragraph to discuss how these equations are related to the average rate of change given by

$$\frac{\Delta y}{\Delta x} = \frac{y_n - y_{n-1}}{x_n - x_{n-1}}.$$

2.2 Investigating Instantaneous Rates of Change

Instantaneous Rate of Change

A difference quotient gives an average rate of change over an interval. An *instantaneous rate of change* represents the rate at which a quantity is changing at an instant of time. A difference quotient calculated over a small interval usually gives a good approximation for an instantaneous rate of change. In Problem 5 of Exercise Set 2.1 we determined the average velocity of a particle over the interval from $t = 1$ to $t = 1.001$. The value of $\frac{\Delta d}{\Delta t}$ between $t = 1$ and $t = 1.001$ gives a good approximation of the particle's instantaneous rate of change, or instantaneous velocity, at $t = 1$. In general, as Δt decreases, an average rate of change becomes a better approximation of an instantaneous rate of change.

instantaneous rate of change

If we imagine driving a car and watching the speedometer, the odometer, and a clock, we can get a good sense of the distinction between average and instantaneous rates of change. To determine an average velocity (that is, an average rate of change of distance with respect to time), we need to make two odometer readings and two clock readings. Suppose at 1:00 PM the odometer reads 30520, and at 2:00 PM the odometer reads 30575. The average velocity over this time interval is equal to

$$\frac{30575 - 30520}{2 - 1} \text{ mph,}$$

or 55 mph. To determine the instantaneous velocity at any particular time, we need to make one speedometer reading and one clock reading. For example, if the speedometer reads 58 mph at 1:35 PM , then the instantaneous velocity of the car is 58 mph at 1:35 PM. If we did not have a speedometer, we could estimate the instantaneous velocity of the car at a particular time by using two odometer readings a short time apart and forming a difference quotient

$$\frac{\text{change in distance}}{\text{change in time}}.$$

Graphing Rate-of-Change Functions

Tien drove his car to school on Friday. The graph in Figure 2.7 shows the relationship between his distance from home d and time t. The distance is zero until he leaves home at 7:00 and increases as he drives toward school. The distance from home remains constant while Tien is at school. (To keep the graph at a reasonable size, we have removed part of this section. The jagged lines in the graph and t axis are used to indicate the removal.) The distance begins decreasing at 3:00 when he leaves school to drive home. Given this graph for distance from home versus time, how can we graph the instantaneous rate of change of distance from home with respect to time? In other words, we want to use information from the graph that shows ordered pairs (time, distance) to create a graph of ordered pairs (time, instantaneous rate of change of distance). To sketch our graph we estimate instantaneous rate of change values by using average rate of change values over small intervals.

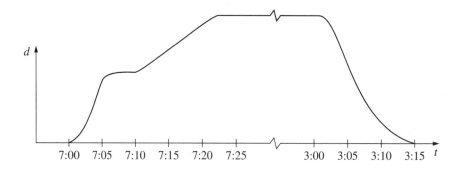

Figure 2.7 Tien's distance from home versus time

From 7:00 to 7:05, Tien's distance from home increases with time, meaning that $\frac{\Delta d}{\Delta t}$ is positive over any interval of time between 7:00 and 7:05. Recall that $\frac{\Delta d}{\Delta t}$ represents Tien's average velocity over a time interval of length Δt. A secant line (whose slope is an average velocity) drawn through the points on the curve with $t = 7{:}04$ and $t = 7{:}05$ is steeper than a secant drawn through the points with $t = 7{:}01$ and $t = 7{:}02$. This increase in steepness indicates that Tien's velocity increases between 7:00 and 7:05, and he appears to achieve a maximum velocity at 7:05. These facts about Tien's velocity are represented graphically in Figure 2.8. Notice in the graph that we have estimated the instantaneous rate of change for every point in the interval from $t = 7{:}00$ to $t = 7{:}05$.

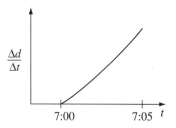

Figure 2.8 Tien's instantaneous rate of change from 7:00 to 7:05

After 7:05, Tien's distance from home is still increasing, but it is not increasing as quickly as it was before; therefore, Tien's velocity decreases from 7:05 to 7:10. A secant line drawn between 7:07 and 7:08 is less steep than a secant drawn between 7:04 and 7:05. From about 7:08 to 7:10, Tien's distance from home does not change. Since his distance remains constant over this time interval, his rate of change is zero. These facts about Tien's instantaneous rate of change are represented graphically in Figure 2.9. As with the graph in Figure 2.8, we have estimated the instantaneous rate of change for every point in the interval from $t = 7:05$ to $t = 7:10$.

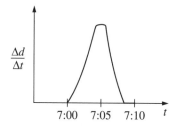

Figure 2.9 Tien's instantaneous rate of change from 7:00 to 7:10

At about 7:10 Tien's distance from home increases again. From 7:10 to 7:20 the graph in Figure 2.7 is linear. Tien's distance changes at a constant rate, which yields a constant value for $\frac{\Delta d}{\Delta t}$. The leveling off of the graph in Figure 2.7 between 7:20 and 7:25 corresponds to Tien slowing down to arrive at school. These facts about Tien's instantaneous rate of change are represented graphically in Figure 2.10.

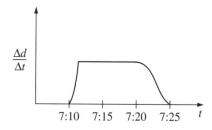

Figure 2.10 Tien's instantaneous rate of change from 7:10 to 7:25

Between the time when Tien arrives at school and the time when he leaves school his distance from home remains constant. His rate of change between 7:25 and 3:00 therefore remains zero. When Tien leaves school and drives toward home, his distance from home decreases, which means that $\frac{\Delta d}{\Delta t}$ has negative values over any interval of time between 3:00 and 3:15. Secant lines over small time intervals on the graph in Figure 2.7 have the steepest negative slope at about 3:05. At about this time the instantaneous rate of change achieves its most negative value. Between 3:10 and 3:15, Tien's distance from home changes very gradually, as indicated by the leveling off of the curve in Figure 2.7. This means that his instantaneous rate of change is approaching zero as time approaches 3:15. These facts about Tien's rate of change are represented graphically in Figure 2.11.

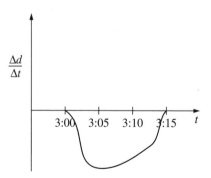

Figure 2.11 Tien's instantaneous rate of change from 3:00 to 3:15

The graphs in Figures 2.9, 2.10, and 2.11 are consolidated in Figure 2.12, thus showing a graph of the instantaneous rate-of-change function.

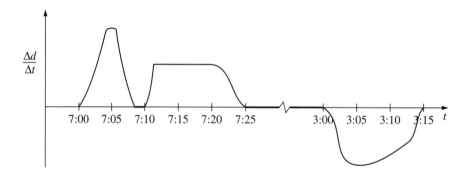

Figure 2.12 Tien's instantaneous rate of change from 7:00 to 3:15

What generalizations can we infer from the previous analysis? When distance is increasing over time, the rate of change is positive. When distance is decreasing over time, the rate of change is negative. When the distance versus time graph is relatively flat (distance is not changing or is changing gradually), the rate of change is zero or close to zero. When the distance is changing at a constant rate (the distance versus time graph is linear), the rate of change has a constant value. Notice also that where the secants to the graph are steep, the rate of change is greater in magnitude.

Notice that we did not indicate a vertical scale on the graph of distance versus time in Figure 2.7 or on any of the corresponding instantaneous rate-of-change graphs. The graphs are approximate representations of the given situation and are not meant to be interpreted exactly. At this time we are interested primarily in the shapes of the graphs, and in particular in how the shapes of the graphs of a function and of its instantaneous rate of change are related. We will use graphical representations of rates of change frequently throughout this course. The following exercises provide some initial practice with graphing rates of change. We will return to these concepts in successive sections.

Exercise Set 2.2.A

1. For each function f graphed below, sketch a graph of the instantaneous rate of change of f.

   a.

   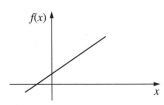

   b.

   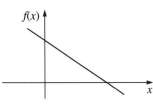

   c.

   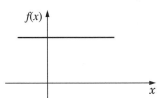

   d.

   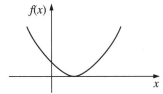

   e.

   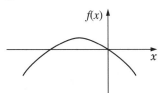

   f.

   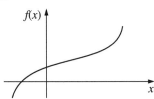

2. In each situation described, draw graphs to represent how the quantity and the rate of change depend on time. Indicating scales on the axes is not necessary; only the general shape of the graphs is required in these exercises.

   a. When an egg is dropped from the roof of a tall building, both its velocity and its height depend on time. Sketch height and velocity graphs.

   b. If you leave school in your car to drive home and you never turn around or drive backwards, your distance from school never decreases. Sketch graphs of your distance from school versus time and the instantaneous rate of change of your distance from school versus time.

   c. Three students leave for the mall. One student walks there, and one rides a bike. The third student leaves late, runs to the bus stop, and catches a bus to ride to the mall. Sketch position and velocity graphs for each student.

d. Sketch a graph of a typical person's height from age 0 to age 30. Sketch a companion graph to show the instantaneous rate of growth.

e. A parachutist jumps from a hovering helicopter. Sketch a graph that shows the parachutist's distance from the ground as a function of time. Sketch a graph of the corresponding velocity as a function of time.

f. Suppose a rumor starts and spreads through the school community. Sketch a graph of the number of students who have heard the rumor as a function of time. Sketch a graph of the instantaneous rate of change of the spread of the rumor.

g. When a cup of hot coffee sits in a room, its temperature decreases until it eventually reaches room temperature. Sketch a graph to show the relationship between temperature and time. Sketch a graph of the instantaneous rate of change of temperature with respect to time.

h. Suppose you take a bath in a bathtub. Sketch a graph that shows the water level in the tub from the time you first enter the bathroom until you have completed your bath. Sketch a graph of the instantaneous rate of change of the water level versus time.

3. For each function f, sketch a graph of the instantaneous rate of change of f.

a.

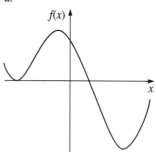

b.

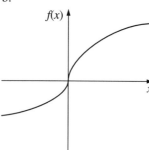

c.

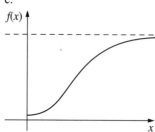

d.

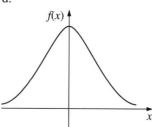

e.

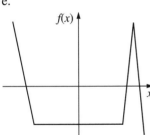

f.

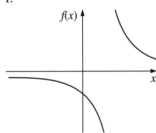

4.   Sketch graphs of the following functions:

$$x^3, \ \sin x, \ \cos x, \ \tan x, \ e^x, \ 2^x, \ \frac{1}{x}, \ \frac{1}{x^2}, \ \ln x, \ \sqrt{x}.$$

For each function, sketch a graph of the instantaneous rate-of-change function. Concentrate on the overall shapes of the graphs rather than the actual function values.

5.   Describe the special characteristics or values of a graph that you identify when you are asked to sketch its instantaneous rate-of-change graph. For each characteristic or value, explain the information it gives about the instantaneous rate of change.

6.   Each graph represents the instantaneous rate of change with respect to time of some unknown function. Use the given information about the rate of change to sketch a graph of the function. Answers are not unique.

a.

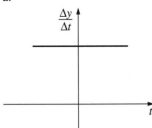

b.

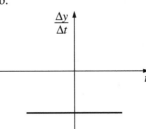

c.

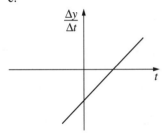

d.

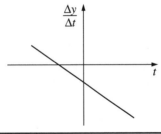

In Lab 1 we will investigate the rate of change of certain elementary functions by looking at difference quotients. The average rate of change of the function with equation $y = f(x)$ over the interval from x to $x + \Delta x$ is

$$\frac{\text{change in } y \text{ values}}{\text{change in } x \text{ values}} = \frac{\Delta y}{\Delta x} = \frac{f(x + \Delta x) - f(x)}{\Delta x}.$$

If Δx is close to zero, then this average rate of change is close to an instantaneous rate of change. In the language of calculus, a function whose values represent instantaneous rates of change is called a **derivative**.

derivative

Lab 1 *Derivatives of Functions*

In this lab we will investigate derivatives of some of the elementary functions that you may have studied in previous courses. We will approximate the derivative using difference quotients over small intervals. Then, we will graph these difference quotients against the x values to create an average rate-of-change function, which gives a good approximation for the instantaneous rate-of-change function, or the derivative, if Δx is sufficiently small.

1. Graph the function $f(x) = x^2$ over the domain $[-3, 3]$. Use a computer or graphing calculator to calculate the difference quotients over this domain. Let Δx, the difference between successive x values in the difference quotients, equal 0.05.

    a. An approximate graph of the derivative can be generated by plotting the difference quotients versus x, which are the ordered pairs $(x, \frac{f(x+\Delta x)-f(x)}{\Delta x})$. Examine the graphs of f and the approximate derivative of f over the interval.

    b. Guess a formula for the derivative. Check your guess by graphing it on the same axes as the difference quotients.

2. Adjusting the domain as necessary, repeat 1.a and 1.b for each of the following functions:

$$x^3, \quad \sin x, \quad \cos x, \quad \tan x, \quad e^x, \quad 2^x, \quad \frac{1}{x}, \quad \frac{1}{x^2}, \quad \ln x, \quad \sqrt{x}.$$

For each of these elementary functions, sketch a graph of the original function and of the approximate derivative of that function. The domain of the trigonometric functions must be in radians, so be sure that your computer or graphing calculator is set for radians. Include scales and coordinate axes for each graph. Make conjectures about equations for the derivative functions and check your conjectures. You may not be able to come up with the exact equations, but the graph of your guess should have the same basic shape as the graph of the difference quotients.

3. List any patterns you see in your guesses for the derivatives. Can you make any generalizations? For instance, what do the derivatives for the three trigonometric functions have in common? What do the derivatives for the even functions have in common?

4. For $f(x) = x^2$ and domain $[-3, 3]$, use a computer or graphing calculator to plot difference quotients for various values of Δx. Try $\Delta x = 1, 0.5, 0.1$, and 0.001. What effect does the value of Δx have on your guess for the derivative? Why does the plot of difference quotients change as Δx decreases? As Δx approaches zero, the graph of the difference quotients approaches a fixed graph. Can you identify the equation of this fixed graph? Try this same investigation with $f(x) = \ln x$ and an appropriate domain.

Summarize your results, interpretations, and conclusions in a written report.

You may have noticed in Lab 1 that derivatives of elementary functions all appear to be elementary functions. Every function we investigated in the lab has a derivative that we recognize, or at least can come close to guessing. The knowledge gained in the lab suggests that working with derivatives does not necessarily require creating new functions, but instead requires working with familiar functions in new ways.

Recall that our broad goal, set forth in Chapter 1, is to find a function that possesses a particular rate of change. For instance, suppose we have some data that show the flow rate of water in gallons per minute out of a water storage tank. Suppose also that we find a function f that models that data, giving us the derivative of the number of gallons at time t. A different function g, which has a derivative equal to f, would then model the total amount of water taken from the tank at time t. To succeed at the task of finding g once we know f, we must be familiar with the relationships between functions and their associated derivatives.

Exercise Set 2.2.B

1. Write a formula for the derivative for each of the functions we investigated in Lab 1, namely:

$$x^2, x^3, \sin x, \cos x, \tan x, e^x, 2^x, \frac{1}{x}, \frac{1}{x^2}, \ln x, \sqrt{x}.$$

2. Sketch a graph of each of the following functions. Next to each graph, sketch a graph of an approximate derivative function.

    a. $y = \sqrt[3]{x}$ b. $y = x^{2/3}$

    c. $y = -x^2$ d. $y = \sec x$

3. Describe what the graph of a function looks like near an x value at which the value of the derivative

   a. is zero;

   b. has a maximum or a minimum;

   c. has a vertical asymptote.

4. Sketch the graph of a function for which the derivative is

   a. constant;   b. linear;

   c. always positive;   d. always negative;

   e. positive and decreasing.

5. An object falls freely after being dropped from a fourth floor window. Assume that acceleration due to gravity is 9.8 m/sec². (Acceleration is the derivative of velocity with respect to time and has units of m/sec per sec, written m/sec².)

   a. Describe how the velocity of the object changes over time.

   b. Use your result from part a to describe how the position of the object changes over time.

2.3 Secants, Tangents, and Local Linearity

learning curve

A student is taking a keyboarding class as part of the computer science curriculum at school. The curve sketched in Figure 2.13 is a model for the number of words per minute N that the student can type as a function of the number of days t in the class. This curve represents a smoothing of the actual data for the student's typing speed. Initially, proficiency is low, and typing speed increases relatively quickly. After a while, progress does not come so easily, and more time is required to achieve increases in typing speed. In the long run, the curve levels off at the upper limit of the student's typing speed. This graph is an example of a *learning curve*.

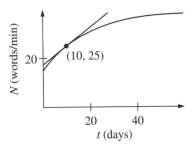

Figure 2.13 Typing speed versus time spent in the class

The point (10, 25) is on this curve, which implies that after 10 days in the class, the student can type 25 words per minute. In Figure 2.13, the line drawn through the point on the curve with coordinates (10, 25) is called a ***tangent line***. This line has two important characteristics: (1) it has the same N value as the learning curve function at $t = 10$; and, (2) it has the same instantaneous rate of change, or derivative, as the learning curve function at $t = 10$. The derivative of this tangent line (that is, its instantaneous rate of change) is simply its slope, which means that the slope of the tangent line is equal to the derivative of the learning curve function at $t = 10$.

tangent line

Near $t = 10$, the shape of the learning curve is approximated by the tangent line. This illustrates an important principle of calculus known as ***local linearity***. A locally linear function, no matter how curved, is one that can be approximated around a point by a tangent line. The word *local* is used to indicate that the tangent line approximation is good only over a limited interval of values of the independent variable. One way to think about local linearity is to imagine zooming in on a graph. If we graph a function on a computer or calculator and zoom in a sufficient amount, then the graph will usually look like a line. (Exceptions to this generalization will be explored in Lab 2.) A sequence of zooms on the learning curve and its tangent line at $t = 10$ is shown in Figure 2.14. Observe that as we zoom in on the curve around the point of tangency, the curve becomes indistinguishable from the tangent line. Hence we use the term local linearity. We will utilize the concept of local linearity many times throughout this course since it forms the basis for much of what we do in calculus. Often, we will want to determine the equation of a tangent line which we can use to approximate values of a function near the point of tangency.

local linearity

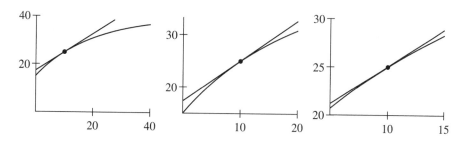

Figure 2.14 Zooms on the learning curve around (10, 25)

Calculating the slope of a line tangent to a curve is sometimes a complex task. If we know a formula for the derivative of the function represented by the curve, then the slope of the tangent line is the value of the derivative at the point of tangency. If we do not know a derivative formula, then we can use slopes of secant lines to approximate the slope of the tangent line. Example 1 illustrates both methods.

Example 1

Investigate the slopes of secant lines drawn near $x = 2$ on the graph of the function with equation $f(x) = \frac{1}{x}$. Determine the slope of the tangent line at the point on the graph with $x = 2$, using the slopes of the secant lines. Then find the slope of the tangent line at $x = 2$ using the derivative of $f(x) = \frac{1}{x}$ that you found in Lab 1. Write the equation of the tangent line at $x = 2$.

Solution   We first look at the secant line drawn through the points on the graph with $x = 2$ and $x = 4$, which is shown in Figure 2.15. Since $f(2) = 0.5$ and $f(4) = 0.25$, the secant line is drawn through the points $(2, 0.5)$ and $(4, 0.25)$. The slope of the secant is $\frac{0.25 - 0.5}{4 - 2}$, or -0.125. The value of Δx, which represents the difference between the x values where the secant line intersects the graph, is 2.

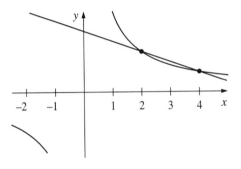

Figure 2.15 Secant line containing the points with $x = 2$ and $x = 4$ on $f(x) = \frac{1}{x}$

If we fix one of the points of intersection at $(2, 0.5)$ and decrease Δx to 1, then the secant line contains the points $(2, 0.5)$ and $(3, 0.3333)$, as shown in Figure 2.16. The slope of this line is approximately -0.1667.

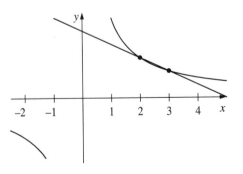

Figure 2.16 Secant line containing the points with $x = 2$ and $x = 3$ on $f(x) = \frac{1}{x}$

If we shrink Δx even more, we obtain the slopes of secant lines shown in Table 2.17. The slopes of the secant lines appear to approach –0.25 as the points of intersection of the secant and the curve move closer together.

Δx	Endpoints of secant line	Slope of secant line
0.5	(2, 0.5) and (2.5, 0.4)	–0.2000
0.1	(2, 0.5) and (2.1, 0.4762)	–0.2381
0.01	(2, 0.5) and (2.01, 0.4975)	–0.2488
0.001	(2, 0.5) and (2.001, 0.4998)	–0.2499

Table 2.17 Slopes of secant lines with decreasing Δx

We also calculate the slopes of the secant lines using negative values for Δx and keeping one of the points of intersection fixed at $x = 2$. The slopes of these secant lines with Δx approaching zero are shown in Table 2.18. The slopes of the secants appear to approach –0.25 for negative values of Δx as well.

Δx	Endpoints of secant line	Slope of secant line
–0.5	(2, 0.5) and (1.5, 0.6667)	–0.3333
–0.1	(2, 0.5) and (1.9, 0.5263)	–0.2632
–0.01	(2, 0.5) and (1.99, 0.5025)	–0.2513
–0.001	(2, 0.5) and (1.999, 0.5003)	–0.2501

Table 2.18 Slopes of secant lines with negative values of Δx

As the two points on the graph that determine the secant line become closer together, the secant line becomes very similar to the tangent line. Fixing one of the points of intersection at $x = 2$ causes the slopes of the secants to approach the slope of the tangent at the point (2, 0.5). The slope of the line tangent to $f(x) = \frac{1}{x}$ at the point (2, 0.5) appears to be –0.25. In Lab 1 you may have found that the derivative of $\frac{1}{x}$ is $-\frac{1}{x^2}$. The value of this derivative at $x = 2$ is –0.25, which is the slope of the line tangent to the graph of $f(x) = \frac{1}{x}$ at $x = 2$. Since this tangent line contains the point (2, 0.5) and has slope –0.25, the equation of this tangent line is

$$y - 0.5 = -0.25(x - 2)$$

or

$$y = -0.25x + 1.$$

A graph of $f(x) = \frac{1}{x}$ and this tangent line is shown in Figure 2.19. ∎

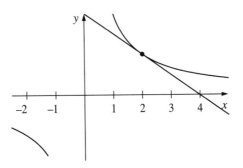

Figure 2.19 Graph of $f(x) = \frac{1}{x}$ and tangent line at $(2, 0.5)$

We can generalize the process used to find the slope of the line tangent to $f(x) = \frac{1}{x}$ at $x = 2$. The slope of the secant line through $(2, f(2))$ and $(2 + \Delta x, f(2 + \Delta x))$ can be expressed as

$$\frac{\text{change in } f(x) \text{ values}}{\text{change in } x \text{ values}} = \frac{f(2 + \Delta x) - f(2)}{(2 + \Delta x) - 2}.$$

Since we are interested in the slope of the tangent line, we want to know what happens to these secant slopes as Δx approaches zero. Do the secant slopes approach some limiting value as Δx gets closer to zero? If so, what value do they approach? A traditional notation for this limiting value is, in this case,

$$\lim_{\Delta x \to 0} \frac{f(2 + \Delta x) - f(2)}{(2 + \Delta x) - 2}, \tag{1}$$

which is read "the limit as Δx approaches zero of the difference quotient." The expression in (1) represents the value that the ratio $\frac{f(2+\Delta x)-f(2)}{(2+\Delta x)-2}$ approaches as Δx gets closer and closer to zero. In Example 1, we found that with $f(x) = \frac{1}{x}$,

$$\lim_{\Delta x \to 0} \frac{f(2 + \Delta x) - f(2)}{(2 + \Delta x) - 2} = -0.25.$$

We also found that the value of this limit is equal to the value of the derivative of $f(x) = \frac{1}{x}$ evaluated at $x = 2$.

Example 2

A particle is moving in such a way that its distance d in centimeters from its original position (the origin) at time t seconds is given by $d = 0.5t^2 + 6t$.

a.   What is the average rate of change of the distance from the origin over the interval from $t = 3$ to $t = 4$?

b.   What is the average rate of change from $t = 2$ to $t = 3$?

c.   Estimate the instantaneous rate of change of the distance from the origin at $t = 3$.

d.   What is the equation for the linear function that will approximate the distance function at the instant when $t = 3$?

e.   Use your answer from part d to approximate the value of d when $t = 4$. Compare your approximation to the actual value of d when $t = 4$.

Solution

a.   When $t = 3$, $d = 22.5$, and when $t = 4$, $d = 32$. The average rate of change over this time interval is $\frac{32 - 22.5}{4 - 3}$, or 9.5 cm/sec.

b.   When $t = 2$, $d = 14$, and when $t = 3$, $d = 22.5$. The average rate of change over this time interval is $\frac{22.5 - 14}{3 - 2}$, or 8.5 cm/sec.

c.   The instantaneous rate of change at $t = 3$ can be estimated by looking at the average rate of change over a very small interval containing $t = 3$. When $t = 3$, $d = 22.5$, and when $t = 3.1$, $d = 23.405$. The average rate of change over this interval of time is $\frac{23.405 - 22.5}{3.1 - 3}$, or 9.05 cm/sec. When $t = 3$, $d = 22.5$, and when $t = 3.01$, $d = 22.59005$. The average rate of change over this interval of time is $\frac{22.59005 - 22.5}{3.01 - 3}$, or 9.005 cm/sec. As the interval of time around $t = 3$ approaches zero, the average rate of change appears to approach 9 cm/sec. Therefore, we estimate the instantaneous rate of change at $t = 3$ to be 9 cm/sec.

d.   To approximate the distance function at $t = 3$, we use the tangent line. We estimated that the tangent line has slope 9 (the instantaneous rate of change at $t = 3$). The equation of the tangent is $d - 225 = 9(t - 3)$, which can be rewritten as $d = 9t - 4.5$. Graphs of the distance function and the tangent line approximation near the point with (3, 22.5) are shown in Figure 2.20.

e.   The equation for the tangent line at the point with $t = 3$ is $d = 9t - 4.5$. We can use this line to approximate the value of d when $t = 4$ by substituting 4 for t, which gives $9(4) - 4.5$, or 31.5. The actual value of d at time $t = 4$ is 32 cm (found in part a). The error in the linear approximation is −0.5 cm. The magnitude of the error relative to the magnitude of the actual distance is $\frac{0.5}{32}$, or approximately 0.016, so the relative error is about 1.6%. ■

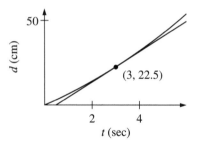

Figure 2.20 Distance versus time graph with a tangent line approximation

The examples we have studied illustrate the connection between rates of change and slopes. If P has coordinates $(x, f(x))$ and Q has coordinates $(x + \Delta x, f(x + \Delta x))$, then as Δx approaches 0, we can state the following generalizations.

The secant line through P and Q approaches the tangent line to the curve at P; therefore, the slope of the secant line through P and Q approaches the slope of the tangent line through P.

The average rate of change of f over the interval from P to Q approaches the instantaneous rate of change of f at P; therefore, the value of $\frac{f(x+\Delta x)-f(x)}{\Delta x}$ approaches the value of the derivative of f at P.

Using these generalizations, we can see that $\lim\limits_{\Delta x \to 0} \frac{f(x+\Delta x)-f(x)}{\Delta x}$ is equal to the slope of the line tangent to the graph of $y = f(x)$ at point P and is also equal to the value of the derivative of f at P.

Exercise Set 2.3

1. Graph $y = \sin x$ and $y = x$ on the same viewing window centered at $(0, 0)$.

    a. Compare the y values of the two functions as you zoom in on the graph. For what values are the curves similar?

    b. When you find $\sin 0.001$ on your calculator, what result do you expect? Why? Test your prediction.

2. Compare the graphs of $y = \ln x$ and $y = x - 1$ near $x = 1$.

    a. When can you use values of $x - 1$ to approximate values of $\ln x$? Explain.

    b. When you find $\ln 1.05$ on your calculator, what result do you expect? Explain your reasoning.

3. Investigate the tangent line approximation for the function with equation $f(x) = \sqrt{x}$ at the point with $x = 9$.

a.   Investigate the slopes of secant lines drawn through the point (9, 3) on the function $f(x) = \sqrt{x}$. Let the other point of intersection have x value $9 + \Delta x$. Begin with $\Delta x = 2$; then decrease the interval by letting $\Delta x = 1, 0.1$, 0.01, and 0.001. What do you observe?

b.   Based on your work in part a, estimate the slope of the line tangent to the curve $f(x) = \sqrt{x}$ at the point (9, 3). Use the derivative of $f(x) = \sqrt{x}$ that you found in Lab 1 to find the slope of the tangent line at $x = 9$. How do your answers from the two methods compare?

c.   Write the equation of the line tangent to $f(x) = \sqrt{x}$ at $x = 9$.

d.   Use your calculator to find $\sqrt{9.1}$. Find the value on the tangent line from part c for $x = 9.1$. How well does the tangent line approximate the value of the function?

e.   Use the tangent line at (9, 3) to approximate $\sqrt{9.5}$, $\sqrt{8.9}$, and $\sqrt{11}$. How well does the tangent line approximate the function values? Explain.

4.   Investigate the tangent line approximation for the function $f(x) = x^2 + 3x + 5$ at the point (0, 5).

a.   Investigate the slopes of secant lines drawn through the point (0, 5) on the function $f(x) = x^2 + 3x + 5$. Let the other point of intersection have x value $0 + \Delta x$. Begin with $\Delta x = 2$; then decrease the interval by letting $\Delta x = 1, 0.1$, 0.01, and 0.001. What do you observe?

b.   Based on your work in part a, estimate the slope of the line tangent to the curve $f(x) = x^2 + 3x + 5$ at the point (0, 5). Write the equation of the tangent line.

c.   Compare the interval over which the tangent line at $x = 9$ is a good approximation for $f(x) = \sqrt{x}$ (see problem 3) to the interval over which the tangent line at $x = 0$ is a good approximation for $f(x) = x^2 + 3x + 5$. Explain why these intervals are different.

5.   Let $f(x) = 4^x$.

a.   Use difference quotients to approximate the slope of the line tangent to $f(x)$ at the point (1, 4).

b.   Approximate the slope of the line tangent to $f(x)$ at the point on $f(x)$ where the second coordinate is twice as big, that is, $(\frac{3}{2}, 8)$. How does this slope compare to the one you found in part a?

c.   Approximate the slope of the line tangent to $f(x)$ at the point on $f(x)$ where the second coordinate is half as big, that is, $(\frac{1}{2}, 2)$. How does this slope compare to the one you found in part a?

d.   Generalize your answers to parts b and c. For the function $f(x) = 4^x$, how does the slope of the tangent line appear to be related to the coordinates of the tangency point? How is this related to what you studied in Chapter 1?

6. Investigate the slopes of secant lines drawn through the point $(0, 0)$ on the graph of $f(x) = x^{2/3}$.

   a. Begin with positive values of Δx, so the secant line is through the points with $x = 0$ and $x = 0 + \Delta x$, and let Δx decrease to zero. What appears to be the slope of the tangent line at $x = 0$?

   b. Investigate the slopes of the secant lines for $f(x) = x^{2/3}$ at $(0, 0)$ using negative values of Δx. What appears to be the slope of the tangent line at $x = 0$?

   c. Is there a line tangent to $f(x) = x^{2/3}$ at $(0, 0)$? If so, what is the slope of the tangent? Support your answer.

7. Let $f(x) = x \ln x$.

   a. What is the average rate of change of $f(x)$ over the interval from $x = 2$ to $x = 4$?

   b. What is the instantaneous rate of change at $x = 3$?

   c. What linear function approximates $f(x)$ at $(3, 3 \ln 3)$?

8. Let $g(x) = \begin{cases} 6x & \text{if } 0 < x < 2, \\ 12 & \text{if } x \geq 2. \end{cases}$

   a. What is the average rate of change of $g(x)$ over the interval from $x = 1$ to $x = 3$?

   b. What is the instantaneous rate of change at $x = 2$?

   c. What is the equation of the line that approximates the function at $(2, 12)$?

9. In Example 2, the average rate of change of $d = 0.5t^2 + 6t$ over the interval from $t = 2$ to $t = 4$ is equal to the instantaneous rate of change at $t = 3$. Sketch a graph showing the function with the corresponding secant line and tangent line. Explain why the rates of change are the same. Is this result unique to this function and interval? Cite specific examples to support your answer.

10. Write a paragraph stating a definition of a line tangent to a curve. Be sure to state under what conditions a tangent line exists and under what conditions a tangent line does not exist. If a tangent line does exist at a particular point, is it unique? How do you find the equation of such a line?

Lab 2 *Secants, Tangents, and Local Linearity*

In this lab we will examine what happens to the slopes of the secant lines for the graph of a function when one of the two points which determine the secant line moves closer to the other. This lab will use certain functions and points on the graphs of these functions to investigate how secants, tangents, and local linearity are related.

1. For each of the functions listed below, investigate the secant lines near the point $(a, f(a))$ according to the directions given in parts a through d.

Function	$(a, f(a))$
$f(x) = \sqrt{x}$	$(4, 2)$
$f(x) = \sqrt{x}$	$(1, 1)$
$f(x) = e^x$	$(3, e^3)$
$f(x) = \frac{1}{x}$	$(-2, -0.5)$
$f(x) = \begin{cases} -x^2 & \text{if } x < 1, \\ (x-1)^2 - 1 & \text{if } x \geq 1 \end{cases}$	$(1, -1)$

a. Find the slope and the equation of the secant line containing the points $(a, f(a))$ and $(a + \Delta x, f(a + \Delta x))$ with $\Delta x = 3$. Graph the secant line and the function on an appropriate domain containing $x = a$ and $x = a + \Delta x$.

b. Now choose a point $(a + \Delta x, f(a + \Delta x))$ such that Δx is less than 3 but still greater than 0. Find the slope and the equation of the secant line containing $(a, f(a))$ and $(a + \Delta x, f(a + \Delta x))$. Graph the secant line and the function on an appropriate domain containing $x = a$ and $x = a + \Delta x$.

c. Continue the process in part b, in each case choosing smaller values of Δx until Δx is less than 0.1. Record your observations about the graphs. (You may want to keep the same viewing window for several values of Δx to help you see what is happening.)

d. Repeat the process outlined in a through c, but start with $\Delta x = -1$, and let Δx approach 0 through negative values.

2. What do you notice about the graphs of the secant line and the function as the Δx values get closer to 0? As Δx gets closer to 0, how is the secant line related to the line tangent to the curve at $(a, f(a))$? What is the slope of the tangent line?

3. Explain the relationship between the slopes of the tangents found in this lab and values of the derivatives investigated in Lab 1.

4. For the last function in part 1 (piecewise defined), compare the values for the slopes of the secants found using Δx greater than 0 with the values found using Δx less than 0. What is the slope of the line tangent to the curve at the point where $a = 1$? Does a tangent exist at $a = 1$? Explain your reasoning.

5. What is the slope of the line tangent to the function $f(x) = |x|$ at $(0, 0)$, if such a line exists? If the tangent does not exist, why not? How does the existence of a tangent relate to local linearity? Explain carefully.

Summarize your results and observations in a written report. Your report should include numerical values from part 1 and thorough discussions of parts 2–5.

2.4   Limit Definition of the Derivative

In the previous sections of this chapter, we approximated the derivative, or instantaneous rate of change, of a function both graphically and numerically. If y is a function of x, so that $y = f(x)$, then the symbol $\frac{dy}{dx}$, which is read "the derivative of y with respect to x," represents the instantaneous rate of change of y with respect to x. This notation indicates that y is the dependent variable and x is the independent variable. Other notations for the derivative of $y = f(x)$ include y', $f'(x)$, $\frac{df}{dx}$, and $\frac{d}{dx}(f(x))$.

In Lab 1 you may have found that the derivative of $f(x) = x^2$ is a linear function with equation $y = 2x$. We can now use derivative notation to write $f'(x) = 2x$ or $\frac{d}{dx}(x^2) = 2x$. If we consider $y = f(x)$, then we can also write $y' = 2x$. This derivative indicates how the instantaneous rate of change of f depends on x. When $x = 3$, the instantaneous rate of change of f is $f'(x)$ evaluated at $x = 3$. This is written $f'(3)$ and is equal to 6. When $x = -0.5$, the instantaneous rate of change of f is $f'(-0.5)$, which equals -1. The value of the derivative at each x value is the slope of the line tangent to the graph of f at the point $(x, f(x))$. The tangent line at $(3, 9)$ has slope $f'(3) = 6$ and the slope of the tangent line at $(-0.5, 0.25)$ is $f'(-0.5) = -1$. The tangent lines at $(3, 9)$ and at $(-0.5, 0.25)$ are shown in Figure 2.21.

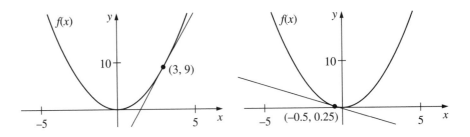

Figure 2.21 Tangent lines to $f(x) = x^2$

How do we know that the derivative of x^2 is exactly $2x$, and not $1.999x$ or $2.001x$? How do we know that the derivative of $\sin x$ is exactly $\cos x$, and not some function that is approximately equal to $\cos x$? You may have seen in Lab 1 that the derivative of 2^x is approximately $(0.7)2^x$, but what is the exact formula for this derivative? The graphical approach used in Lab 1 does not enable us to answer these questions easily, if at all. The derivatives we found in Lab 1 are qualitatively correct; that is, each derivative has the required overall qualities when its graph is compared to the graph of the difference quotients computed using a small Δx. Approximations are all that the graphical approach can give us. We need to use algebra to find an exact representation of a formula for a derivative.

Using algebra to find a formula for a derivative, we begin with a secant line determined by the points $(x, f(x))$ and $(x + \Delta x, f(x + \Delta x))$, where x is any value in the domain of f and Δx is some increment to the value of x. The slope of the secant line containing these two points is $\frac{f(x+\Delta x)-f(x)}{\Delta x}$. This slope corresponds to an average rate of change. If f behaves nicely (we'll say more about what that means, and about what happens if f does not, later) over the interval between x and $x + \Delta x$, then as Δx approaches 0, the slope of the secant approaches the slope of the tangent at the point $(x, f(x))$. The process of secants approaching a tangent is illustrated in Figure 2.22.

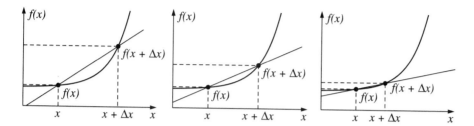

Figure 2.22 Secant lines approaching a tangent line

Likewise, as Δx approaches 0, the average rate of change of f over the interval from x to $x + \Delta x$ approaches the instantaneous rate of change of f at $(x, f(x))$. The value of the derivative f' at each x in the domain of f is defined as the limiting value of the average rate of change as Δx approaches zero, provided the limiting value exists. This limit definition of the derivative function is written as

$$f'(x) = \lim_{\Delta x \to 0} \frac{f(x+\Delta x)-f(x)}{\Delta x}. \tag{2}$$

The domain of the derivative of a function f is the set of all values in the domain of f for which the limit on the right side of (2) exists. (Recall that in Lab 2 we saw values in the domain of a function for which the slope of the tangent, *i.e.* the derivative, could not be uniquely determined.)

We can now use the definition in equation (2) to find the derivatives of specific functions. When we ***differentiate*** a function (that is, find its derivative), we find the value of the limit expression on the right side of (2) for all x in the domain of f for which this limit exists.   *differentiate*

Example 1

Use the limit definition of the derivative to find $f'(x)$ if $f(x) = mx + b$.

Solution   We begin with the limit definition in (2),

$$f'(x) = \lim_{\Delta x \to 0} \frac{f(x+\Delta x)-f(x)}{\Delta x}$$

and substitute using $f(x) = mx + b$ so that

$$f'(x) = \lim_{\Delta x \to 0} \frac{m(x + \Delta x) + b - (mx + b)}{\Delta x}.$$

Simplifying this expression yields

$$f'(x) = \lim_{\Delta x \to 0} \frac{m \Delta x}{\Delta x}$$

$$= \lim_{\Delta x \to 0} m \frac{\Delta x}{\Delta x}.$$

When we write the symbols $\lim_{\Delta x \to 0}$ we imply that Δx approaches zero through both positive and negative values. Since Δx is only close to zero, but not actually equal to zero, we can see that $\frac{\Delta x}{\Delta x} = 1$. Now we have

$$f'(x) = \lim_{\Delta x \to 0} m.$$

The constant m does not vary as Δx approaches zero, so $\lim_{\Delta x \to 0} m = m$, and

$$f'(x) = m. \qquad \blacksquare$$

In your previous work, you may have seen that the derivative of a linear function is constant and equal to its slope. Example 1 confirms this observation. The fact that $f'(x) = m$ when $f(x) = mx + b$ means that at any x value, the instantaneous rate of change of $f(x)$ is simply m.

In Chapter 1 we learned that when the relationship between two variables exhibits a constant rate of change then the variables are related in a linear way. Now we know the converse: if y is a linear function of x, then $\frac{dy}{dx}$ equals a constant.

Example 2

Use the limit definition of the derivative to verify that $\frac{dy}{dx} = 2x$ if $y = x^2$.

Solution We substitute into the limit definition of the derivative (2) using $f(x) = x^2$, so that

$$\frac{dy}{dx} = \lim_{\Delta x \to 0} \frac{(x + \Delta x)^2 - x^2}{\Delta x}. \qquad (3)$$

We expand the right side of equation (3) to get

$$\frac{dy}{dx} = \lim_{\Delta x \to 0} \frac{x^2 + 2x\Delta x + (\Delta x)^2 - x^2}{\Delta x}. \qquad (4)$$

Simplifying equation (4) yields

$$\frac{dy}{dx} = \lim_{\Delta x \to 0} \frac{2x\Delta x + (\Delta x)^2}{\Delta x},$$

which can be factored as

$$\frac{dy}{dx} = \lim_{\Delta x \to 0} \frac{(2x + \Delta x)\Delta x}{\Delta x}. \tag{5}$$

As in Example 1, for non-zero values of Δx, $\frac{\Delta x}{\Delta x} = 1$. It is important to remember that Δx is not equal to zero, but becomes arbitrarily close to zero through both positive and negative values. Equation (5) simplifies to

$$\frac{dy}{dx} = \lim_{\Delta x \to 0} 2x + \Delta x.$$

As Δx approaches zero, the quantity $2x + \Delta x$ approaches $2x$, so the derivative of $y = x^2$ is

$$\frac{dy}{dx} = 2x. \qquad\qquad \blacksquare$$

Exercise Set 2.4 includes additional problems using the limit definition of the derivative to find exact derivative formulas. Verification of the exact derivative formula for the function $\sin x$ will follow in Section 2.6 of this chapter. We now return to the problem of finding equations of tangent lines using what we have learned in this section.

Example 3

Write the equations of the two tangent lines to the graph of $f(x) = x^2$ shown here.

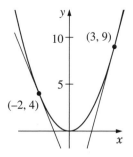

Figure 2.23 Graph of $f(x) = x^2$ and two tangent lines

Solution First we focus on the tangent to the curve at the point $(-2, 4)$. The slope of the tangent is the value of the derivative at the point $(-2, 4)$. The derivative of $f(x) = x^2$ is $f'(x) = 2x$. At $(-2, 4)$, $f'(-2) = -4$; therefore, the tangent line contains $(-2, 4)$ and has slope -4. The equation of the tangent line is thus

$$y - 4 = -4(x + 2)$$

or

$$y = -4x - 4.$$

Now we use the same procedure to find the equation for the tangent at (3, 9). The slope of the tangent is $f'(3)$, which is 6. The equation of the tangent line is

$$y - 9 = 6(x - 3)$$

or

$$y = 6x - 9. \qquad\qquad \blacksquare$$

We can write a general equation of a tangent line to the graph of $y = f(x)$ at the point with $x = a$. The slope of the tangent at $x = a$ is the derivative evaluated at $x = a$, which we symbolize as $f'(a)$. The tangent contains the point $(a, f(a))$. Using this point and slope we can write the equation of the tangent line as

$$y - f(a) = f'(a)(x - a)$$

or

$$y = f'(a)(x - a) + f(a). \qquad\qquad (6)$$

linear approximation

We will call this line the ***linear approximation*** of the function $f(x)$ at $x = a$.

In this section, we have introduced the limit definition of the derivative. Perhaps our work seems redundant, merely verifying derivative formulas that we already know. The limit definition of the derivative and the accompanying algebra, however, allow us to know these derivative formulas with greater certainty than is possible with the graphical method used previously. The limit definition of the derivative allows us to prove these derivative formulas.

Exercise Set 2.4

1. For each of the following functions:

    i.   sketch a graph of the function and sketch a graph of the derivative;

    ii. use the limit definition to find the derivative; and

    iii. check that the result in ii is consistent with the graph in i. If it is not, correct your work.

    a. $f(x) = 3x$ b. $f(x) = k$

    c. $f(x) = 5x^2 - 3$ d. $f(x) = x^2 + x$

    e. $f(x) = \sqrt{x}$ Hint: You will need to rationalize the numerator.

    f. $f(x) = |x|$ Hint: Use the definition $|a| = \begin{cases} a & \text{if } a \geq 0, \\ -a & \text{if } a < 0. \end{cases}$

2. Use parts a–c to develop a rule for the derivative of a power function $f(x) = x^n$.

   a. Use the limit definition of the derivative to find $f'(x)$ for $f(x) = x^3$.

   b. Use the limit definition of the derivative to find $f'(x)$ for $f(x) = x^4$.

   c. Write a general formula for the derivative of $f(x) = x^n$, where n is a positive integer.

3. Use parts a–d to develop a rule for the derivative of a reciprocal power function $f(x) = \frac{1}{x^n}$.

   a. Use the limit definition of the derivative to find $f'(x)$ for $f(x) = \frac{1}{x}$.

   b. Use the limit definition of the derivative to find $f'(x)$ for $f(x) = \frac{1}{x^2}$.

   c. Guess the derivative of $f(x) = \frac{1}{x^3}$. Check the reasonableness of your conjecture for $f'(x)$ by analyzing the graphs of $f(x)$ and $f'(x)$.

   d. Write a general formula for the derivative of $f(x) = \frac{1}{x^n}$, where n is a positive integer.

4. Use the fact that $f(x) = \frac{1}{x^n}$ can be written as $f(x) = x^{-n}$ to help you compare your derivative formulas from problems 2 and 3. Write one general result for the derivative of $f(x) = x^n$, for integer values of n.

5. Use the limit definition of the derivative to verify that $\frac{d}{dx}(x^2 + 3) = 2x$. Use a graphical argument to explain why $f(x) = x^2 + 3$ has the same derivative as $g(x) = x^2$.

6. Use the definition of the derivative to verify that $\frac{d}{dx}(4x^2) = 8x$. How does this result compare with the derivative of $f(x) = x^2$? How does a comparison between the derivatives of x^2 and $4x^2$ relate to a comparison of the graphs of $y = x^2$ and $y = 4x^2$?

7. The slope of the line tangent to the graph of $f(x) = \sin x$ at $x = 0$ is 1. In a–d, make conjectures for the slopes of the tangents to the functions at the point where $x = 0$.

   a. $y = \sin 2x$ b. $y = 5\sin x$

   c. $y = \sin\left(\frac{x}{3}\right)$ d. $y = \sin x + 6$

   e. At what x value is the slope of the line tangent to the graph of $y = \sin(x+1)$ equal to 1?

8. What conditions will prevent a function from having a derivative at a particular point? Explain your ideas and include examples.

9. Suppose $f'(x) = 0$ at $x = a$. Describe how the graph of $f(x)$ might behave near $x = a$. Be as specific as possible.

10. Does having a derivative at $x = a$ guarantee that $f(x)$ is locally linear at $x = a$? Does being locally linear at $x = a$ guarantee that $f(x)$ has a derivative at $x = a$? Explain and give examples.

2.5 Techniques of Differentiation

Based on work in preceding sections, exercises, and labs, you may have observed the derivative formulas listed in Table 2.24, where b, c, m, and n are constants.

Function	Derivative	Function	Derivative
$y = c$	$y' = 0$	$y = \sin x$	$y' = \cos x$
$y = mx + b$	$y' = m$	$y = \cos x$	$y' = -\sin x$
$y = x^2$	$y' = 2x$	$y = \tan x$	$y' = \sec^2 x$
$y = \sqrt{x}$	$y' = \dfrac{1}{2\sqrt{x}}$	$y = e^x$	$y' = e^x$
$y = x^n$	$y' = nx^{n-1}$ (n is an integer)	$y = \ln x$	$y' = \dfrac{1}{x}$

Table 2.24 A toolkit of derivative formulas

We can think of many other functions built from the functions in Table 2.24 for which we would like to find derivatives. For example, we might like to add, multiply, or compose two or more of these functions to form a new function. Fortunately, to find the derivative of this new function, we have options other than guessing the derivative using a graphical approach or using the limit definition of the derivative. There are many general rules for differentiation that simplify the process of finding derivatives. We will develop several of these rules in this section.

Vertical Shifts and Stretches

The graph in Figure 2.25 shows several functions whose equations are of the form $f(x) = x^2 + k$. Each parabola is simply a vertical shift of $y = x^2$. How do the derivatives for these shifted parabolas compare with the derivative of $y = x^2$? Since the vertical shifts change all y values by the same amount, they do not change the ratios $\frac{\Delta y}{\Delta x}$ over a fixed interval. (Convince yourself this is true). The limit of the ratio $\frac{\Delta y}{\Delta x}$ as Δx approaches zero defines the derivative; therefore, the vertical shifts do not change the derivatives. Geometrically, we can see that the tangent lines drawn in Figure 2.25 at the points with $x = 1$ are all parallel to each other. The derivative of each vertically shifted parabola is $2x$, so all the tangent lines drawn in Figure 2.25 have slope 2.

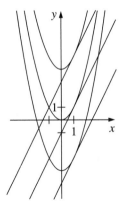

Figure 2.25 Graphs of $f(x) = x^2 + k$ and tangent lines at points with $x = 1$

The fact that shifting the graph of a function vertically does not change the derivative can be expressed as

$$\frac{d}{dx}[f(x) + k] = f'(x). \tag{7}$$

Applying the statement in (7), we know that, for example:

$$\frac{d}{dx}(\cos x + 7) = -\sin x$$

$$\frac{d}{dx}(\ln x - 9) = \frac{1}{x}.$$

How do other transformations influence the derivative? That is, how do the derivatives of $k \cdot f(x)$, $f(x + k)$, and $f(k \cdot x)$ compare to the derivative of $f(x)$?

The k in $k \cdot f(x)$ stretches or shrinks the graph of f vertically by a factor of k. Graphs for two values of k are shown in Figure 2.26. For the same x values, the y values are all k times as big. This means that the ratios $\frac{\Delta y}{\Delta x}$ for $y = k \cdot f(x)$ are k times as big as they are for $y = f(x)$.

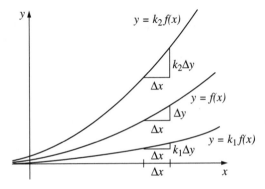

Figure 2.26 Comparison of $\frac{\Delta y}{\Delta x}$ for $y = f(x)$ and $y = k \cdot f(x)$ for two values of k

For a symbolic representation of this, consider the x values x_1 and x_2, and the corresponding function values. Using $y = k \cdot f(x)$,

$$\frac{\Delta y}{\Delta x} = \frac{k \cdot f(x_2) - k \cdot f(x_1)}{x_2 - x_1}$$

$$= \frac{k \cdot [f(x_2) - f(x_1)]}{x_2 - x_1}$$

$$= k \cdot \frac{f(x_2) - f(x_1)}{x_2 - x_1}.$$

Since $\lim\limits_{\Delta x \to 0} \frac{\Delta y}{\Delta x}$ defines the derivative, a vertical stretch or shrink changes the derivative by a factor of k. As a consequence, we write

$$\frac{d}{dx}[k \cdot f(x)] = k \cdot f'(x). \tag{8}$$

Applying the statement in (8), we know that

$$\frac{d}{dx}(7 \cdot x^3) = 7 \cdot \frac{d}{dx}(x^3),$$

which means that

$$\frac{d}{dx}(7 \cdot x^3) = 7 \cdot 3x^2,$$

or simply,

$$\frac{d}{dx}(7 \cdot x^3) = 21x^2.$$

What is the effect on the derivative of a combined vertical shift and vertical stretch? Suppose the function f is stretched vertically by a factor m and shifted vertically by an amount b to create the function expressed by $m \cdot f(x) + b$. We have seen in (7) that a vertical shift does not affect the derivative, so only the vertical stretch needs to be considered. Using this fact with (8), we see that the derivative of the transformed function is therefore given by the equation

$$\frac{d}{dx}[m \cdot f(x) + b] = m \cdot f'(x). \tag{9}$$

So, for example, we know that

$$\frac{d}{dx}(2 \sin x + 3) = 2 \cos x.$$

Horizontal Shifts and Stretches

How does the k in $f(x+k)$ influence the derivative? The graph in Figure 2.27 shows three functions whose equations are of the form $y = (x+k)^3$ for $k = 3, 0$, and -3. The graph of each cubic function is a horizontal shift of $y = x^3$. The tangent line at the point $(4, 1)$ on the right-hand graph is parallel to the tangent line at $(1, 1)$ on the middle graph, and they are both parallel to the tangent line at $(-2, 1)$ on the left-hand graph. This is to be expected, since the change in y for $y = x^3$ over an interval of length Δx is the same as the change in y for $y = (x+k)^3$ over the corresponding shifted interval of length Δx. Shifting the entire graph horizontally by k units does not change the shape of the graph, so $\frac{\Delta y}{\Delta x}$ values remain the same. The fact that the three tangent lines have the same slope means that the derivatives of the three functions have the same value. The derivatives with the same value occur, however, at different x values because of the horizontal shifts.

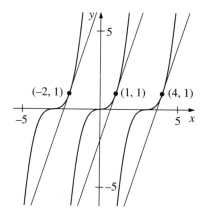

Figure 2.27 Graphs of $f(x) = (x+k)^3$ and tangent lines

These observations support the fact that the derivative of $g(x) = f(x+k)$ is the derivative of $f(x)$ evaluated at $x + k$. In other words, a horizontal shift in a function shifts the derivative function by the same amount, as expressed by the equation

$$g'(x) = \frac{d}{dx}[f(x+k)] = f'(x+k). \tag{10}$$

Applying the result in (10), we know that

$$\frac{d}{dx}\left(e^{x-9}\right) = e^{x-9}$$

and

$$\frac{d}{dx}\sqrt{x+5} = \frac{1}{2\sqrt{x+5}}.$$

Now, let's consider how the k in $y = f(k \cdot x)$ influences the derivative. We know that this k compresses or stretches the graph of f horizontally by a factor of k. As we will see, the effect of a horizontal stretch is not as simple as the effects described with the previous transformations. To better see this effect, it will be helpful for us to keep the Δy value equal to one, rather than keeping the Δx value equal to one. For the same y values, the x values in the graph of $y = f(k \cdot x)$ are $\frac{1}{k}$ times as big as the x values in the graph of $y = f(x)$. Consider for example the function $f(x) = \sqrt{x}$ and the transformation of this function $g(x) = \sqrt{4x}$. If we look at the points $(4, 2)$ and $(9, 3)$ on the graph of f, we notice that Δy is 1. We locate points $(1, 2)$ and $(2.25, 3)$ on the graph of g with the same Δy. Figure 2.28 shows these pairs of points on the two graphs. Because of the horizontal compression, the Δx associated with the graph of $g(x) = \sqrt{4x}$ is one-fourth the length of the Δx associated with the graph of $f(x) = \sqrt{x}$. If we keep the Δy's associated with f and g equal to each other but let them get smaller, the Δx's for $g(x)$ will remain one-fourth as big as the Δx's for $f(x)$. Because the derivative is determined by the ratio $\frac{\Delta y}{\Delta x}$, the derivative of g at $x = 1$ will be 4 times the derivative of f at $x = 4$.

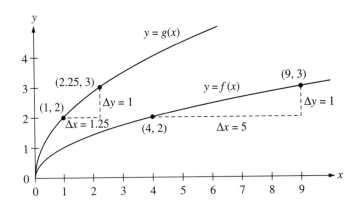

Figure 2.28 Graphs of $f(x) = \sqrt{x}$ and $g(x) = \sqrt{4x}$

In general, the value of Δx that yields a change of Δy in the function for $y = f(k \cdot x)$ is $\frac{1}{k}$ times as big as the Δx that yields the corresponding change Δy for $y = f(x)$. Thus, the denominators of the ratios $\frac{\Delta y}{\Delta x}$ for $y = f(k \cdot x)$ are $\frac{1}{k}$ times as big as the denominators of the ratios $\frac{\Delta y}{\Delta x}$ for $y = f(x)$. As a consequence, the ratios $\frac{\Delta y}{\Delta x}$ are k times as big for $y = f(k \cdot x)$ as they are for $y = f(x)$, as illustrated in Figure 2.29.

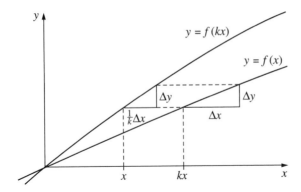

Figure 2.29 Comparison of $\frac{\Delta y}{\Delta x}$ for $y = f(x)$ and $y = f(k \cdot x)$

Though the slopes being compared in Figure 2.29 occur at the same y value, they do not occur at the same x value for the two functions because of the horizontal stretch or compression. As a consequence, the derivative of $y = f(k \cdot x)$ is k times the derivative of f evaluated at $k \cdot x$. That is,

$$\frac{d}{dx}[f(k \cdot x)] = f'(k \cdot x) \cdot k. \tag{11}$$

The notation $f'(k \cdot x)$ represents the derivative function f' evaluated at $k \cdot x$.

Applying the statement in (11), we know that

$$\frac{d}{dx}[\sin(5x)] = [\cos(5x)] \cdot 5$$

and

$$\frac{d}{dx}(9x)^7 = 7(9x)^6 \cdot 9.$$

What is the effect on the derivative of a combined horizontal shift and horizontal stretch? Suppose the function f is stretched and shifted horizontally to create the function expressed by $f(mx + b)$. The transformed function is equivalent to $f(m[x + \frac{b}{m}])$, a form that illustrates that f is compressed by a factor m and shifted by $\frac{b}{m}$. We have seen that a horizontal shift merely moves the derivative horizontally by the same amount for all x values. You can see from (11) that the horizontal stretch affects the derivative both vertically and horizontally—vertically because the derivative of f is multiplied by k and horizontally because the derivative is evaluated at $k \cdot x$ rather than at x. The derivative of the transformed function $f(m[x + \frac{b}{m}])$ is therefore given by the equation

$$\frac{d}{dx}\left[f\left(m\left[x + \frac{b}{m}\right]\right)\right] = f'\left(m\left[x + \frac{b}{m}\right]\right) \cdot m,$$

which is equivalent to

$$\frac{d}{dx}[f(m \cdot x + b)] = f'(m \cdot x + b) \cdot m.$$

So, for example, we know that

$$\frac{d}{dx}[\sin(2x + 3)] = 2\cos(2x + 3).$$

Example 1

Use a horizontal stretch to show that if $y = 2^x$, $\frac{dy}{dx} = \ln 2 \cdot 2^x$.

Solution First we write $y = 2^x$ as $y = (e^{\ln 2})^x$, which is equivalent to $y = e^{\ln 2 \cdot x}$. This function is a horizontal stretch of $y = e^x$ by a factor of $\ln 2$. We know that the derivative of the re-expressed function is $\frac{dy}{dx} = \ln 2 \cdot e^{\ln 2 \cdot x}$, and this is precisely $\frac{dy}{dx} = \ln 2 \cdot 2^x$. ∎

Combining Horizontal and Vertical Transformations

We will now use the four principles we have developed to find derivatives of functions which involve combinations of transformations.

Example 2

Find $\frac{dy}{dx}$ for $y = 3\sqrt{2x} + 11$.

Solution This function is a transformation of $f(x) = \sqrt{x}$. The coefficient "3" corresponds to a vertical stretch, the "2" to a horizontal compression, and the "11" to a vertical shift. All together, this function can be expressed as

$$y = 3f(2x) + 11,$$

where $f(x) = \sqrt{x}$. We know that $f'(x) = \frac{1}{2\sqrt{x}}$. Based on our observations of how transformations affect the derivative, we can calculate the derivative as follows:

$$\frac{dy}{dx} = 3 \cdot f'(2x) \cdot 2$$

$$= 3 \cdot \frac{1}{2\sqrt{2x}} \cdot 2$$

$$= \frac{3}{\sqrt{2x}}.$$ ∎

Example 3

Find $\frac{dy}{dx}$ for $y = \frac{1}{6}\cos(x + \pi) - 5$.

Solution   This function is a transformation of $g(x) = \cos x$. The "$\frac{1}{6}$" corresponds to a vertical compression, the "π" to a horizontal shift, and the "5" to a vertical shift. Thus,

$$\frac{dy}{dx} = \frac{1}{6}g'(x + \pi),$$

and since $g'(x) = -\sin x$,

$$\frac{dy}{dx} = -\frac{1}{6}\sin(x + \pi). \qquad \blacksquare$$

Example 4

Find $\frac{dy}{dx}$ for $y = 2e^{0.1x-10}$.

Solution   This function is a transformation of $g(x) = e^x$. The "2" corresponds to a vertical stretch, the "0.1" to a horizontal stretch, and the "−10" to a horizontal shift. Thus,

$$\frac{dy}{dx} = 2 \cdot g'(0.1x - 10) \cdot 0.1,$$

and since $g'(x) = e^x$,

$$\frac{dy}{dx} = 0.2e^{0.1x-10}. \qquad \blacksquare$$

Exercise Set 2.5.A

1. Find $\frac{df}{dx}$ for each function.

    a. $f(x) = e^x + 2$ b. $f(x) = 3 + \sin x$

    c. $f(x) = x^2 - 3$ d. $f(x) = 5x^2$

    e. $f(x) = -x^2$ f. $f(x) = 3\sin x$

    g. $f(x) = 3\sqrt{x}$ h. $f(x) = 2\ln x$

2. For each function given, write an expression for the derivative. Check the reasonableness of your derivative by viewing the graphs of $f(x)$ and $f'(x)$.

    a. $f(x) = e^{-x}$ b. $f(x) = -e^x$

    c. $f(x) = 2\sin x$ d. $f(x) = \sin 2x$

    e. $f(x) = (x + 3)^2$

3. Find $\frac{dy}{dx}$ for each function.

    a. $y = \sin(x - \frac{\pi}{4})$ b. $y = \frac{1}{4x+5}$

    c. $y = (4x)^5$ d. $y = \sqrt{x+3}$

    e. $y = \sqrt{6x}$ f. $y = e^{x-4}$

    g. $y = \sin(x + 3)$ h. $y = (7x - 2)^5$

    i. $y = (-6x + 7)^3$ j. $y = \sqrt{6x - 5} + 2$

    k. $y = \cos(0.5x - 1)$ l. $y = e^{3x}$

    m. $y = \sqrt{5x+3}$ n. $y = \sin(3x + \frac{\pi}{2})$

    o. $y = \ln(2x - 5)$ p. $y = \frac{1}{(3x-1)^2}$

4. If $y = e^{kx}$, then $\frac{dy}{dx} = k \cdot e^{kx}$, which can also be written as $\frac{dy}{dx} = k \cdot y$. Write a short paragraph to explain how this is related to what you studied about proportional rate of change in Chapter 1.

5. Show how $f(x) = x^2$ and $g(x) = x^3$ can be used to show that $\frac{d}{dx}(f \cdot g) \neq \frac{df}{dx} \cdot \frac{dg}{dx}$.

6. Use the limit definition of the derivative to prove that:

    a. $\frac{d}{dx}\left[f(x) + k\right] = f'(x)$ b. $\frac{d}{dx}\left[k \cdot f(x)\right] = k \cdot f'(x)$

7. Use the limit definition of the derivative to show that:

    a. $\frac{d}{dx} f(kx) = k \cdot f'(kx)$ Hint : First let $u = kx$ and find $\frac{df}{du}$.

    b. $\frac{d}{dx} f(x + k) = f'(x + k)$ Hint : First let $u = x + k$ and find $\frac{df}{du}$.

The Addition Rule for Derivatives

In Exercise Set 2.4, you may have seen that the derivative of $f(x) = x^2 + x$ is $f'(x) = 2x + 1$. Notice that the derivative of x^2 is $2x$, the derivative of x is 1, and the derivative of $x^2 + x$ is the sum of the two derivatives $2x + 1$. Is it just a lucky coincidence that the derivative of the sum is the sum of the derivatives? Not at all. Several times in this course we will use arguments based on local linearity to verify that certain relationships are true. We will use such an argument here to show that the derivative of a sum of two functions is the sum of the derivatives.

First let us review what we know about local linearity. In previous sections of this chapter we have used a tangent line to approximate the values of a function f near the point of tangency with $x = a$. We use a tangent line because the slope of the tangent is equal to $f'(a)$, the instantaneous rate of change of the function at $x = a$. Provided the function is locally linear (see Section 2.3) around the point with $x = a$, a tangent line is a good approximation for $f(x)$ near $x = a$. The equation of the tangent line at $(a, f(a))$ is

$$y - f(a) = f'(a)(x - a)$$

or

$$y = f'(a)(x - a) + f(a). \tag{12}$$

Equation (12) gives the linear approximation for $f(x)$ near $x = a$, so we can write

$$f(x) \approx f'(a)(x - a) + f(a)$$

for x values close to a.

Suppose the function h is the sum of functions f and g, that is $h(x) = f(x) + g(x)$. We are interested in the derivative of h at $x = a$. We must first assume that both f and g have derivatives at $x = a$. With this assumption, we can approximate both of these functions near $x = a$ by their linear approximations:

$$f(x) \approx f'(a)(x - a) + f(a);$$

$$g(x) \approx g'(a)(x - a) + g(a).$$

The function h, being the sum of f and g, can be approximated near $x = a$ by the sum of the two linear approximations, so that

$$h(x) \approx [f'(a)(x - a) + f(a)] + [g'(a)(x - a) + g(a)]. \tag{13}$$

Simplifying the right side of (13) yields

$$h(x) \approx [f'(a) + g'(a)](x - a) + f(a) + g(a). \tag{14}$$

The derivative of the right side of (14) is an approximation for the derivative of h near $x = a$. Since $f'(a)$, $g'(a)$, $f(a)$, and $g(a)$ are all constants, the derivative of (14) can be written as

$$h'(x) \approx f'(a) + g'(a). \tag{15}$$

The linear approximations for $f(x)$ and $g(x)$ are exact at $x = a$, thus the approximation for $h'(x)$ in (15) is exact at $x = a$, so we write

$$h'(a) = f'(a) + g'(a). \tag{16}$$

Our original choice of $x = a$ was arbitrary and depended only on both functions f and g being defined and having derivatives at $x = a$; therefore, the relationship in equation (16) is true for any x where the functions f and g are defined and have derivatives. Substituting x for a in (16), we have the ***addition rule for derivatives:***

addition rule for derivatives

If $h(x) = f(x) + g(x)$, then

$$h'(x) = f'(x) + g'(x)$$

for all x such that f and g are defined and have derivatives.

As you might expect, this rule can be extended to more than two functions. For example, suppose $h(x) = x^3 + x^2 + 6x + 2$. To determine $h'(x)$, we consider h as the sum of the four simpler functions whose derivatives we already know. So, $h'(x)$ is equal to the derivative of x^3 plus the derivative of x^2 plus the derivative of $6x$ plus the derivative of 2; therefore, $h'(x) = 3x^2 + 2x + 6$.

The Product Rule for Derivatives

Suppose the function P models the U.S. population over time t, and suppose that E models the per capita energy consumption in the U.S. over time. In any given year, the product of population and per capita energy consumption will give the total energy consumed in the U.S. Therefore, we can multiply P and E to obtain a new function T that represents total energy consumption, which is expressed by the following:

$$T(t) = P(t) \cdot E(t).$$

If P and E are both polynomials then T is also a polynomial. In this case, to find the rate of change of total energy consumption in the U.S., we can easily differentiate the function T. For example, if $P(t) = t^2$ and $E(t) = t^3 - 2t$, then $T(t) = t^5 - 2t^3$. The derivative of T is $T'(t) = 5t^4 - 6t^2$.

If P and E, however, are not both polynomials, then their product T may not be a function that we can easily differentiate. For example, if $P(t) = 10e^{0.2t}$ and $E(t) = 0.2t + 1.4$, then their product is given by $T(t) = (0.2t + 1.4)(10e^{0.2t})$. We do not yet know how to differentiate such a product. The derivative of a product is *not* the product of the derivatives, as can be seen in the example in the previous paragraph and in problem 5 of Exercise Set 2.5.A.

product rule　　　　There is a technique called the ***product rule*** that allows us to find the derivative of a product of two functions. We will develop a justification for the product rule that is based on local linearity.

Suppose the function h is the product of functions f and g, that is $h(x) = f(x) \cdot g(x)$. We are interested in the derivative of h at some point $x = a$. We must first assume that both f and g have derivatives at $x = a$. With this assumption, we can approximate both of these functions near $x = a$ by their linear approximations:

$$f(x) \approx f'(a)(x - a) + f(a);$$

$$g(x) \approx g'(a)(x - a) + g(a).$$

The function h, being the product of f and g, can be approximated near $x = a$ by the product of the two linear approximations, so that

$$h(x) \approx [f'(a)(x - a) + f(a)] \cdot [g'(a)(x - a) + g(a)]. \tag{17}$$

Expanding the right side of (17) yields

$$h(x) \approx f'(a)g'(a)(x - a)^2 + f(a)g'(a)(x - a) + g(a)f'(a)(x - a) + f(a)g(a). \tag{18}$$

The derivative of the right side of (18) is an approximation for the derivative of $h(x)$ near $x = a$, which implies

$$h'(x) \approx 2f'(a)g'(a)(x - a) + f(a)g'(a) + g(a)f'(a). \qquad (19)$$

The value of the right side of (19) at $x = a$ is simply $f(a)g'(a) + g(a)f'(a)$. The linear approximations for $f(x)$ and $g(x)$ are exact at $x = a$, thus the approximation for $h'(x)$ in (19) is exact at $x = a$, so we write

$$h'(a) = f(a)g'(a) + g(a)f'(a). \qquad (20)$$

Our original choice of $x = a$ was arbitrary and depended only on both functions f and g being defined and having derivatives at $x = a$; therefore, the relationship in equation (20) is true for any x where the functions f and g are defined and have derivatives. Substituting x for a in (20), we have the ***product rule for derivatives:***

product rule for derivatives

If $h(x) = f(x) \cdot g(x)$, then

$$h'(x) = f(x)g'(x) + g(x)f'(x)$$

for all x such that f and g are defined and have derivatives.

Returning to the formula for total energy $T(t) = P(t) \cdot E(t)$, the product rule tells us that $T'(t) = P(t)E'(t) + E(t)P'(t)$. This equation expresses the relationship between the rate of change of total energy consumption and the rates of change of population and per capita energy consumption.

If we use the polynomial functions $P(t) = t^2$ and $E(t) = t^3 - 2t$, as stated earlier in the section, then the product is $T(t) = t^2(t^3 - 2t)$. Using the product rule we see that the derivative of the product is

$$\begin{aligned} T'(t) &= (t^2)(3t^2 - 2) + (t^3 - 2t)(2t) \\ &= 3t^4 - 2t^2 + 2t^4 - 4t^2 \\ &= 5t^4 - 6t^2, \end{aligned}$$

which agrees with our earlier work.

Example 5

Find $\frac{dy}{dx}$ for $y = \sqrt{x} \cdot \sin x$.

Solution   This function is the product of $f(x) = \sqrt{x}$ and $g(x) = \sin x$. We know that $f'(x) = \frac{1}{2\sqrt{x}}$ and $g'(x) = \cos x$, so the derivative of y with respect to x is

$$y' = \sqrt{x} \cdot \cos x + \sin x \cdot \frac{1}{2\sqrt{x}}. \qquad \blacksquare$$

Example 6

Find $\dfrac{dy}{dx}$ if $y = \dfrac{e^{2x}}{x^4}$.

Solution   This function is the product of $f(x) = e^{2x}$ and $g(x) = x^{-4}$. We can find the derivative as follows:

$$\dfrac{dy}{dx} = e^{2x}(-4x^{-5}) + (x^{-4})2e^{2x}$$

$$= -4e^{2x}x^{-5} + 2e^{2x}x^{-4}$$

$$= \dfrac{-4e^{2x}}{x^5} + \dfrac{2e^{2x}}{x^4}$$

$$= \dfrac{-4e^{2x} + 2xe^{2x}}{x^5}.$$ ∎

Exercise Set 2.5.B

1. Find $\dfrac{dy}{dx}$ for each function.

 a. $y = 3\sin x - 2\cos x$ b. $y = -6x^4 + x^3 - 7x + 1$

 c. $y = \dfrac{1}{x} + \sqrt{x}$ d. $y = 5x^3 - 7x + \dfrac{2}{x}$

 e. $y = x + e^{-x}$ f. $y = (x-1)^2 + \cos 2x$

2. Find $\dfrac{dy}{dx}$ for each function.

 a. $y = x\sin x$ b. $y = x^2 \sin x$

 c. $y = \dfrac{1}{x} \cdot e^x$ d. $y = (x-3)e^x$

 e. $y = (\cos x)(5x^2 + 3x - 1)$ f. $y = x^5 \cdot \sqrt{x}$

 g. $y = \sqrt{x} \cdot 2^x$ h. $y = \dfrac{1}{x}\cos 2x$

 i. $y = \dfrac{4}{x}(1 + \sqrt{x} + e^x)$ j. $y = 5x^4 \cdot \sqrt{3x - 1}$

 k. $y = \dfrac{7x - 4}{e^{2x}}$ l. $y = \dfrac{(3 - 7x)^5}{(x - 4)^2}$

3. Use the limit definition of the derivative to prove that the derivative of the sum of two functions is the sum of the derivatives. If it is helpful, you may use the assumption that

$$\lim_{x \to a} [r(x) + s(x)] = \lim_{x \to a} r(x) + \lim_{x \to a} s(x).$$

This assumption can be shown to be true, but we will not do so in this text.

4. A cable company is installing new cable at the rate of 1000 new subscribers per year. Their charge for basic services has been going up by $1 per year. Discuss why this is not enough information to determine the rate of change of their revenue with respect to time. (**Note:** revenue is number of subscribers times charge per subscriber.)

5. For $f(x) = \sqrt{x} + 4$ and $g(x) = x^2 + 4$, find the equation of the line tangent to $p(x) = f(x) \cdot g(x)$ at $x = 1$.

6. Graph the function $f(x) = x \ln x$. For what value of a is $(a, f(a))$ the lowest point on the graph? What is the slope of the tangent line at the lowest point on the graph? Use a derivative formula to find an exact value for a.

7. Suppose the function $P(t) = 12 + (\frac{25}{t})(\sin t)$ models the number of printers in use at a computer facility, where t represents the number of hours since the facility opened.

    a. Find the instantaneous rate of change of P at $t = 1$ hour, $t = 2$ hours, $t = 3$ hours, $t = 4$ hours, and $t = 5$ hours.

    b. What is happening to the number of printers in use at the computer facility during the first five hours? (Calculate more values of $P'(t)$ as needed.)

8. Suppose the function $P(t) = 25e^{0.34t}$ models the population (in millions) of a certain country t years after 1990. Also suppose that $E(t) = 0.42t + 7.1$ models the per capita energy consumption (in 10^7 British thermal units, or BTUs) in that country. At what rate do you expect the total energy consumption to be changing when $t = 10$?

Lab 3 *Energy Consumption and the Product Rule*

In this lab we will look at two sets of data—one for the U.S. population and one for the per capita energy consumption in the U.S. We will find an exponential model for the population and a linear model for the per capita energy consumption. Then we will look at the derivative of the product of the two models so that we can have a model for the rate of growth of total energy consumption.

Let P be a function that models the total U.S. population, and let E be a function that models the per capita energy consumption. The product $T(t) = P(t) \cdot E(t)$ models the total energy consumption in the U.S. We will explore how the derivative $\frac{dT}{dt}$ is related to $\frac{dP}{dt}$ and $\frac{dE}{dt}$.

We will start with the data for population and per capita energy use given in the following table. These are mid-year figures with population reported in millions and energy consumption reported in 10^8 BTUs.

Year	Population	Per Capita Energy Use
1982	232.2	3.05
1983	234.3	3.01
1984	236.4	3.13
1985	238.5	3.10
1986	240.7	3.08
1987	242.8	3.16
1988	245.1	3.27
1989	247.3	3.29
1990	249.9	3.26

1. Use a computer or graphing calculator to make a scatter plot of each data set.

2. Find a linear model $E(t) = mt + b$ for the per capita energy use as a function of time.

3. Find an exponential model $P(t) = Ca^t$ for the population data.

4. Using your models, generate numerical values for the instantaneous rate of change of population with respect to time for each year. Do the same for per capita energy consumption with respect to time. You can generate these numbers by taking the derivative of each function and then evaluating the derivatives for each year.

5. Consider the function $T(t) = P(t) \cdot E(t)$, which represents the total energy consumption in the United States. Use difference quotients over small time intervals to approximate the instantaneous rate of change of this function with respect to time for each year. How do these approximations for the values of the derivative compare to the values generated by the product of the two derivatives of the individual functions E and P? Why are they not equal?

6. Use the numbers you generated in parts 4 and 5 together with function values of P and E to confirm what the product rule tells you about the rate of change of T. You should use your models for these functions and their derivatives, not the data points.

Write up your results and include an intuitive explanation of what the product rule tells you about $\frac{dT}{dt}$.

Reference

U.S. Department of Commerce, *Statistical Abstract of the United States 1992*, 112th Edition, Washington, D.C., U.S. Government Printing Office, 1992.

The Chain Rule for Derivatives

Some scientists claim that in recent decades the mean temperature of the earth's atmosphere has increased. Much of this global warming is thought to be caused by the increasing levels of carbon dioxide (CO_2) in the atmosphere. Other gases generated by human activity, the so-called greenhouse gases, also may be contributing to global warming, but CO_2 is thought to have the most significant effect on temperature.

Because of the potential negative consequences of global warming, scientists are concerned about how the temperature T is changing with respect to recent changes in CO_2 levels. This change is represented by the derivative $\frac{dT}{dCO_2}$, where T is a function that models ordered pairs of the form (CO_2 level, temperature). Such ordered pairs are not directly available to us, so the function T is not directly available to us either. We do, however, have data for CO_2 levels over time and for temperature levels over time. These data allow us to find two models: $C(t)$ models ordered pairs (time, CO_2) and $Temp(t)$ models ordered pairs (time, temperature). How is T related to the functions C and $Temp$? The keys to studying the relationship between the functions C, $Temp$, and T are composition and inverses of these functions.

If we think of these functions in terms of input (the values supplied to the functions) and output (the values that the functions produce), we have the following situation.

Function	Input	Output
C	time	CO_2
$Temp$	time	temperature
T	CO_2	temperature

To obtain T, we need an input CO_2 and an output temperature. The functions C and $Temp$ cannot be composed to yield such a function. Since the function C that models ordered pairs (time, CO_2) is an increasing function over the domain in question, it is a one-to-one function. This guarantees that the inverse relation C^{-1} is also a function. We have the following situation.

Function	Input	Output
C^{-1}	CO_2	time
$Temp$	time	temperature
T	CO_2	temperature

This shows that the composition $Temp \circ C^{-1}$ has CO_2 as input and temperature as output, which is precisely the characteristic we seek for T. We therefore define T as

$$T = Temp \circ C^{-1}$$

or

$$T(CO_2) = Temp[C^{-1}(CO_2)].$$

The derivative $\frac{dT}{dCO_2}$ measures how the temperature T is changing with respect to changes in CO_2. To determine $\frac{dT}{dCO_2}$ we need to know how to find the derivative of a composition because $T(CO_2) = Temp\big(C^{-1}(CO_2)\big)$. We will use local linearity to develop a mathematical formula for the derivative of a composition.

Suppose we have a function h which is defined by the composition $h(x) = g(f(x))$, and we are interested in the derivative of h at $x = a$. The first thing we notice is that $f(a)$ must be defined for $h(a)$ to be defined. In addition, we also assume that $f'(a)$ exists; therefore, we can use the linear approximation for $f(x)$ at $x = a$, which is $f(x) \approx f'(a)(x-a) + f(a)$.

We use a linear approximation for $f(x)$ because earlier we learned how to differentiate functions of the form $g(mx+b)$. So if g is composed with a linear function, then we know that the derivative of this composition is given by

$$\frac{d}{dx}\big[g(mx+b)\big] = g'(mx+b) \cdot m. \tag{21}$$

Substituting the linear approximation for $f(x)$, we have an approximation for $h(x) = g(f(x))$ given by

$$h(x) = g(f(x)) \approx g(f'(a)(x-a) + f(a)). \tag{22}$$

These statements are made under the assumption that when values close together are used as input values to the function g, the resulting output values are also close together. In effect, we are assuming that g is a *continuous* function, a concept that will be discussed in more detail later in this chapter.

Though the right side of (22) may appear complicated, it is actually the composition of g with a linear function of x. The slope of this linear function is $f'(a)$, and the other terms in the linear function only contribute a horizontal shift. Differentiating both sides of (22) and applying the result from (21), we know that near $x = a$,

$$h'(x) \approx g'(f'(a)(x-a) + f(a)) \cdot f'(a). \tag{23}$$

The right side of (23) is an approximation of $h'(x)$ for x values close to a. The approximation is exact for $x = a$ (since the linear approximation for $f(x)$ is exact for $x = a$), so we can write the equation

$$h'(a) = g'(f'(a)(a-a) + f(a)) \cdot f'(a)$$
$$= g'\big(f(a)\big) \cdot f'(a). \tag{24}$$

Our choice of a is arbitrary (as long as the assumptions stated above are true), so the relationship given in equation (24) is true for all x for which f is defined and has a derivative and for which g is defined and has a derivative at $f(x)$. Writing equation (24) using x instead of a yields the derivative function

$$h'(x) = g'(f(x)) \cdot f'(x). \tag{25}$$

This result is known as the ***chain rule***. It is called that because it tells us how to differentiate "chains" of functions, which are compositions of functions. The chain rule also can be expressed in terms of "inner" and "outer" functions. The derivative of a composition is equal to the product of the derivative of the outer function (evaluated at the inner function) and the derivative of the inner function.

chain rule

Sometimes it is helpful if we think of the chain rule in a different form. Let $y = g(f(x))$, and $u = f(x)$, so that $y = g(u)$. Suppose we want to know $\frac{dy}{dx}$ at some fixed number x. If we change x by a small amount Δx, then Δu is the corresponding change in u. Since y can be written as a function of u, then y has a change Δy that corresponds to the change Δu. Assuming that Δx and Δu are not zero, we can write

$$\frac{\Delta y}{\Delta x} = \frac{\Delta y}{\Delta u} \cdot \frac{\Delta u}{\Delta x}. \tag{26}$$

The two fractions $\frac{\Delta y}{\Delta u}$ and $\frac{\Delta u}{\Delta x}$ on the right side of (26) are average rates of change that approximate the instantaneous rates of change, or derivatives, $\frac{dy}{du}$ and $\frac{du}{dx}$. The approximations all get better as Δx (and thus Δu and Δy) approaches zero. So, in the limit as Δx approaches zero, the equation in (26) becomes the chain rule written in the form

$$\frac{dy}{dx} = \frac{dy}{du} \cdot \frac{du}{dx}. \tag{27}$$

To further develop our intuition about the chain rule as written in (27), suppose that $\frac{du}{dx} = 2$. This means that if x changes 1 unit, then u changes 2 units. Suppose also that $\frac{dy}{du} = 4$, which means that if u changes 1 unit, then y changes 4 units. Since y changes 4 units for each unit change in u, and u changes 2 units for each unit change in x, we conclude that y changes, $4 \cdot 2$, or 8, units for each unit change in x. This example illustrates the multiplication of derivatives found in the chain rule.

We now return to the global warming example where

$$T(CO_2) = Temp\big(C^{-1}(CO_2)\big).$$

The chain rule applied to these functions can be written as

$$\frac{dT}{dCO_2} = \frac{dTemp}{dC^{-1}} \cdot \frac{dC^{-1}}{dCO_2}. \tag{28}$$

At this point we notice that if the expressions in equation (28) were to represent fractions, we could note that $\frac{dC^{-1}}{dC^{-1}} = 1$, and the right side of the equation would be identical to the left side. While this is a good way to ensure that we have the chain rule set up properly, "canceling" the dC^{-1} does not represent a legitimate mathematical operation, since symbols like $\frac{dy}{du}$ represent functions, not ratios of two finite quantities.

Example 7

Use the chain rule to find the derivative of each function.

a. $y = (2x^2 + 5x)^7$ b. $y = \sin(x^3)$

c. $y = (\sin x)^2$ d. $y = e^{x^3 + 7x}$

Solution

a. Using notation consistent with (25), this function can be expressed $h(x) = g(f(x))$, where $g(x) = x^7$ and $f(x) = 2x^2 + 5x$. We know that $g'(x) = 7x^6$ and $f'(x) = 4x + 5$. Applying the chain rule, $h'(x) = g'(f(x))f'(x)$, we get

$$\frac{dy}{dx} = 7(2x^2 + 5x)^6(4x + 5).$$

b. Using notation consistent with (27), this function can be expressed as the composition $y = \sin u$ where $u = x^3$. We know that $\frac{dy}{du} = \cos u$ and $\frac{du}{dx} = 3x^2$, so applying the chain rule $\frac{dy}{dx} = \frac{dy}{du} \cdot \frac{du}{dx}$ yields

$$\frac{dy}{dx} = \cos(x^3) \cdot 3x^2.$$

c. This function is a composition of $m(x) = x^2$ and $k(x) = \sin x$, so $y = m(k(x))$. Therefore, $\frac{dy}{dx} = m'(k(x)) \cdot k'(x)$, or

$$\frac{dy}{dx} = 2 \sin x \cdot \cos x.$$

d. This function is equal to $f(g(x))$, where $f(x) = e^x$ and $g(x) = x^3 + 7x$. Since $f'(x) = e^x$ and $g'(x) = 3x^2 + 7$, $\frac{dy}{dx} = f'(g(x)) \cdot g'(x)$, so that

$$\frac{dy}{dx} = e^{x^3 + 7x} \cdot (3x^2 + 7). \qquad \blacksquare$$

Exercise Set 2.5.C

1. Find $\frac{dy}{dx}$ for each function.

a. $y = (x^2 + 5x - 1)^5$ b. $y = \sin \sqrt{3x + 1}$

c. $y = e^{(x^2 + 7)}$ d. $y = \ln(\sin x)$

e. $y = \frac{2x-1}{x+1}$ f. $y = e^{2x^3}$

g. $y = 7 \ln x + \ln 7x$ h. $y = \sin(4x^2) + \sin^2(4x)$

i. $y = x^3 \cos(x^2)$ j. $y = e^{\sin x} + \sin(e^x)$

k. $y = \dfrac{1}{(1 + \ln x)^2}$ l. $y = x \cdot e^{-x^2}$

m. $y = \sqrt{e^x \cdot \cos x}$ n. $y = \dfrac{x}{(x^2 + 1)}$

2. Find the equation of the line tangent to $f(x) = e^{\sin x}$ at $x = \pi$.

3. We know that the derivative of a function $y = f(x)$, denoted by $\dfrac{dy}{dx}$ or $f'(x)$, gives the instantaneous rate of change of y with respect to x. Sometimes we are interested in the rate of change of the rate of change, which is called the **second derivative**. The second derivative is denoted by $f''(x)$ or by

second derivative

$$\frac{d}{dx}\left(\frac{dy}{dx}\right) = \frac{d^2y}{dx^2}.$$

a. For $f(x) = x^n$, find $f'(x)$ and $f''(x)$.

b. For $f(x) = \sin x$, find $f'(x)$ and $f''(x)$.

c. For $f(x) = e^{-x^2}$, find $f'(x)$ and $f''(x)$.

4. We know that velocity is the derivative of position with respect to time and that acceleration is the derivative of velocity with respect to time. If acceleration is constant, what kind of function must position be? Explain how your answer relates to the heights of objects thrown into the air.

5. a. Find the coordinates of the highest point on the graph of $y = \frac{x}{1+x^2}$ for $x \geq 0$. What is the value of the derivative at this point?

b. Find the greatest tangent line slope to the graph of $y = \frac{x}{1+x^2}$ for $x \geq 0$.

6. The oxygen content of the water in a lake changes when organic waste is added to the lake. If the normal oxygen content of a lake is taken to be 1, then the fraction of the normal oxygen content in the lake t days after waste is dumped into the lake is given by the function $C(t) = \frac{1}{1200}(t^3 - 15t^2 + 1200)$ where t is time in days and $0 \leq t \leq 15$.

a. When is the oxygen content lowest?

b. When is the oxygen content changing most rapidly?

7. Your lab partner is having trouble deciding how the product rule and the chain rule can be used to find the derivative of the function

$$f(x) = \frac{(4x^2)\ln(x+1)}{\sin^2 x}.$$

Write some advice to help your partner think about this question.

8. Many calculus books present what is called the **quotient rule**, which is written as

quotient rule

$$\frac{d}{dx}\left[\frac{f(x)}{g(x)}\right] = \frac{g(x)f'(x) - f(x)g'(x)}{[g(x)]^2}.$$

Use the product and chain rules to show that this rule is correct.

Lab 4 *Global Warming and the Chain Rule*

Many scientists believe that recent increases in the mean global temperature are a result, in part, of the accumulation of greenhouse gases, especially carbon dioxide (CO_2). In this lab we will look at data for the concentration of CO_2 in the atmosphere over a period from 1965 to 1990 (measured in parts per million) and the yearly mean global temperature over the same time period (measured in degrees Celsius deviation from the mean of the 1950–1979 mean temperatures).

Increases in the concentration of CO_2 and increases in the mean global temperature have been measured and can be modeled mathematically. A related quantity currently being studied is the rate at which temperature is increasing in correlation with increases in the CO_2 level. In this lab, we will find models relating temperature and time and relating CO_2 concentration and time. We then will find a relationship for temperature as a function of CO_2 levels. This will allow us to determine the rate at which the temperature has changed with respect to CO_2 levels. This information is of critical importance as the world increases in population, uses more energy, and clears more rain forests (which produce oxygen and consume CO_2).

The following data and graph represent the deviation from the 1950–1979 mean of the average daily temperature for each year from 1965 through 1990.

Average Temperature Deviation from 1950–1979 Mean

Year	Temp Deviation	Year	Temp Deviation
1965	−0.12	1978	+0.04
1966	−0.04	1979	+0.10
1967	−0.01	1980	+0.19
1968	−0.06	1981	+0.27
1969	+0.06	1982	+0.09
1970	+0.05	1983	+0.30
1971	−0.08	1984	+0.12
1972	+0.02	1985	+0.09
1973	+0.14	1986	+0.17
1974	−0.10	1987	+0.30
1975	−0.09	1988	+0.34
1976	−0.22	1989	+0.25
1977	+0.11	1990	+0.39

Reference

Jones, P.D.; T.M.L. Wigley; and K.R. Briffa. Global and hemispheric temperature anomalies—land and marine instrumental records. pp. 603–608. In T.A. Boden, D.P. Kaiser, R.J. Sepanski, and F.W. Stoss (eds.), *Trends '93: A Compendium of Data on Global Change*. ORNL/CDIAC-65. Carbon Dioxide Information Analysis Center, Oak Ridge National Laboratory, Oak Ridge, Tenn. 1994.

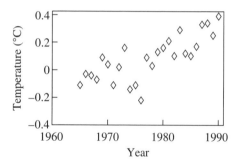

Scatter plot for global mean temperature deviation

1. Find a linear model *Temp(t)* for the temperature deviation as a function of time.

Now we can look at the data and scatter plot for the carbon dioxide levels in the atmosphere for the same time period. The average daily CO_2 levels, measured in parts per million at Mauna Loa Observatory, are given in the table below.

Average CO_2 Levels for 1965–1990

Year	CO_2 level	Year	CO_2 level
1965	319.87	1978	335.34
1966	321.21	1979	336.68
1967	322.02	1980	338.52
1968	322.89	1981	339.76
1969	324.46	1982	340.96
1970	325.52	1983	342.61
1971	326.16	1984	344.25
1972	327.29	1985	345.73
1973	329.51	1986	346.99
1974	330.08	1987	348.79
1975	330.99	1988	351.34
1976	331.98	1989	352.75
1977	333.73	1990	353.99

Reference

Keeling, C.D., and T.P. Whorf. Atmospheric CO_2 records from sites in the SIO air sampling network. pp. 16–26. In T.A. Boden, D.P. Kaiser, R.J. Sepanski, and F.W. Stoss (eds.), *Trends '93: A Compendium of Data on Global Change*. ORNL/CDIAC-65. Carbon Dioxide Information Analysis Center, Oak Ridge National Laboratory, Oak Ridge, Tenn. 1994.

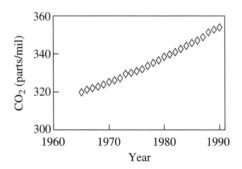

Scatter plot for carbon dioxide concentrations

2. The scatter plot for CO_2 levels appears only slightly curved. If we include data from years prior to 1965, a more pronounced curvature appears. We therefore will fit a curve to the data, rather than a line. Find an exponential model $C(t)$ for the level of CO_2 in the atmosphere as a function of time.

3. Use $C(t)$ to find an expression for t as a function of CO_2. Compose this result with $Temp(t)$ to write a function $T(CO_2)$ for temperature in terms of CO_2 levels.

4. Use your models to generate numerical values for the rate at which temperature is changing with respect to time for each year from 1965 to 1990. Do the same for the rate at which time is changing with respect to CO_2. Finally do the same for the rate at which temperature is changing with respect to CO_2 levels. Technology would be helpful for this problem.

5. Use the values you generated in part 4 to confirm what the chain rule tells you about the rate of change of temperature with respect to CO_2.

6. What conclusions can you draw about the relationship between carbon dioxide levels and the mean global temperature based on your models and calculations? Explain your reasoning.

7. According to the models you have found, how much would the CO_2 level need to change to cause the temperature to rise 1 degree? Does this depend on the current CO_2 level or the current temperature?

In a written report, summarize your observations and conclusions, and generalize your findings to explain the rule for differentiating functions that are compositions.

2.6   Indeterminate Forms

Based on what we observed in Lab 1, we believe that the derivative of $\sin x$ is $\cos x$, and that the derivative of e^x is e^x. Our knowledge of these derivative formulas comes from approximating the derivative values with difference quotients in which we use a small value of Δx. Though we are confident that our conjectures for the derivatives of the elementary functions are correct, nevertheless, how do we know that the difference quotients for $\sin x$ really behave like $\cos x$ as we shrink Δx? Could it actually be $1.0001 \cos x$ or $\cos x + 0.0001$? We could strive for greater accuracy in the difference quotients by shrinking the interval between successive points even further; however, the issue still is not resolved. If our results have 10-digit accuracy, could we be in error in the 11th digit? One way to deal with this issue is to use the limit definition of the derivative to try to verify our conjectures.

Before we begin to look at the derivative of the sine function, recall that in Lab 1 we cautioned you to use the radians unit for the domain of the trigonometric functions. There are several reasons for this. The radian, by definition, is the measure of an arc of a circle divided by the measure of the radius of the circle (using the same units for each). Regardless of the units used in measuring these lengths, the division causes the units to "cancel" each other, making the radian unitless, or simply a real number. If we use radians and draw graphs of the sine and cosine functions using equal scaling on each axis, then the graphs are recognizable as sine waves. Suppose we use degrees on the horizontal axis and real numbers with no units on the vertical axis, keeping the scales the same for both axes. (That is, one degree on the horizontal axis is the same distance from the origin as one on the vertical axis.) Over only one full period, the graphs of the sine and cosine are nearly indistinguishable from the horizontal axis.

A second reason for working with radians arises from the definition of the derivative. With all of the elementary functions, units are not assumed, so the change in the dependent variable is in terms of real numbers, as is the change in the independent variable. This results in real numbers with no units for the derivatives. If we use degrees for the domain of $y = \sin x$ or $y = \cos x$, then the units of the derivative will be in units of change in y per degrees of change in x. This would be inconsistent with all our other derivatives.

The limit definition of the derivative of $\sin x$ is given by

$$\frac{d}{dx}(\sin x) = \lim_{\Delta x \to 0} \frac{\sin(x + \Delta x) - \sin x}{\Delta x}. \tag{29}$$

As Δx approaches zero, both the numerator and denominator in (29) approach 0, so we cannot tell easily what $\frac{\sin(x+\Delta x)-\sin x}{\Delta x}$ approaches. We might be able to evaluate this limit better if we use the trigonometric identity for expanding $\sin(a + b)$:

$$\sin(a + b) = \sin a \cos b + \cos a \sin b.$$

This tells us that

$$\sin(x + \Delta x) = \sin x \cos \Delta x + \cos x \sin \Delta x.$$

Substituting this result into (29) yields

$$\frac{d}{dx}(\sin x) = \lim_{\Delta x \to 0} \frac{\sin x \cdot \cos \Delta x + \cos x \cdot \sin \Delta x - \sin x}{\Delta x}$$

$$= \lim_{\Delta x \to 0} \frac{\sin x \cdot (\cos \Delta x - 1) + \cos x \cdot \sin \Delta x}{\Delta x}.$$

Distributing the division by Δx through each term allows us to write (29) as

$$\frac{d}{dx}(\sin x) = \lim_{\Delta x \to 0} \left(\frac{\sin x \cdot (\cos \Delta x - 1)}{\Delta x} + \frac{\cos x \cdot \sin \Delta x}{\Delta x} \right).$$

As Δx approaches zero, the values of $\sin x$ and $\cos x$ in the previous limit expression remain unchanged. This means that the derivative of $\sin x$ can be written as

$$\frac{d}{dx}(\sin x) = \sin x \cdot \lim_{\Delta x \to 0} \frac{(\cos \Delta x - 1)}{\Delta x} + \cos x \cdot \lim_{\Delta x \to 0} \frac{\sin \Delta x}{\Delta x}.$$

To complete the work of finding the derivative of $\sin x$, we need to evaluate two limits,

$$\lim_{\Delta x \to 0} \frac{\cos \Delta x - 1}{\Delta x} \quad \text{and} \quad \lim_{\Delta x \to 0} \frac{\sin \Delta x}{\Delta x}.$$

indeterminate forms

In each limit, both the numerator and the denominator approach 0 as Δx approaches zero. Limits that approach $\frac{0}{0}$ are examples of expressions called ***indeterminate forms***. The word *indeterminate* does not mean that the limit does not exist, but rather implies that evaluating the limit presents a dilemma. Can we make an educated guess about the value of $\lim_{\Delta x \to 0} \frac{\sin \Delta x}{\Delta x}$? We know that $\sin \Delta x$ and Δx both approach 0 as Δx approaches 0. What does this mean about the limit of $\sin \Delta x$ divided by Δx? Does the limit equal 0? Does the ratio approach infinity? Does the limit equal some non-zero number? Does it simply not exist? If we focus on the numerator, we might think that the limit is zero because 0 divided by anything other than zero is 0. If we focus on the denominator, we might think the ratio approaches infinity because a ratio often increases without bound as the denominator approaches 0. We will investigate these questions in the next lab.

Similar questions arise when we look at the derivative of e^x. How can we be sure that the derivative of e^x is exactly e^x, and not $(2.72)^x$, or some other function? Again, the limit definition of the derivative allows us to analytically determine $\frac{d}{dx}(e^x)$. The algebraic steps associated with the limit definition of the derivative follow.

$$\frac{d}{dx}\left(e^x\right) = \lim_{\Delta x \to 0} \frac{e^{(x+\Delta x)} - e^x}{\Delta x}$$

$$= \lim_{\Delta x \to 0} \frac{e^x e^{\Delta x} - e^x}{\Delta x}$$

$$= \lim_{\Delta x \to 0} \frac{e^x(e^{\Delta x} - 1)}{\Delta x}$$

$$= e^x \cdot \lim_{\Delta x \to 0} \frac{e^{\Delta x} - 1}{\Delta x}.$$

Once again, we encounter a limit in which both the numerator and denominator approach zero as Δx approaches zero. In the next lab we will investigate this limit both numerically and graphically.

How unusual are these indeterminate forms? When we use the limit definition to find the derivative of any function, we must evaluate the limit

$$\lim_{\Delta x \to 0} \frac{f(x + \Delta x) - f(x)}{\Delta x}.$$

For any function f, this limit is indeterminate since the numerator and the denominator both approach zero as Δx approaches zero. The dilemma presented by this indeterminate form can be approached in several ways. For example, the limit that arises in finding the derivative of $f(x) = x^2$ can be evaluated by first simplifying algebraically, which allows us to factor Δx from the numerator and denominator, as we did in Example 2 of Section 2.4:

$$\frac{d}{dx}\left(x^2\right) = \lim_{\Delta x \to 0} \frac{(x + \Delta x)^2 - x^2}{\Delta x}$$

$$= \lim_{\Delta x \to 0} \frac{x^2 + 2x\Delta x + (\Delta x)^2 - x^2}{\Delta x}$$

$$= \lim_{\Delta x \to 0} \frac{2x\Delta x + (\Delta x)^2}{\Delta x}$$

$$= \lim_{\Delta x \to 0} (2x + \Delta x) \cdot \frac{\Delta x}{\Delta x}$$

$$= \lim_{\Delta x \to 0} (2x + \Delta x)$$

$$= 2x.$$

If we use algebra to simplify $\lim_{\Delta x \to 0} \frac{\sin(x + \Delta x) - \sin x}{\Delta x}$, we still have to evaluate $\lim_{\Delta x \to 0} \frac{\sin \Delta x}{\Delta x}$ and $\lim_{\Delta x \to 0} \frac{\cos \Delta x - 1}{\Delta x}$. When we use algebra with $f(x) = e^x$ we still have to evaluate the limit of the ratio $\frac{e^{\Delta x} - 1}{\Delta x}$ as $\Delta x \to 0$. We will examine these limits in the next lab.

Lab 5 *Limits of the Form $\frac{0}{0}$*

In this lab we will investigate limits of the form $\lim\limits_{x \to a} \frac{f(x)}{g(x)}$ where $f(x)$ and $g(x)$ both approach zero as x approaches a. We will examine tables of values for $f(x)$, $g(x)$, and $\frac{f(x)}{g(x)}$ near $x = a$ to estimate the value of the limit. We will also look at graphs of $f(x)$ and $g(x)$ near $x = a$ to help us understand why these limits behave as they do.

1. Define $f(x) = \sin x$ and $g(x) = x$. Let $a = 0$.

2. Construct a table of values for x, $f(x)$, $g(x)$, and $\frac{f(x)}{g(x)}$ for x values near a, but greater than a. What are $\lim\limits_{x \to a} f(x)$ and $\lim\limits_{x \to a} g(x)$? Make a conjecture about the value of $\lim\limits_{x \to a} \frac{f(x)}{g(x)}$.

3. Modify the table so that it will show x values near a, but less than a. If necessary, modify your conjecture about the value of $\lim\limits_{x \to a} \frac{f(x)}{g(x)}$.

4. Graph $f(x)$ and $g(x)$ on the same axes near $x = a$. What information about the ratio $\frac{f(x)}{g(x)}$ can be obtained from the graph? Both f and g are locally linear, so if you zoom in around $x = a$, the graphs of f and g look like lines. What are the equations of those lines? What is the limit of the ratio of the linear approximations of f and g as x approaches a? How does the value of this limit compare with your conjectures for the value of $\lim\limits_{x \to a} \frac{f(x)}{g(x)}$ in problems 2 and 3?

5. Repeat parts 1–4 for the following functions and values of a.

   a. $f(x) = \sin x$ $g(x) = x^2 + 6x$ $a = 0$

   b. $f(x) = \sin 3x$ $g(x) = 5x$ $a = 0$

   c. $f(x) = \sin(2x - 2)$ $g(x) = x - 1$ $a = 1$

   d. $f(x) = \cos x - 1$ $g(x) = x$ $a = 0$

   e. $f(x) = e^x - 1$ $g(x) = x$ $a = 0$

   f. $f(x) = 2^x - 1$ $g(x) = x$ $a = 0$

   g. $f(x) = e^{x-1} - 1$ $g(x) = x - 1$ $a = 1$

   h. $f(x) = \ln(x + 1)$ $g(x) = x^2 + 2x$ $a = 0$

6. Use the concept of local linearity to help you describe how the derivatives of locally-linear functions f and g are related to the graphs of f and g near $x = a$. How is the value of $\lim\limits_{x \to a} \frac{f(x)}{g(x)}$ related to the derivatives of f and g?

Summarize your observations and answers in a written report. Be sure to include a conjecture about limits of the form $\lim\limits_{x \to a} \frac{f(x)}{g(x)}$ where both $f(x)$ and $g(x)$ approach zero as x approaches a. You should explain why your conclusions are reasonable and how local linearity relates to your conclusions.

Exercise Set 2.6

1. Use the limit definition of the derivative and the results of Lab 5 to justify the statements $\frac{d}{dx}(\sin x) = \cos x$ and $\frac{d}{dx}\left(e^x\right) = e^x$.

2. Justify the statement $\frac{d}{dx}(\cos x) = -\sin x$ by writing $\cos x$ as a transformation of $\sin x$ and using rules for differentiation.

3. Justify the statement $\frac{d}{dx}\left(2^x\right) = \ln 2 \cdot 2^x$ by writing 2^x as a transformation of e^x and using rules for differentiation. (Hint: Use the fact that $2 = e^{\ln 2}$.)

4. Use the product rule and the chain rule to justify the statement $\frac{d}{dx}(\tan x) = \sec^2 x$.

5. Evaluate each limit. Use results of Lab 5 where appropriate.

   a. $\displaystyle\lim_{x \to 3} \frac{(x-3)^2}{(x-3)^7}$

   b. $\displaystyle\lim_{x \to 3} \frac{(x-3)^{11}}{(x-3)^5}$

   c. $\displaystyle\lim_{x \to 3} \frac{(x-3)^m}{(x-3)^n}$

   d. $\displaystyle\lim_{x \to -3} \frac{x^3 + 2x + 3}{x + 6}$

   e. $\displaystyle\lim_{x \to \frac{\pi}{2}} \frac{1 - \sin x}{\cos x}$

   f. $\displaystyle\lim_{x \to 0} \frac{1 - \cos x}{x^2}$

   g. $\displaystyle\lim_{x \to 0} \frac{e^x - 1}{x^3}$

   h. $\displaystyle\lim_{x \to 5} \frac{5 - x}{e^{5-x}}$

   i. $\displaystyle\lim_{x \to 5} \frac{x - 5}{e^{x-5} - 1}$

   j. $\displaystyle\lim_{x \to 0} \frac{e^x}{x}$

   k. $\displaystyle\lim_{x \to 0} \frac{\ln(x+1)}{x^2 + 7x + 4}$

There are indeterminate forms of limits other than $\frac{0}{0}$. In problems 6 and 7, we introduce the indeterminate form $\frac{\infty}{\infty}$, meaning the limit of a ratio in which the numerator and denominator both increase without bound. This form is indeterminate because of the dilemma it presents. If we focus on the numerator, we might think the limit of the ratio is infinity, since a number approaching infinity divided by any non-zero finite number approaches infinity. On the other hand, if we focus on the denominator, we might think the limit of the ratio is zero, since any finite number divided by a number approaching infinity goes to zero. Does a ratio of this type approach infinity, zero, or something in between?

6. If we allow x to increase without bound, rather than approaching some value a, we say that x is approaching ∞. Use your calculator or a computer to evaluate each limit numerically.

   a. $\displaystyle\lim_{x \to \infty} \frac{\ln x}{x}$

   b. $\displaystyle\lim_{x \to \infty} \frac{\ln x}{x^7}$

c. $\lim\limits_{x\to\infty} \dfrac{e^x}{x}$ d. $\lim\limits_{x\to\infty} \dfrac{e^x}{x^5}$

e.   When function values $f(x)$ increase without bound as x approaches a or ∞, we write $\lim\limits_{x\to a} f(x) = \infty$ or $\lim\limits_{x\to\infty} f(x) = \infty$. Does it appear that

$$\lim_{x\to\infty} \frac{f(x)}{g(x)} = \lim_{x\to\infty} \frac{f'(x)}{g'(x)}$$

when $\lim\limits_{x\to\infty} f(x) = \infty$ and $\lim\limits_{x\to\infty} g(x) = \infty$?

7.   Evaluate $\lim\limits_{x\to\infty} xe^{-x}$ numerically. Rewrite xe^{-x} so that the limit is in the indeterminate form $\dfrac{\infty}{\infty}$, and use the result in problem 6 to evaluate the limit.

2.7   Limits, Continuity, and Differentiability

The concepts of limits and continuity play a central role in calculus. We have defined the derivative as a limit, and many of the rules that we have developed for derivatives depend upon functions being continuous. Our use of the concepts of limits and continuity has been informal, even though much mathematical theory has been developed about limits and continuity. In this section, we will touch upon this theory.

Limits

We say that a function with equation $y = f(x)$ has a limit at $x = a$ if, as x values get close to a, the corresponding y values are getting close to some finite number. The expression $\lim\limits_{x\to a} f(x)$ represents the value that $f(x)$ approaches as x gets closer and closer to a. If we write

$$\lim_{x\to a} f(x) = D, \tag{30}$$

we mean that as x approaches a, $f(x)$ approaches D. How close to D does $f(x)$ get? As close as we like. The limit statement in (30) means that no matter how close we would like $f(x)$ to get to D, we can make that happen by using x close enough to a.

To evaluate the expression $\lim\limits_{x\to a} f(x)$, we must determine what happens to values of $f(x)$ as x values get closer and closer to a. For instance, we write

$$\lim_{x\to 2} (x^2 + 3) = 7,$$

because as the values of x get close to 2, the values of $x^2 + 3$ get close to 7. Although it may appear that we are simply substituting $x = 2$ into the expression $x^2 + 3$, that is not what the limit notation represents. The notation $x \to 2$ means that x gets arbitrarily close to 2, but does not equal 2. If we evaluate the expression $x^2 + 3$ for values of x near 2 (such as 2.1, 2.01, 2.001, ... , or 1.9, 1.99, 1.999, ...), we see that the values of $x^2 + 3$ are approaching 7, as shown in Table 2.30.

x	$f(x)$	x	$f(x)$
1.9	6.61000	2.1	7.41000
1.99	6.96010	2.01	7.04010
1.999	6.99600	2.001	7.00400
1.9999	6.99960	2.0001	7.00040
1.99999	6.99996	2.00001	7.00004

Table 2.30 Values of $f(x) = x^2 + 3$

Figure 2.31 shows the graph of $f(x) = x^2 + 3$. The graph shows that when t (on the horizontal axis) is near 2, then $f(t)$ (on the vertical axis) is near 7.

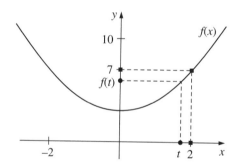

Figure 2.31 Graph of $f(x) = x^2 + 3$

Notice that in this example, $\lim_{x \to 2} f(x) = f(2)$. Though many limits of the form $\lim_{x \to a} f(x)$ can be evaluated by substituting a into the expression for $f(x)$, simple substitution does not suffice always. For instance, to evaluate $\lim_{x \to 2} \frac{x^2-4}{x-2}$ we cannot substitute 2 for x. If we try substituting $x = 2$ into $f(x) = \frac{x^2-4}{x-2}$, we end up with $\frac{0}{0}$, which is an indeterminate form. As we have seen before, limits of the form $\frac{0}{0}$ may actually equal some finite number. Table 2.32 shows what happens if we substitute values close to 2 into the function $f(x) = \frac{x^2-4}{x-2}$. We see that $f(x)$ approaches 4 as x approaches 2.

x	$f(x)$	x	$f(x)$
1	3	3	5
1.5	3.5	2.5	4.5
1.75	3.75	2.25	4.25
1.9	3.9	2.1	4.1
1.99	3.99	2.01	4.01
1.999	3.999	2.001	4.001

Table 2.32 Values of $f(x) = \dfrac{x^2 - 4}{x - 2}$

Even though $f(2)$ is undefined, $\lim\limits_{x \to 2} f(x)$ seems to equal 4. While our numerical approach of substituting numbers close to 2 leads us to believe that the limit is 4, we have even more evidence that this limit is exactly 4, as algebra can help us simplify this limit expression. The expression $\frac{x^2-4}{x-2}$ can be factored as $\frac{(x+2)(x-2)}{(x-2)}$. Since x approaches 2, but is not equal to 2, the expression $\frac{(x+2)(x-2)}{(x-2)}$ simplifies to $x + 2$. The quantity $x + 2$ approaches 4 as x approaches 2, so we can conclude that $\lim\limits_{x \to 2} \frac{x^2-4}{x-2} = 4$.

The graph of $f(x)$, shown in Figure 2.33, consists of the line $y = x + 2$ with a hole, represented by an open circle, at the point $(2, 4)$. Notice that when x values are near 2, the values of $f(x)$ are near 4.

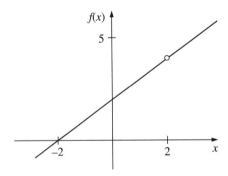

Figure 2.33 Graph of $f(x) = \dfrac{x^2 - 4}{x - 2}$

Example 1

Find the value of $\lim\limits_{x \to 4} \dfrac{x^3 - 64}{x - 4}$.

Solution   The expression given produces $\frac{0}{0}$ if we try to substitute $x = 4$. However, we find that the value of the limit is 48 by substituting numbers close to 4 and seeing that the values of $\frac{x^3-64}{x-4}$ seem to approach 48. We can also verify (by factoring the numerator and observing that $\frac{x-4}{x-4} = 1$) that $\frac{x^3-64}{x-4} = x^2 + 4x + 16$, if $x \neq 4$. The value of $x^2 + 4x + 16$ clearly approaches 48 as x approaches 4. ∎

Example 2

Find the value of $\displaystyle\lim_{x\to 0}\frac{\sin 2x}{\sin 3x}$.

Solution   This expression also produces $\frac{0}{0}$ when we substitute $x = 0$. Substituting x values close to zero shows that $\frac{\sin 2x}{\sin 3x}$ seems to approach $\frac{2}{3}$. Neither direct substitution nor factoring can help us find this limit. We are relying solely on numerical substitution when we say that $\displaystyle\lim_{x\to 0}\frac{\sin 2x}{\sin 3x}=\frac{2}{3}$. ■

The limits in Examples 1 and 2 can also be evaluated using the concepts developed in Lab 5. What we observed in Lab 5 is that a limit of the form $\frac{0}{0}$ (both numerator and denominator approaching 0 as x approaches a) has a value that depends upon the relative rates at which the numerator and denominator approach 0. We express this concept in symbolic form as

$$\lim_{x\to a}\frac{f(x)}{g(x)}=\lim_{x\to a}\frac{f'(x)}{g'(x)}, \tag{31}$$

assuming that both $\displaystyle\lim_{x\to a} f(x) = 0$ and $\displaystyle\lim_{x\to a} g(x) = 0$. The rule embodied in equation (31) is known as *l'Hôpital's rule*.

l'Hôpital's rule

Applying l'Hôpital's rule to the limit in Example 1, we see that

$$\lim_{x\to 4}\frac{x^3-64}{x-4}=\lim_{x\to 4}\frac{3x^2}{1},$$

which equals 48, as we expected. Applying l'Hôpital's rule to the limit in Example 2, we find that

$$\lim_{x\to 0}\frac{\sin 2x}{\sin 3x}=\lim_{x\to 0}\frac{2\cos 2x}{3\cos 3x}.$$

Since $\displaystyle\lim_{x\to 0}\cos 2x = 1$ and $\displaystyle\lim_{x\to 0}\cos 3x = 1$, this limit equals $\frac{2}{3}$, which is what we found in Example 2.

Earlier in this chapter we tried to use the limit definition of the derivative to find $\frac{d}{dx}(\sin x)$. The initial work is repeated here.

$$
\begin{aligned}
\frac{d}{dx}(\sin x) &= \lim_{\Delta x\to 0}\frac{\sin(x+\Delta x)-\sin x}{\Delta x} \\
&= \lim_{\Delta x\to 0}\frac{\sin x\cdot\cos\Delta x+\cos x\sin\Delta x-\sin x}{\Delta x} \\
&= \lim_{\Delta x\to 0}\frac{\sin x(\cos\Delta x-1)+\sin\Delta x\cos x}{\Delta x} \\
&= \sin x\cdot\lim_{\Delta x\to 0}\frac{\cos x\Delta-1}{\Delta x}+\cos x\cdot\lim_{\Delta x\to 0}\frac{\sin\Delta x}{\Delta x}.
\end{aligned}
$$

We then investigated the limits as $\Delta x \to 0$ of $\frac{\sin \Delta x}{\Delta x}$ and $\frac{\cos \Delta x - 1}{\Delta x}$ in Lab 5. Our numerical substitutions convinced us that $\lim_{\Delta x \to 0} \frac{\sin \Delta x}{\Delta x} = 1$ and $\lim_{\Delta x \to 0} \frac{\cos \Delta x - 1}{\Delta x}$. These results now allow us to continue our algebraic simplification of the limit involved in finding $\frac{d}{dx} (\sin x)$.

$$\frac{d}{dx} (\sin x) = \sin x \cdot \lim_{\Delta x \to 0} \frac{\cos \Delta x - 1}{\Delta x} + \cos x \cdot \lim_{\Delta x \to 0} \frac{\sin \Delta x}{\Delta x}$$

$$= \sin x \cdot 0 + \cos x \cdot 1$$

$$= \cos x.$$

We now have further evidence that the derivative of the sine function is exactly the cosine function. Prior to completing this demonstration, we had used the fact that the derivative of $\sin x$ was $\cos x$ based on graphical evidence from Lab 1.

The demonstration completed above depends upon the fact that $\lim_{\Delta x \to 0} \frac{\sin \Delta x}{\Delta x} = 1$. We could also evaluate $\lim_{\Delta x \to 0} \frac{\sin \Delta x}{\Delta x}$ by using l'Hôpital's rule, but to use l'Hôpital's rule in this context would involve circular reasoning. We cannot use the derivative of the sine function in the process of trying to show that $\frac{d}{dx} (\sin x) = \cos x$. This circularity is not particular to the sine function. For any function $f(x)$, the derivative function is defined to be $\lim_{\Delta x \to 0} \frac{f(x + \Delta x) - f(x)}{\Delta x}$, provided this limit exists. Every limit of this form is a limit in which both numerator and denominator approach zero. It would be circular reasoning to try to evaluate such a limit by taking the derivative of the numerator and denominator, since this would require knowing $f'(x)$.

Exercise Set 2.7.A

1. Let $f(x) = x^2 + 3x + 1$.
    a. Find the value of $\lim_{x \to 2} f(x)$ and discuss how this value is related to the graph of $f(x)$.
    b. Find the value of $\lim_{\Delta x \to 0} \frac{f(2 + \Delta x) - f(2)}{\Delta x}$ and discuss how this value is related to the graph of $f(x)$ and the derivative of f.

2. Let $f(x) = 2x + 3$.
    a. If $\lim_{x \to 1} f(x) = L$, what is the value of L?
    b. If x is within 0.1 of 1 (between 0.9 and 1.1), how close is $f(x)$ to L?
    c. If $f(x)$ is within 0.1 of L, how close is x to 1?

3. Let $f(x) = x^2 - 3x + 5$.
    a. If $\lim_{x \to 2} f(x) = L$, what is the value of L?
    b. When x is within 0.1 of 2, how close is $f(x)$ to L?
    c. When $f(x)$ is within 0.1 of L, how close is x to 2?
    d. When $f(x)$ is within 0.01 of L, how close is x to 2?

4. Let $q(x) = \frac{\sqrt{x}-2}{x-4}$, $x \geq 0$, $x \neq 4$.

   a. Use a graphing calculator or computer to graph $q(x)$ with a viewing window showing $0 \leq x \leq 10$.

   b. Use your graph to find $\lim\limits_{x \to 4} \frac{\sqrt{x}-2}{x-4}$.

   c. If x is within 0.05 of 4, how close is $q(x)$ to the limit you found in part b?

   d. Factor the denominator using $\sqrt{x} - 2$ as one factor. Take the limit and show that your answer to part b is correct.

5. Let $p(x) = \frac{\sin(x-2)}{x-2}$.

   a. Use a graphing calculator or computer to graph $p(x)$ with a viewing window showing $0 \leq x \leq 4$.

   b. Evaluate $\lim\limits_{x \to 2} \frac{\sin(x-2)}{x-2}$.

6. For the function $f(x) = x^2 - 3$, find an interval about $x = 3$ such that $f(x)$ is between 5.9 and 6.1.

7. A classmate missed calculus class on the day that limits were discussed and is trying to find $\lim\limits_{x \to 3} \frac{x^2+x-12}{x-3}$. Your classmate has asked you to explain this concept. What would you say?

8. If $g(x) = x^3$, then the ratio $\frac{x^3-8}{x-2}$ represents the slope of the secant line containing the points $(2, 8)$ and (x, x^3). As x approaches 2, the secant line becomes a tangent line. Note that when $x = 2$ there is no secant line between the points $(2, 8)$ and (x, x^3).

   a. What does the expression

   $$\lim_{x \to 2} \frac{x^3 - 8}{x - 2}$$

   represent about the graph of $g(x)$?

   b. Use your knowledge of the derivative of $g(x)$ to help you evaluate the expression in part a.

9. In section 2.6, we gave some reasons why units for the domains of the trigonometric function are usually radians. Suppose we felt these arguments were not convincing. What is the limit of $\frac{\sin x}{x}$ as x approaches 0 if x is in degrees?

Continuity

Imagine a skydiver who jumps from a plane, is in freefall for a period of time, and then opens the parachute. Even though the skydiver's velocity will change rather abruptly when the parachute opens, the velocity and altitude change in a continuous way.

Neither the parachutist's altitude nor velocity can change from one value to another without going through all intermediate values. Both altitude and velocity are thus continuous functions of time.

Most functions that we use in calculus are continuous for all values in their domains. For example, polynomial, exponential, and sinusoidal functions are continuous for all real numbers. Other functions, such as some rational and trigonometric functions, are continuous for all values except where vertical asymptotes occur.

Though we have an intuitive understanding of the concept of continuity, we need to be more precise in what we mean by continuity, and we need some theoretical notions which we can use in our occasional encounters with discontinuity. We will develop the concept of continuity by looking at some examples first.

In the previous section, we investigated the limit of the function $f(x) = x^2 + 3$ as x approaches 2. We determined that this limit is 7. Since there are no sudden breaks or jumps in the graph of this function, the limit of the function as x approaches 2 is equal to the value of the function at $x = 2$, or in symbols $\lim_{x \to 2} f(x) = f(2)$. We say that f is continuous at $x = 2$.

The second function we investigated in our discussion of limits is the function $f(x) = \frac{x^2-4}{x-2}$. We noted that the graph of this function has a "hole" where $x = 2$. We also noted that the limit of $f(x)$ as x approaches 2 exists and is equal to 4, but because $f(2)$ is undefined, we say that f is discontinuous at $x = 2$.

The graphs of the functions $g(x) = \frac{1}{x}$ and $h(x) = \frac{1}{x^2}$ are shown in Figure 2.34. We see that both functions are undefined at $x = 0$, and therefore, both are discontinuous at $x = 0$. Neither of these functions has a limit as x approaches 0. (Although we say that $\lim_{x \to 0} h(x) = \infty$, remember that the symbol ∞ does not represent a real number, and therefore, the limit does not exist.)

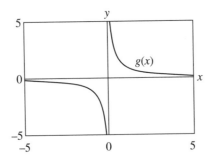
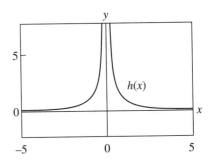

Figure 2.34 Graphs of $g(x) = \frac{1}{x}$ and $h(x) = \frac{1}{x^2}$

The discontinuities of g and h at $x = 0$ are different from the discontinuity of $f(x) = \frac{x^2-4}{x-2}$ at $x = 2$. The discontinuity in f can be removed by changing the definition of f at the point $x = 2$, so that

$$f(x) = \begin{cases} \dfrac{x^2 - 4}{x - 2} & \text{if } x \neq 2, \\ 4 & \text{if } x = 2. \end{cases}$$

Defining $f(2)$ to be 4 removes the hole in the graph of f. This definition for $f(2)$ also makes f a continuous function at $x = 2$. Since $\lim\limits_{x \to 2} f(x) = 4$, defining $f(2)$ to be 4 makes it true that $\lim\limits_{x \to 2} f(x) = f(2)$. If $f(2)$ were defined to be any value other than 4, $\lim\limits_{x \to 2} f(x)$ would still be 4, but $f(2)$ would not equal 4. This would mean that as x approaches 2, $f(x)$ values would not approach $f(2)$.

No assignment of a value for $g(0)$ or $h(0)$ can be made to remove the discontinuities in g and h, since the limits of g and h as x approaches 0 do not exist (*i.e.* do not equal a finite, real number). We can define values for $g(0)$ and $h(0)$; however, these values cannot possibly make either function continuous at $x = 0$.

Figure 2.35 shows the graph of the piecewise-defined function

$$f(x) = \begin{cases} -1 & \text{if } x < 0, \\ 1 & \text{if } x > 0. \end{cases}$$

While we could define $f(0)$ to be any value we want, no value would make this function continuous since the limit as x approaches 0 does not exist. When we approach zero from the positive side, the limit is 1; whereas, when we approach zero from the negative side, the limit is -1. The sudden jump in the graph at $x = 0$ makes this function discontinuous regardless of the value of the function at $x = 0$.

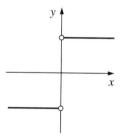

Figure 2.35 Graph of a piecewise function

The behavior of the functions examined in this section suggests three different ways that a function can fail to be continuous at a particular x value:

1. f is discontinuous at $x = a$ if $f(a)$ is undefined;
2. f is discontinuous at $x = a$ if $\lim\limits_{x \to a} f(x)$ does not exist;
3. f is discontinuous at $x = a$ if $\lim\limits_{x \to a} f(x)$ exists but is not equal to $f(a)$.

This leads us to a criterion for continuity of f at $x = a$:

f is continuous at $x = a$ if and only if $\lim\limits_{x \to a} f(x) = f(a)$.

In other words, a function is continuous at $x = a$ if as x approaches a, $f(x)$ approaches $f(a)$. Exercises at the end of this section investigate the implications of this statement for other functions.

If a function is continuous for all values in some interval, the function is said to be continuous over that interval. For example, the function $f(x) = \tan x$ is continuous for all x for which $-\frac{\pi}{2} < x < \frac{\pi}{2}$. The notation $\left(-\frac{\pi}{2}, \frac{\pi}{2}\right)$ is used to represent the open interval $-\frac{\pi}{2} < x < \frac{\pi}{2}$. The parentheses denote that the endpoints are not included in the interval. If a function g is continuous for all x for which $-1 \leq x \leq 1$, then g is continuous on the closed interval $[-1, 1]$. While functions can also be continuous on half-open intervals (one endpoint not included and the other included), we will see few of these in this course.

Differentiability

differentiable

If the derivative of a function exists at a point, on an interval, or for all x values, then $f(x)$ is said to be ***differentiable*** at that point, on that interval, or for all x, respectively. If $\lim_{\Delta x \to 0} \frac{f(x+\Delta x)-f(x)}{\Delta x}$ fails to exist at a particular x value, then f is not differentiable at that x value.

You have seen in Lab 2 that $f(x) = |x|$ is not locally linear at $x = 0$. Is this function differentiable at $x = 0$? The answer is no, because $\lim_{\Delta x \to 0} \frac{|x+\Delta x|-|x|}{\Delta x}$ does not exist when $x = 0$. When Δx approaches 0 through positive values, which we write as $\Delta x \to 0^+$, the value of $\lim_{\Delta x \to 0^+} \frac{|0+\Delta x|-|0|}{\Delta x}$ is 1. When Δx approaches zero through negative values, which we write as $\Delta x \to 0^-$, the value of $\lim_{\Delta x \to 0^-} \frac{|0+\Delta x|-|0|}{\Delta x}$ is -1. Therefore, because the limits as we approach zero from the left and the right are not the same, $\lim_{\Delta x \to 0} \frac{|0+\Delta x|-|0|}{\Delta x}$ does not exist, and $f(x) = |x|$ is not differentiable at $x = 0$. The graph of the piecewise function in Figure 2.35 is actually the graph of the derivative of the absolute value function.

The function $f(x) = \frac{1}{x}$ is not differentiable at $x = 0$ because the function is not defined at that x value. The function $f(x) = \sqrt{x}$ is not differentiable for x values that are less than or equal to zero. For negative x values the square root function is not differentiable because the function is undefined. At $x = 0$, the function $f(x) = \sqrt{x}$ is not differentiable for another reason. As x approaches 0 through positive values, slopes of the lines tangent to f approach infinity, not a finite value, so the derivative fails to exist at $x = 0$. Note that $f'(x) = \frac{1}{2\sqrt{x}}$, which is not defined for $x = 0$. While there are many other functions that fail to be differentiable at one or more points, most of the functions commonly used to model real world phenomena are differentiable for all x in their domains.

Continuity and differentiability are closely linked. Suppose a function $f(x)$ is differentiable at $x = a$. This means that $\lim_{\Delta x \to 0} \frac{f(a+\Delta x)-f(a)}{\Delta x}$ exists. One implication of this limit's existence is that $f(a)$ must be defined. Another implication has to do with continuity. If the limit exists, is it possible for f to be discontinuous at $x = a$? Suppose that f is discontinuous at $x = a$, even though $f(a)$ exists. That would mean that as x values get close to a, values of $f(x)$ do not get close to $f(a)$. This implies that as the x value $a + \Delta x$ gets close to a (that is, Δx approaches 0) the corresponding value $f(a + \Delta x)$ does not get close to $f(a)$, and thus $f(a + \Delta x) - f(a)$ does not approach zero. If $f(a + \Delta x) - f(a)$ does not approach zero, the ratio $\frac{f(a+\Delta x)-f(a)}{\Delta x}$ will not have a limit as Δx approaches 0. But this means that the derivative of f at $x = a$ does not exist. Because this contradicts our original assumption that f is differentiable at $x = a$, our supposition that f is discontinuous at $x = a$ must be false. Therefore, we can conclude that if a function is differentiable at a particular x value, it must also be continuous at that x value.

This important relationship between continuity and differentiability can be summarized as follows:

> If a function has a derivative that is defined at a particular x value, the function must be continuous at that x value; therefore, differentiability implies continuity. The converse is not necessarily true—a function can be continuous at a particular x value but not be differentiable at that x value.

Exercise Set 2.7.B

1. The function f is graphed below. For what values of x is the function discontinuous? Explain each answer.

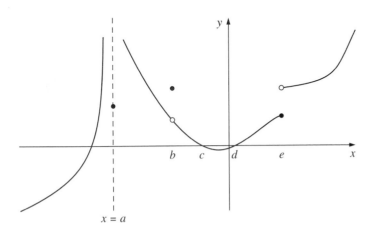

2. What value for the constant h would make the following function continuous?

$$F(x) = \begin{cases} \dfrac{x^2 - 9}{x - 3} & \text{if } x \neq 3, \\ h & \text{if } x = 3. \end{cases}$$

3. For this problem, use the function

$$g(t) = \begin{cases} t^2 & \text{if } t \geq 0, \\ t + 1 & \text{if } t < 0. \end{cases}$$

a. Find $\lim\limits_{t \to 0^+} g(t)$ and $\lim\limits_{t \to 0^-} g(t)$.

b. What is $\lim\limits_{t \to 0} g(t)$? Describe the behavior of this function around $t = 0$.

4. For this problem, use the function

$$f(x) = \begin{cases} 2x^2 - 1 & \text{if } x \geq 3, \\ 13x - 22 & \text{if } x < 3. \end{cases}$$

a. Find $\lim\limits_{x \to 3^+} f(x)$ and $\lim\limits_{x \to 3^-} f(x)$. What is $\lim\limits_{x \to 3} f(x)$?

b. Is this function continuous at $x = 3$? Why or why not?

c. Is this function differentiable at $x = 3$? Why or why not?

5. Let $f(x)$ be defined as

$$f(x) = \frac{\sin x}{x}, \quad x \neq 0.$$

Is it possible to define $f(x)$ at $x = 0$ so that f is continuous at $x = 0$? Explain.

6. Suppose f and g are both discontinuous at $x = a$. Give examples to show that the function $h(x) = f(x) + g(x)$ could be either continuous or discontinuous at $x = a$.

7. Give an example of a function g such that $\lim\limits_{x \to 1} g(x)$ exists but g is not continuous at $x = 1$.

8. Is it possible for a function h to be continuous at $x = 1$, but not to have a limit as x approaches 1? Explain.

9. Graph the following function and discuss its continuity.

$$f(x) = \begin{cases} x + \dfrac{2}{x} & \text{if } x < 1, \\ -x^3 + 4 & \text{if } 1 \leq x \leq 2, \\ -1 & \text{if } 2 < x < 3, \\ 1 & \text{if } x = 3, \\ x - 4 & \text{if } x > 3. \end{cases}$$

10. Find values for the constants a and b which would make

$$f(x) = \begin{cases} ax^2 + bx & \text{if } x \le 2, \\ 3x + 1 & \text{if } x > 2 \end{cases}$$

both continuous and differentiable at $x = 2$.

11. For this problem, use the function

$$F(x) = \begin{cases} 1 & \text{if } x > 0, \\ -1 & \text{if } x < 0, \text{ and} \\ 0 & \text{if } x = 0. \end{cases}$$

a. What is the value of $\lim\limits_{x \to 0} F(x)$?

b. What is the value of $\lim\limits_{\Delta x \to 0} \frac{F(0 + \Delta x) - F(0)}{\Delta x}$?

c. Discuss the differentiability and continuity of this function.

12. The descent of a roller coaster starts at the maximum point of a parabola whose equation is $y = 100 - \frac{1}{4}x^2$ and ends along the horizontal line $y = 0$. In order to make the ride smooth, a section of track following the graph of the function $y = a(x - 40)^2$ is added as a transition from the parabola to the line. Determine the value for a which makes the ride from the top to the end along the horizontal line smooth, that is, continuous and differentiable. At what point do the two parabolas meet? What is the slope at this point?

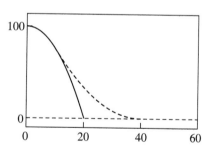

13. Discuss how graphing calculators and computers treat discontinuities when graphing a function. What pitfalls should you be aware of when graphing functions with a calculator or computer?

END OF CHAPTER EXERCISES

Review and Extensions

1. Write a paragraph to describe how average rate of change, instantaneous rate of change and the derivative are related.

2. Find the equation of the line tangent to $f(x) = (\sin x)^2$ at $x = \frac{\pi}{4}$.

3. For each of the following functions find $\frac{dy}{dx}$.

   a. $y = x(2^x)$

   b. $y = \dfrac{\tan x}{e^x}$

   c. $y = \dfrac{x-4}{x^2 - 8}$

   d. $y = \sqrt[3]{x^2 + 4x + 7}$

   e. $y = \dfrac{1}{\sin x^2 + x}$

4. Why is the tangent line important in our study of calculus?

5. Evaluate the following limits.

   a. $\displaystyle\lim_{x \to 0} \frac{\sin 2x}{x^2}$

   b. $\displaystyle\lim_{x \to 2} \frac{x^3 - 8}{x + 2}$

   c. $\displaystyle\lim_{x \to \infty} \frac{2x^2 + 4}{3x^2 + x + 1}$

6. a. The graph of the function $y = f(x)$ is shown below. Sketch $y = f'(x)$.

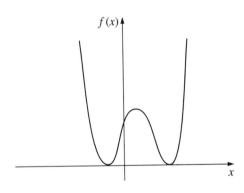

b.   The graph of the function $y = f'(x)$ is shown below. Sketch $y = f(x)$.

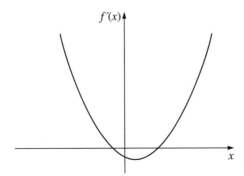

7.   What do we mean when we say $\lim\limits_{x \to 2} x^2 + x + 3 = 9$?

8.   Suppose $\lim\limits_{x \to a} f(x) = 0$ and $\lim\limits_{x \to a} g(x) = \infty$. What is $\lim\limits_{x \to a} \dfrac{f(x)}{g(x)}$?

9.   Make a summary of elementary functions and their derivatives.

10. Use your understanding of the definition of derivative to find $f'(1)$ if $f(x) = \tan^{-1} x$. Use the definition to graph $f'(x)$.

CHAPTER 3

Interpreting and Applying the Derivative

3.1 Investigating the First and Second Derivatives

The following article appeared in the April 6, 1991 issue of *The Economist*.[1] While reading it, consider how the author uses the language and concepts of calculus.

The Tyranny of Differential Calculus
Rates of change and the pace of decay in all around we see

"The pace of change slows," said a headline on the Financial Times' survey of world paints and coatings last week. Growth has been slowing in various countries—slowing quite quickly in some cases. Employers were invited recently to a conference on "techniques for improving performance enhancement." It's not enough to enhance your performance, Jones, you must improve your enhancement.

   Suddenly, everywhere, it is not the rate of change of things that matters, it is the rate of change of rates of change. Nobody cares much about inflation; only whether it is going up or down. Or rather, whether it is going up fast or down fast. "Inflation drops by disappointing two points," cries the billboard. Which, roughly translated, means that prices are still rising, but less fast than they were, though not quite as much less fast as everybody had hoped.

   No respectable American budget director has discussed reducing the national debt for decades; all talk sternly about the need to reduce the budget deficit, which is, after all, roughly the rate at which the national debt is

---

increasing. Indeed, in recent years it is not the absolute size of the deficit that has mattered so much as the trend; is the rate of change of the rate of change of the national debt positive or negative?

Blame Liebniz, who invented calculus (yes, so did Newton, but he called it fluxions, and did his best to make it incomprehensible). Rates of change of rates of change are what mathematicians call second-order differentials or second derivatives. Or blame Herr Daimler. Until the motor car came along, what mattered was speed, a first-order differential. The railway age was an era of speed. With the car came the era of acceleration, a second-order differential: the fact that a car can do 0-60 mph in eight seconds is a far more important criterion for the buyer than that it can do 110 mph. Acceleration limits are what designers of jet fighters, rockets and racing cars are chiefly bound by.

Politicians, too, are infected by ever higher orders of political calculus. No longer is it necessary to have a view on abortion or poll taxes. Far better to commission an opinion poll to find out what the electorate's view is, then adopt that view. Political commentators, and journalists, make a living out of the second-order differential: predicting or interpreting what politicians think the electorate thinks. Even ethics has become infected ("you did nothing wrong, Senator, and even the appearance of it does not stink, but people might think it will appear to stink").

Bring back the integral

It boils down to biology. There is virtually nothing in the human brain or the nerves that measures a steady state. Everything responds to change. Try putting one hand in hot water and the other in cold water. After a minute put both hands in tepid water. It will feel hot to one, cold to the other. The skin's heat sensors measure changes in temperature.

The visual system of the frog is a blank screen on which only things that move (delicious flies or dangerous enemies) show up, like shooting stars in the night sky. Human eyes have "edge detectors" to catch second-order differentials. They respond to the places where the rate of change of the rate of change of light hits zero—these mark the edges of things.

Soon second-order differentials will be passé, and the third order will be all the rage, with headlines reading "inflation's rate of increase is leveling off", or "growth is slowing quite quickly". Frenzy will then be a steady state. It will be high time to reverse the slide into perpetual differentiation. Workers of the world, integrate!

As we have seen in Chapter 2, the derivative of a function gives us a way of describing the behavior of a function. While the formula for a function will tell us the value of a function at any point, the derivative will tell us how the function is changing at any point. In addition, as we have seen in the preceding article, the second derivative tells us how the derivative is changing.

Class Exercises

These exercises refer to the article from *The Economist*.

1. Write a definition of the second derivative of a function. What information about the behavior of a function does it provide?

2. Discuss how the phrase from the second paragraph, "Nobody cares about inflation; only whether it is going up or down," might translate into the language of calculus.

3. Discuss how the last sentence of paragraph three might translate into the language of calculus.

4. Find three other phrases in the article that refer to how things change and translate them into the language of calculus.

5. Summarize the main point of the article using the terminology of first and second derivatives.

Up to now, the derivative of a function f has been symbolized by a new function f', $\frac{d}{dx}f$, or $\frac{df}{dx}$, representing the instantaneous rate of change of $f(x)$ with respect to x. If $y = f(x)$, then the derivative is also symbolized by the notation $\frac{dy}{dx}$ or y'. The second derivative, a function that is the derivative of the derivative, is symbolized by f'' or $\frac{d}{dx}\left[\frac{d}{dx}f\right]$. The brackets in the latter expression are usually dropped, and we replace the symbols dd with d^2 and $dxdx$ with dx^2, resulting in the notation $\frac{d^2}{dx^2}(f)$, which is equivalent to $\frac{d^2y}{dx^2}$ or y'' if $y = f(x)$. The 2's in these expressions are not exponents in the usual sense. They do not mean the squaring of a number; rather, they mean that the derivative operation is performed twice in succession.

The purpose of the following lab is to discover relationships between the graph of a function and its first and second derivatives. We will investigate intervals in which the function increases or decreases and locate maximum and minimum points. We will also investigate concavity and points where concavity changes. A graph that turns upward is called ***concave up***, while a graph that turns downward is called ***concave down***. A point on a graph where the concavity changes is called a ***point of inflection***. The graph shown in Figure 3.1 is concave up for $x < A$, for $B < x < C$, and for $x > D$. The graph shown in Figure 3.1 is concave down for $A < x < B$ and for $C < x < D$. Points of inflection occur at $x = A$, $x = B$, $x = C$, and $x = D$.

concave up
concave down
point of inflection

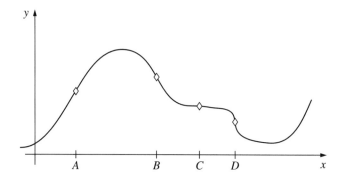

Figure 3.1 Illustration of concave up, concave down, and points of inflection

Lab 6 *Investigating Graphs and First and Second Derivatives*

The purpose of this laboratory activity is to explore relationships between graphs of functions and their first and second derivatives. In particular, we will study how positive, negative, and zero values of the first and second derivatives are related to some features of the original function, such as intervals of increase and decrease, turning points, concavity, and points of inflection.

We will begin the investigation in the lab with a simple polynomial function, move on to a more complicated polynomial, and then include rational and exponential functions. The final function we will investigate is one that does not have a derivative at every point in its domain.

1. Let $f(x) = x^3 - x$. Find $f'(x)$ and $f''(x)$.

2. Graph $y = f(x)$ and $y = f'(x)$ on the same axes using the intervals $x \in [-2, 2]$ and $y \in [-5, 5]$.

    a. What do you observe about the graph of $f(x)$ at those values of x where the derivative $f'(x)$ is positive?

    b. What do you observe about the graph of $f(x)$ at those values of x where the derivative $f'(x)$ is negative?

    c. What is happening on the graph of $f(x)$ at those values of x where the derivative $f'(x)$ changes sign?

3. Graph $f(x)$ and $f''(x)$ on the same axes.

    a. Describe the shape of the graph of $f(x)$ at those values of x where $f''(x)$ is positive.

b. Describe the shape of the graph of $f(x)$ at those values of x where $f''(x)$ is negative.

c. Describe the shape of the graph of $f(x)$ at those values of x where $f''(x)$ changes sign.

4. Repeat the analysis in parts 1–3 for the following functions over the indicated intervals:

a. $f(x) = x^4 - x^3 - 7x^2 + x + 5$ $x \in [-4, 4]$ $y \in [-25, 25]$

b. $f(x) = e^{-x^2/2}$ $x \in [-4, 4]$ $y \in [-1, 1]$

c. $f(x) = x + \frac{1}{x}$ $x \in [-4, 4]$ $y \in [-10, 10]$

d. $f(x) = xe^{-x}$ $x \in [-1, 4]$ $y \in [-1, 1]$

e. $f(x) = x^4$ $x \in [-2, 2]$ $y \in [-10, 20]$

5. For any function, $f(x)$, what information do the graphs of $f'(x)$ and $f''(x)$ give about the behavior of the graph of $f(x)$? If you were given just the graphs of $f'(x)$ and $f''(x)$, what features of f could you sketch?

6. All of the functions in part 4 are differentiable at all points in their domains. Now consider the function $f(x) = x^{2/3} + x$ where x is a real number. Note that f is defined at all points in its domain.

a. Find $f'(x)$ and $f''(x)$. Identify the domain of each of these derivatives.

b. Use $f'(x)$ and $f''(x)$ to describe the graph of the original function f. Include intervals in which the function increases or decreases, locations of maximum and minimum points, concavity, and points where the concavity changes. (Note that the fractional exponents in the equations for $f(x)$ and its derivatives sometimes cause inaccuracies in a graph produced using graphing calculators or computers. Use your understanding of functions to confirm that your graphs drawn with technology are correct.)

c. Use information from $f'(x)$ and $f''(x)$ to make a paper-and-pencil graph of $f(x)$.

d. If necessary, modify the generalizations you made in part 5 to account for the behavior of $f(x)$.

7. Describe a specific procedure for using a function's first and second derivatives to determine the function's intervals of increase and decrease, locations of maximum and minimum points, concavity, and inflection points. Your procedure should allow you to reach conclusions about the graph of $f(x)$ based on information from $f'(x)$ or $f''(x)$.

Summarize your observations and comments in a report. As usual, your observations are important, but your explanations and generalizations of those observations are even more important.

Exercise Set 3.1

1. Find $\frac{dy}{dx}$ and $\frac{d^2y}{dx^2}$ for each function.

   a. $y = e^{x^2}$

   b. $y = \sin(x^3)$

   c. $y = \ln(1 - x)$

   d. $y = 7x^2 + 6x + 9$

2. We are not necessarily limited to just first and second derivatives. We can also define the third derivative as the derivative of the second derivative, the fourth derivative as the derivative of the third derivative, and so on. The notation $\frac{d^n y}{dx^n}$ is used to symbolize the *n*th derivative of y with respect to x. Prime notation begins to become cumbersome after the third derivative [as in $f'''(x)$], so the notation

   $$f^{(n)}(x)$$

   nth derivative   is used to denote the **nth derivative** of f with respect to x. Find formulas for the *n*th derivatives indicated below. Your results should depend on n in each case. You may discern patterns by first finding $f'(x)$, $f''(x)$, $f^{(3)}(x)$ and so on.

   a. $\dfrac{d^n}{dx^n}(x^n)$

   b. $\dfrac{d^n}{dx^n}(e^{kx})$

   c. $\dfrac{d^n}{dx^n}(\ln x)$

   d. $\dfrac{d^n}{dx^n}[\ln(1 - x)]$

   e. $\dfrac{d^n}{dx^n}(\sin x)$

   f. $\dfrac{d^n}{dx^n}(\cos x)$

3. What do you know about the values of the first and second derivatives of a continuous function at a turning point of the function?

3.2   Optimization

Using Calculus to Solve Optimization Problems

We now have some understanding of the information that the first and second derivatives give us about the behavior of functions. We will look next at several more examples that illustrate how derivatives can be used. The following examples are all variations on the theme of optimization. Many of these problems involve finding the most efficient or best way of doing something, be it the most economical way to store inventory, the way to make the biggest box from a given size of cardboard, or the most efficient way to route telephone calls. Before we investigate these optimization problems, we summarize what we learned about derivatives in Lab 6.

1. If the first derivative is positive on an interval, then the function is increasing on that interval.

2. If the first derivative is negative on an interval, then the function is decreasing on that interval.

3. If the first derivative is zero or fails to exist at $x = a$ then the function *might* have a turning point at $x = a$.

4. If the first derivative changes from positive to negative, or vice versa, at $x = a$ and the function is continuous, then the function has a turning point at $x = a$.

5. If the second derivative is positive on an interval, then the function is concave up on that interval.

6. If the second derivative is negative on an interval, then the function is concave down on that interval.

7. If the second derivative is zero or fails to exist at $x = a$ then the function *might* have a point of inflection at $x = a$.

8. If the second derivative changes from positive to negative, or vice versa, at $x = a$ and the function is continuous, then the function changes concavity $x = a$, that is, the function has an inflection point at $x = a$.

In the following examples and discussion, we will show how these ideas can help solve optimization problems.

Example 1

Suppose we start with a piece of cardboard that measures 64 cm by 36 cm. We then cut identical squares from each corner and fold up the flaps to make an open box, as shown in Figure 3.2. Determine the dimensions of the box with the maximum possible volume.

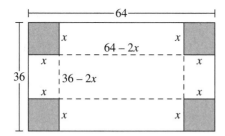 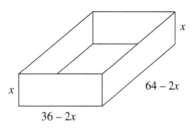

Figure 3.2 Cardboard for an open rectangular box

Solution The volume of a rectangular box is the product of the length, the width, and the height. By letting the variable x represent the length of a side of the squares to be cut from each corner of the cardboard, the length of the box can be expressed as $64 - 2x$, the width can be expressed as $36 - 2x$, and the height is simply x. Note that x can have values only between 0 and 18. The volume is

$$V(x) = (64 - 2x)(36 - 2x)x$$
$$= 2304x - 200x^2 + 4x^3.$$

From the graph of this function shown in Figure 3.3 we observe that the maximum volume occurs at an x value between 5 and 10, which we have labeled x^*.

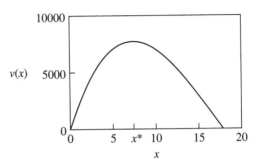

Figure 3.3 Graph of $V(x) = 2304x - 200x^2 + 4x^3$

Observe from the graph of $V(x)$ that for x values less than x^*, the lines tangent to the curve at these points have positive slope. It is a bit unwieldy to continue to talk about the slope of a line tangent to a curve, so we often shorten this expression to the ***slope of a curve***. In Figure 3.3 the slope of the curve is positive for $x < x^*$, and the value of $V(x)$ increases as x increases. For these x values, increasing the size of the square cut out of the cardboard causes an increase in the volume of the box constructed. For x values greater than x^*, the slope of the curve is negative (tangent lines have negative slope) and the value of $V(x)$ decreases as x increases. For these x values, increasing the size of the square cut out of the cardboard decreases the volume of the box constructed. The line tangent to the graph at $x = x^*$ is horizontal, so the slope of the line tangent to the graph at $x = x^*$ is zero.

Since the maximum value of V occurs at a point on the graph with a horizontal tangent line, we can determine the x value at the maximum by finding the derivative of the volume function and then finding the x value at which the derivative is equal to zero. The derivative of volume with respect to x is

$$V'(x) = 2304 - 400x + 12x^2.$$

To find where the derivative is equal to zero we must solve the equation

$$0 = 2304 - 400x + 12x^2.$$

The solutions to this quadratic equation are approximately 7.4 and 25.9. (The exact solutions are irrational numbers. It is impossible to attain such precision in the

slope of a curve

cutting process, so there is no need to use exact values here. All values given in the rest of this example will be approximate.) Since the context of this problem restricts the domain of $V(x)$ so that $0 \le x \le 18$, we can eliminate 25.9 as a solution. Referring to Figure 3.3, we see that over the domain of interest $V'(x)$ is zero at about 7.4, is positive for x values between 0 and 7.4, and is negative for x values between 7.4 and 18. The information about the derivative tells us that the volume function is increasing to the left of $x \approx 7.4$ and decreasing to the right of $x \approx 7.4$, so the point with $x \approx 7.4$ at which the function changes from increasing to decreasing gives the maximum volume. The maximum volume is approximately $V(7.4)$ or 7718.5 cubic centimeters. This analysis based on the derivative agrees with the graph in Figure 3.3. ■

The maximum volume in Example 1 also can be determined by graphing $V(x)$ and zooming in on the maximum point on the graph. This method of solution is perfectly legitimate and allows us to determine the maximum volume without using any calculus. (This method also can provide us with an approximate value only. As we noted earlier, exact answers are not needed, so this is acceptable.) So where do we take advantage of the power of the derivative? Are there optimization problems that require the use of calculus? We will address these questions as we look at another example.

Example 2

Water and wind have been used for centuries to generate mechanical and electrical energy. One type of machine that is used to extract energy from a stream of water or air has surfaces or paddles that move in the same direction as the flow of water or air. These machines are called drag machines, or vanes. The main goal when making a machine to extract energy from a stream of water or air is to maximize the amount of energy extracted. In this example we will look at the amount of energy this type of machine can extract from a stream of air.

When air with a density of ρ (*rho*) and velocity V strikes a surface of a vane with area A, the force F exerted on that surface is given by

$$F(x) = 0.5C\rho A(V - x)^2,$$

where C is a coefficient of pressure and x is the velocity of the moving surface. By a law of physics, the power P is force times velocity. This can be expressed by

$$P(x) = 0.5C\rho A(V - x)^2 x. \qquad (1)$$

If we replace the product of the constants with a new constant k, we can rewrite the power function as

$$P(x) = k(V - x)^2 x.$$

This theoretical equation expresses the power in terms of one variable x that can be controlled by changing the characteristics of the machine, and parameters k and V that are beyond our control. In terms of these two parameters, what velocity x will produce the maximum power?

Solution   For any positive values of the parameters k and V, $P(x)$ is a cubic polynomial that has a positive leading coefficient. Its graph has an x-intercept at $x = 0$ and an x-intercept of multiplicity two at $x = V$. Since the machine cannot turn faster than the wind, the domain of P is the interval $0 \le x \le V$. A graph of $P(x)$ over this domain is shown in Figure 3.4.

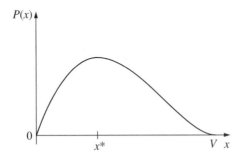

Figure 3.4 Graph of power model $P(x) = k(V - x)^2 x$

From the graph, we see that a velocity of $x = x^*$ produces the maximum power. If specific values of k and V are known, we can find an approximate value for x^* by using technology to graph $P(x)$ and then zooming in on the highest point on the graph. If specific values for k and V are not known, or if we want to generalize the relation between x^*, k, and V, then calculus provides a tool for finding the value of x^* in terms of k and V.

As in Example 1, the maximum value of P occurs at a point on the graph with a horizontal tangent line. We can find the exact location of this point by setting the derivative of $P(x)$ equal to 0 and solving for x. We can do this without knowing specific values of the parameters k and V.

We can find the derivative of $P(x)$ by using the product rule or by first performing the indicated multiplication. The second option is slightly easier. Multiplying the terms in $P(x)$ gives us

$$P(x) = k(V^2 x - 2Vx^2 + x^3),$$

and the derivative is

$$P'(x) = k(V^2 - 4Vx + 3x^2).$$

The right side of this equation can be factored so that

$$P'(x) = k(V-x)(V-3x).$$

Solving $P'(x) = 0$ gives us the values $x = V$ and $x = \frac{V}{3}$, which are precisely the x values of the points on the graph of $P(x)$ that have a horizontal tangent line. A velocity of V would correspond to the machine turning as fast as the wind. While this may sound ideal, it really means that there is no force on the surface of the machine; hence, no power is being generated. Substituting $x = V$ into equation (1), we find that $P(V) = 0$.

Provided k and V are positive, the derivative $P'(x) = k(V-x)(V-3x)$ is positive for $0 < x < \frac{V}{3}$ and negative for $\frac{V}{3} < x < V$, so we know that the value $x = \frac{V}{3}$ produces the maximum power. ∎

As we have already mentioned, there is an important difference between the two previous examples. Since the box problem of Example 1 has no parameters, we can find the maximum volume by graphing and tracing, without using calculus. In Example 2, because the model contains parameters, we could not answer the question by a graph and trace method. In this situation we must use the derivative.

The First Derivative Test

In the preceding two examples we located turning points of a function by finding where the derivative is equal to zero. However, the relationship between the first derivative and the location of turning points is not as simple as finding where a derivative is equal to zero. We are not guaranteed a turning point where the derivative is zero. For instance, $f(x) = x^3$ does not have a turning point at $x = 0$ even though $f'(0) = 0$. Similarly, $g(x) = x + \sin x$ does not have a turning point at $x = \pi$ even though $g'(\pi) = 0$.

To decide if a function really has a turning point, we must also check that the function changes from increasing to decreasing or vice versa at the potential turning point. In addition, we must remember that a function can have a turning point where the derivative does not exist. The functions $f(x) = |x|$ and $f(x) = x^{2/3}$ both have a turning point at $x = 0$, where the derivative does not exist. In both of these cases, $f'(x)$ changes sign at $x = 0$. However, $f(x) = x^{1/3}$ does not have a turning point at $x = 0$ even though f' fails to exist at $x = 0$. In this case, f' does not change sign at $x = 0$.

This discussion is summarized in what is known as the ***first derivative test***:   *first derivative test*

1. If $f'(a) = 0$ and f' changes sign at $x = a$, then f has a turning point at $x = a$.

2. If $f(x)$ is continuous, $f'(a)$ does not exist, and f' changes sign at $x = a$, then f has a turning point at $x = a$.

Concavity and the Second Derivative Test

Now we will examine the relationship between concavity and the second derivative. Why does the fact that a function has a positive second derivative guarantee that the function's graph is concave up? Since the second derivative is the derivative of the first derivative, a positive second derivative means the first derivative is increasing. If the first derivative is increasing, then the curve must have an increasing slope.

The graphs in Figure 3.5 show two possibilities for a curve with an increasing slope. In the graph on the left, the slope is positive and increasing as we move along the curve from $x = a$ to $x = b$. Notice that the graph gets progressively steeper as the slope increases. In the graph on the right, the slope is negative and increasing (which means it is getting closer to zero) as we move along the curve from $x = a$ to $x = b$. Notice that the graph gets progressively flatter as the slope increases. In both cases the graph is concave up.

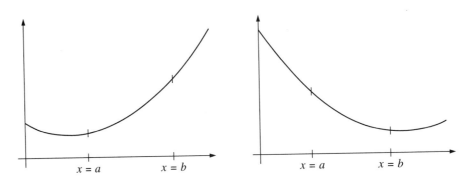

Figure 3.5 Graphs with increasing slope

Why is a function's graph concave down where the function has a negative second derivative? The second derivative being negative means that the first derivative is decreasing, and since the first derivative is decreasing the graph of the original function must have a decreasing slope. The graphs in Figure 3.6 show two possibilities for a curve with a decreasing slope. In the graph on the left, the slope is positive and decreasing as we move along the curve from $x = a$ to $x = b$. Notice that the graph is getting progressively flatter as the slopes decrease. In the graph on the right, the slope is negative and decreasing (getting farther from zero) as we move along the curve from $x = a$ to $x = b$. Notice that the graph is getting progressively steeper as the slopes decrease. In both cases, the graph is concave down.

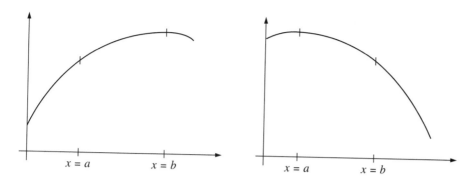

Figure 3.6 Graphs with decreasing slope

Finally, why does the graph of a function have a point of inflection when the second derivative changes sign? When the second derivative changes sign at $x = a$, the concavity must be changing from up to down or from down to up, so by definition the graph of the original function has a point of inflection at $x = a$. Convince yourself that a value of zero for the second derivative does not guarantee a point of inflection, just as a value of zero for the first derivative does not guarantee a turning point.

In the following example, we will see how the concept of concavity can help solve optimization problems.

Example 3

Let us return to the open box problem of Example 1, but now let the length of the cardboard used to form a box be L, and let the width be W. Determine the dimensions of the box with the maximum possible volume.

Solution   If we let x represent the length of a side of the squares to be cut from each corner, the equation for the volume is

$$V(x) = (L - 2x)(W - 2x)x.$$

We can find the maximum volume by finding where the graph of V versus x has a tangent line with slope of zero. Thus, we need to take the derivative of V with respect to x and find where $V'(x) = 0$. It is easier to find the derivative if the volume equation is first rewritten as

$$V(x) = LWx - 2(L + W)x^2 + 4x^3.$$

Differentiating with respect to x and remembering that L and W are constants gives

$$V'(x) = LW - 4(L + W)x + 12x^2.$$

To find where $V'(x)$ is zero, we need to solve the equation

$$0 = LW - 4(L + W)x + 12x^2.$$

This is a quadratic equation in x that has solutions

$$x_1 = \frac{L + W + \sqrt{L^2 - LW + W^2}}{6}$$

and

$$x_2 = \frac{L + W - \sqrt{L^2 - LW + W^2}}{6}.$$

The facts that $V'(x_1) = 0$ and $V'(x_2) = 0$ tell us that the graph of $V(x)$ has horizontal tangent lines at $x = x_1$ and at $x = x_2$. The second derivative $V''(x)$ can help us determine whether these horizontal tangents correspond to a maximum or a minimum value of V. The second derivative of the volume function is

$$V''(x) = -4(L + W) + 24x.$$

When we evaluate $V''(x)$ at $x = x_1$ and at $x = x_2$, we get

$$V''(x_1) = 4\sqrt{L^2 - LW + W^2},$$

which is greater than 0, and

$$V''(x_2) = -4\sqrt{L^2 - LW + W^2},$$

which is less than 0. At x_1, the first derivative is zero and the second derivative is positive; therefore, the graph of $V(x)$ is concave up and has a horizontal tangent line. We have located a minimum. At x_2, the first derivative is zero and the second derivative is negative. The graph of $V(x)$ is concave down and has a horizontal tangent line, and thus we have located a maximum. The maximum volume occurs at $x = x_2$, which means that the square to be cut from the corner has sides of length

$$x = \frac{L + W - \sqrt{L^2 - LW + W^2}}{6}. \qquad \blacksquare$$

Example 3 again shows the power of the derivative to solve a general problem involving parameters. The solution tells us how to achieve the maximum box volume for any values of L and W. In Example 1, we began with a rectangle with length 64 cm and width 36 cm. Using the general solution from Example 3, we maximize the volume of this box by cutting corner squares with sides of length

$$x = \frac{64 + 36 - \sqrt{64^2 - 64 \cdot 36 + 36^2}}{6} \approx 7.4,$$

which matches our solution to Example 1.

The technique of using the second derivative to determine whether a turning point is a maximum or minimum value, as we did in Example 3, is summarized in what is known as the *second derivative test*, which is stated as follows:

second derivative test

1. If $f'(a) = 0$ and $f''(a) > 0$, then f has a local minimum at $x = a$.

2. If $f'(a) = 0$ and $f''(a) < 0$, then f has a local maximum at $x = a$.

3. If $f'(a) = 0$ and $f''(a) = 0$, then the test is inconclusive.

Example 4

Suppose a closed cylindrical can has fixed volume V. What dimensions of the can minimize the amount of material used to construct this can?

Solution   Let the can have radius r and height h. The surface area A consists of the area of the top and bottom and the area of the side walls as shown in Figure 3.7 and is given by the formula $A = 2\pi r^2 + 2\pi rh$. This equation expresses the surface area as a function of the two variables r and h.

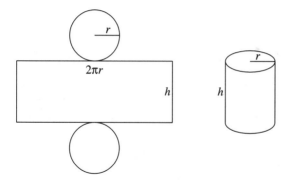

Figure 3.7 Surface area of a closed cylinder

Minimizing the amount of material used to construct the can requires that we find the least possible surface area of the can. In order to use the derivative to minimize the surface area, we first need to express the surface area in terms of a single variable. We could use either r or h. We know that the volume V of a cylindrical can is given by

$$V = \pi r^2 h.$$

For a fixed volume V we will solve for h in terms of V and r, yielding

$$h = \frac{V}{\pi r^2},$$

which then can be substituted into the surface area equation to give

$$A = 2\pi r^2 + \frac{2V}{r}. \tag{2}$$

Since we are using a fixed volume, V is a constant; thus, equation (2) expresses the surface area of the can as a function of a single variable r. The surface area is the sum of the quadratic term $2\pi r^2$ and the reciprocal term $\frac{2V}{r}$. The quadratic term adds little to the sum when r is small, while the reciprocal term adds little to the sum when r is large; therefore, we expect the graph of A to resemble $\frac{2V}{r}$ for small values of r and $2\pi r^2$ for large values of r. Figure 3.8 shows the graph of A versus r for the relationship given in equation (2) with $V = 1000$.

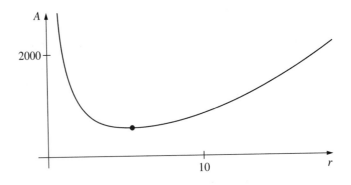

Figure 3.8 Graph of $A = 2\pi r^2 + \dfrac{2V}{r}$ with $V = 1000$

As we can see in Figure 3.8, the minimum surface area occurs at the point where the graph of A versus r has a horizontal tangent line. The derivative of A with respect to r is given by

$$\frac{dA}{dr} = 4\pi r - \frac{2V}{r^2}.$$

We can set $\frac{dA}{dr}$ equal to zero and solve for r to determine where this horizontal tangent line occurs.

$$0 = 4\pi r - \frac{2V}{r^2}$$

$$4\pi r = \frac{2V}{r^2}$$

$$4\pi r^3 = 2V$$

$$r^3 = \frac{V}{2\pi}$$

We find that $\frac{dA}{dr} = 0$ at the point where $r = \left(\frac{V}{2\pi}\right)^{1/3}$.

We can also use the second derivative test to confirm that $r = \left(\frac{V}{2\pi}\right)^{1/3}$ does indeed produce a minimum surface area. The second derivative of A with respect to r is given by

$$\frac{d^2A}{dr^2} = 4\pi + \frac{4V}{r^3}.$$

Since V is a positive constant and r is also positive, $\frac{d^2A}{dr^2}$ is always positive. This means that the graph of A versus r is concave up over the entire domain $r > 0$. Since we have found the r value for which $\frac{dA}{dr}$ is zero, we can be sure that we have found the r value that minimizes A.

What are the dimensions of the can that minimize the surface area? If $r = \left(\frac{V}{2\pi}\right)^{1/3}$ and $h = \frac{V}{\pi r^2}$, by substitution we find that $h = \left(\frac{4V}{\pi}\right)^{1/3}$. The can with volume V that uses the least amount of material must have a radius of $\left(\frac{V}{2\pi}\right)^{1/3}$ and a height of $\left(\frac{4V}{\pi}\right)^{1/3}$. Notice that the ratio of r to h is given by

$$\frac{r}{h} = \sqrt[3]{\left(\frac{V}{2\pi}\right)\cdot\left(\frac{\pi}{4V}\right)}$$

which simplifies to $\frac{1}{2}$. The radius is one-half the height. This means the can should have a square shape when viewed from the side. The large food cans used in school cafeterias often have this square shape. ∎

Whenever we have an optimization problem in which we want to find the maximum or minimum value of a function f that involves no parameters, we have the choice of solving the problem using derivatives or using a graph-and-trace strategy. Which choice is better and more efficient depends on the problem and on the technology available to us. In some situations, the function f includes one or more unknown parameters in addition to the independent variable x. In these situations, calculus allows us to find an expression that gives the maximum or minimum value of $f(x)$ and the x value where it occurs in terms of the parameters. This expression enables us to optimize $f(x)$ for all values of the parameters. By substituting values for these parameters, we might be able to solve all specific examples of our optimization problem with essentially no additional work. Furthermore, the expressions derived using the tools of calculus often give us insight into the general relationships that produce the optimal values.

Exercise Set 3.2

1. Find where the following functions have maximum, minimum, and inflection points. Support your answers.

    a. $f(x) = x^3 - 3x - 4$

    b. $g(x) = Ax^4 - Bx^2 - C,\ A > 0,\ B > 0,\ C > 0$

    c. $y = x^3 - x + 1$

    d. $y = x^4 - 2x^3 + x^2 - p$

    e. $m(x) = e^{-x^2}$

    f. $y = x + \frac{p}{x},\ p > 0$

    g. $r(x) = x + \sin x$

    h. $y = |A - x^2|,\ A > 0$

    i. $f(x) = \frac{4}{x^2 + 1}$

    j. $g(t) = t \cdot e^{Ct},\ C > 0$

k. $q(x) = \ln(x^2 + 3x + 4)$ l. $f(x) = \sin^2 x$

m. $g(t) = |\sin t|$ n. $y = e^x \cos x$

2. Suppose you know that $f'(a) > 0$ and $f''(a) < 0$. Sketch a graph of $f(x)$ for x values near $x = a$.

3. Suppose you know that $f'(a) < 0$ and $f''(a) = 0$. Sketch a graph of $f(x)$ for x values near $x = a$.

4. For what values of x is the quantity $x^3(15 - x^2)$ increasing most rapidly?

5. Let f be the function that has domain $[-1, 4]$ and range $[-1, 2]$. Let $f(-1) = -1$, $f(0) = 0$, and $f(4) = 1$. Also let f have the derivative function f' that is continuous and that has the graph shown in Figure 3.9.

    a. Find all values of x for which f has a turning point that is a maximum. Justify your answer.

    b. Find the value of x for which f assumes its minimum value on its entire domain. Justify your answer.

    c. Find the intervals on which f is concave downward.

    d. Give all values of x for which f has a point of inflection.

    e. Sketch a graph of f.

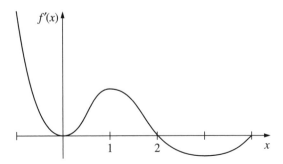

Figure 3.9 Graph of the derivative function, f'

6. Sketch graphs of the following functions. Locate the turning points using calculus. How do the turning points compare with the points where $f(x)$ and $g(x)$ touch the envelope curves, which are given after the function?

    a. $f(x) = x \cos x$ $y = x, y = -x$

    b. $g(x) = \frac{\sin x}{x}$ $y = \frac{1}{x}, y = -\frac{1}{x}$

7. A silo is built in the shape of a cylinder topped by a hemisphere. The top of the silo costs twice as much per unit area to build as does the side of the silo. What dimensions will be most cost effective for a silo that holds a volume V? (Note: Assume the silo is not filled beyond the top of the cylinder.)

8. If you want to send a package through the U.S. mail you will find instructions at the post office indicating that the length plus the girth cannot exceed 108 inches. This regulation is designed to keep the packages a reasonable size for shipping and handling. What size package with a rectangular base has the largest possible volume and meets the condition listed above?

9. A particle moves along the x-axis so that at time t its position is given by $x(t) = \sin(\pi t^2)$ for $-1 \le t \le 1$. Assume that when the particle is moving to the right the velocity is positive.

   a. Find the velocity at time t. When is the particle moving the fastest?

   b. Find the acceleration at time t. When is the acceleration greatest?

   c. For what values of t does the particle change direction?

   d. Find all values of t for which the particle is moving to the left.

10. A rectangle $ABCD$ with sides parallel to the coordinate axes is inscribed in the region enclosed by the graph of $y = -4x^2 + 4$ and the x-axis, as shown in Figure 3.10. Find the x- and y-coordinates of C so that the area of the rectangle is a maximum.

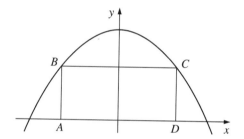

Figure 3.10 Rectangle inscribed in a parabola

11. What are the dimensions of the largest cylinder that can be inscribed in a sphere of radius R?

12. Suppose that $f(t)$ measures the level of oxygen in a pond, where $f(t) = 1$ is the normal level and the time t is measured in weeks. When $t = 0$, some organic waste is dumped into the pond, and as the waste material oxidizes, the amount of oxygen in the pond is given by

$$f(t) = \frac{t^2 - t + 1}{t^2 + 1}, \quad 0 \le t < \infty.$$

   a. Find $f'(t)$ and interpret its meaning.

   b. When is the oxygen level lowest?

   c. When is the oxygen level highest?

13. In an effort to decrease tooth decay and also allow people to eat candy, Mercury Candy Company hires you to study acid levels in the mouth after eating candy. Your lab assistant finds that the function below describes the pH level in the mouth t minutes after eating candy. The pH level is used to measure acidity or alkalinity. On a scale from 0 to 14, 0 represents the highest level of acid and 14 the most basic or alkaline. Neutral is designated as 7.

$$A(t) = 6.5 - \frac{20.4t}{t^2 + 36}$$

Assume that the lower the pH level the more likely tooth decay. The dentist would prefer you eat no candy.

a. Sketch a graph of the function. Label scales on the axes and important points.

b. Using the graph, estimate the normal pH level in the mouth. How many minutes after candy is eaten does it take the mouth to get back to within 0.1 of normal pH level?

c. On the basis of this model and knowing that Mercury Candy's goal is for us all to eat candy, what would you advise Mercury to suggest to people about brushing teeth after eating candy? Give your reasoning.

d. When is the acid level in the mouth increasing after eating the candy? When is it decreasing? Justify your answers.

e. At what time after eating the candy does the mouth begin to recover and begin the change back to normal pH levels? Justify your answer.

Reference
Lancaster, Ron, presented at the 7th Annual Conference on Secondary School Mathematics and Technology, Exeter, NH, 1991.

14. A second type of wind machine used to extract energy from a stream of air is the rotor-type, or the traditional windmill. The function for the power which this type of machine can extract from the air is $P(x) = 2kx^2(V - x)$, where k is a constant, V is the velocity of the wind, and x is the average of the wind speeds in front and behind the windmill. Using this formula for the power that the windmill can extract from the wind, Albert Betz, a German scientist, theorized in 1927 that the maximum power that a windmill can extract from the wind is $\frac{16}{27}$ times the power in the wind stream, which is given by $P = \frac{k}{2}V^3$. Verify Betz's conclusion.

Reference
Betz, Albert, *Introduction to the Theory of Flow Machines*, Pergamon Press, Oxford, England, 1966.

15. A dealer in sports cards has a rare card for sale. Naturally, the idea is to optimize the profits from the sale. The dealer can either sell the card today for $\$C$ or hold onto the card and sell it later, hopefully for more money. Because of the price history, rarity and condition of the card, and other factors, the dealer believes the card will always increase in value. The value of the card (and many other collectibles) typically grows according to the rule

$$V(t) = Ce^{k\sqrt{t}},$$

where C is the value at $t = 0$ and k is a positive constant. The dealer can store the card at no cost. What would be the best time to sell the card?

Notice that the function $V(t) = Ce^{k\sqrt{t}}$ has no maximum value. This seems to mean that the dealer should hold the card forever. But there is another factor to be considered—the interest that could be earned on the money obtained from the sale. Selling the card for $120 now and investing the money so that it earns an additional 6% is one option; another is holding the card for 5 years and then selling it for $135 dollars. Even though $135 is greater than $120, the $120 earning 6%, compounded continuously, would be worth $161.98 in 5 years.

Assume the proceeds of any sale is placed into an account with an interest rate of r and the interest is compounded continuously. The *present value* of V is given by $P(t) = V(t) \cdot e^{-rt}$. For example, if a card is estimated to have a value in 5 years of $V(5) = \$100$, the present value would be $P(5) = V(5) \cdot e^{-5r}$, which equals $74.08 with $r = 0.06$. What this means is that $74.08 today will be worth $100 in 5 years if the $74.08 is deposited in an account that earns 6% annual interest compounded continuously. Thus, present values can be compared, and the dealer wishes to maximize the present value of the investment.

a. Determine the optimum waiting time t in terms of k and r for the present value function P.

b. If $k = 0.5$ and $r = 0.085$, when should the card be sold?

c. Suppose a rare coin can be sold for $2500 today and the proceeds placed in an account that pays 9.7% annual interest compounded continuously. Using the same function for $V(t)$ with $k = 1$, how long should the coin be held before it is sold? What price should the coin command at the time of sale?

d. The value of timber already planted grows according to the equation $V(t) = 250\,000\,e^{2\sqrt{t}}$. How long should the timber grow before being cut and sold if r is the annual interest rate on the continuous-compounding account in which the money can be deposited? What price will the timber bring when it is sold?

Reference

Chiang, Alpha C., *Fundamental Methods of Mathematical Economics*, McGraw-Hill Inc., New York, 1967.

16. The *Extreme Value Theorem* states that if a function f is continuous on the closed interval $[a, b]$, then f has a maximum and minimum value on $[a, b]$. Furthermore, the maximum and minimum values occur where the derivative of f is zero or does not exist, or at an endpoint of the interval. Draw graphs that demonstrate this theorem. Then draw graphs that indicate why continuity is a necessary hypothesis in this theorem. Also use graphs to show why the theorem is only guaranteed on a closed interval.

Extreme Value Theorem

3.3   Derivatives Applied to Economics

Maximum Profit

Many questions in economics are formulated in terms that make calculus a natural tool to use in finding answers. In this section, we will use calculus to investigate how a firm maximizes profit.

First we define cost, revenue, and profit functions as

$C(x) =$ total cost for producing x items (includes marketing, maintaining inventory, general and administrative costs),

$R(x) =$ revenue generated by selling quantity x, and

$P(x) =$ profit for quantity x produced and sold.

Profit is the difference between revenue and cost, so

$$P(x) = R(x) - C(x). \tag{3}$$

If, for example, $R(x) = 4x$ and $C(x) = 0.1x^2 + 20$, then $P(x) = 4x - 0.1x^2 - 20$. The graph of P is a parabola that is concave down. The maximum profit occurs at the vertex of this parabola, which is where $P'(x) = 0$. Since $P'(x) = 4 - 0.2x$, the maximum profit occurs at the point where $x = 20$. Maximizing profit is also equivalent to finding the value of x that gives the greatest difference between $R(x)$ and $C(x)$, assuming $R(x)$ is above $C(x)$. This corresponds to finding the greatest vertical distance between the graphs of $R(x)$ and $C(x)$, as illustrated in Figure 3.11.

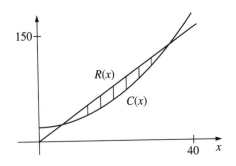

Figure 3.11 Graphs of cost and revenue functions

Applications of the Second Derivative to Economics

The graph of $P(t)$ versus t shown in Figure 3.12 illustrates the profits for a business over time. In this example, the profits are increasing at first, then they begin to decrease, but eventually the business recovers and profits once again rise. At what time might an astute business owner sense that the business was beginning to recover?

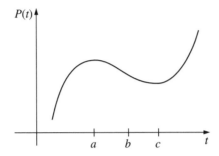

Figure 3.12 Graph of profits over time

We might refer to the turning point at $t = c$ after which profits begin to rise as the beginning of the recovery; however, the increase in profits also might have been anticipated before this turning point is reached. Certainly, after passing the turning point at $t = a$, the outlook for the business is rather gloomy. Profits are not only falling, but they are falling more steeply as time goes on. In other words, the slope of the curve is negative, and the slope is moving farther from zero. In the language of calculus, $P'(t)$ is negative and $P'(t)$ is decreasing. Since $P'(t)$ is decreasing, $P''(t)$ is negative, which gives the downward concavity of the graph for values of t near a.

As we move past the point of inflection at $t = b$ on the graph, the profits still decrease, but not as quickly as to the left of $t = b$. To the left of $t = b$ the curve is getting steeper over time, and to the right of $t = b$ the curve begins to level off. This means that although $P'(t)$ is still negative, $P'(t)$ begins increasing, which implies that $P''(t)$ is positive. A positive second derivative is connected with the upward concavity of the graph to the right of $t = b$. The point with $t = b$ marks the beginning of what eventually becomes an upturn (a term suggested by the shape of the graph) in the profitability of the business.

We now examine another economic concept related to the second derivative. Suppose a parcel of farmland is used for growing corn. Today's farmer has many factors to consider in order to maximize the land's yield. In addition to the initial investment in land, purchased or leased, decisions must be made regarding factors such as types of seed; pest and weed control; and, most importantly, the amount of fertilizer. Each of these factors has an associated cost and benefit. To maximize return, the farmer must make decisions about how much money to allocate to each resource.

Suppose that an existing farm initially spends about $180 per acre for fertilizer, harvesting, and so on, and the expected yield of the harvest is 110 bushels of corn per acre. If the farmer spends an additional $20 per acre on resources for the same amount of land, the expected yield might rise to 125 bushels per acre. Now, if another $20 is spent on resources, the expected yield might rise to 150 bushels per acre. Add another $20, and the yield increases to 165 bushels per acre. The next investment increases the yield to 170 bushels per acre. What do these figures imply?

Initially, the extra return from additional capital is increasing from 15 for the first investment to 25 for the second investment. Then the extra return from the third declines to 15 bushels. This decline in extra returns continues as the fourth adds an extra 5 bushels. Although the extra returns for each additional investment are initially increasing, eventually the extra returns begin to diminish.

law of diminishing returns   In general, the ***law of diminishing returns*** refers to the observation that adding equal increments of a variable input (investment, in our example) to a constant amount of a fixed input (such as land) eventually results in successively decreasing extra outputs (in this case, bushels of corn). This diminish in extra returns is a result of the fact that the increments of the variable input have a progressively decreasing amount of the fixed resources with which to work. The law is summarized below.

> The Law of Diminishing Returns: An increase in some inputs, when other inputs are fixed or held constant, will cause total output to increase; but after a point the same additions of extra inputs is likely to result in less and less extra outputs.

Let x be a variable input to a production process, and let $Q(x)$ be the amount produced for a given value of x. Assume that for an input amount x, an increase in the input by Δx yields an increased output of ΔQ. Suppose the graph of Q versus x has the shape shown in Figure 3.13. The point to which the law of diminishing returns refers is *point of diminishing* called the ***point of diminishing returns***, which is located at $x = a$ on the graph in *returns* Figure 3.13.

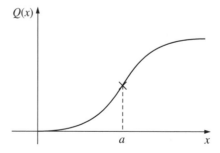

Figure 3.13 Point of diminishing returns

Up to the point where $x = a$, ΔQ increases for each increase Δx. This means that the values of $\frac{\Delta Q}{\Delta x}$ are increasing for $x < a$. For $x > a$, ΔQ decreases for each increase Δx, or in other words $\frac{\Delta Q}{\Delta x}$ is decreasing for $x > a$. For sufficiently small Δx, the derivative $Q'(x)$ is approximately equal to $\frac{\Delta Q}{\Delta x}$. The law of diminishing returns implies that

$$Q'(x) \text{ is increasing if } x < a, \text{ and}$$
$$Q'(x) \text{ is decreasing if } x > a. \tag{4}$$

If the function $Q'(x)$ is increasing, then its derivative $Q''(x)$ is positive, and if the function $Q'(x)$ is decreasing, then its derivative $Q''(x)$ is negative. The statements in (4) imply that

$$Q''(x) > 0 \text{ if } x < a, \text{ and}$$
$$Q''(x) < 0 \text{ if } x > a. \tag{5}$$

The sign of the second derivative relates to concavity, so the statements in (5) imply that $Q(x)$ is concave up for $x < a$, and $Q(x)$ is concave down for $x > a$. The change in concavity at $x = a$ means that the graph of $Q(x)$ has a point of inflection at $(a, Q(a))$. We conclude that the point of diminishing returns occurs at this point of inflection.

Exercise Set 3.3

1. The location of the greatest difference between $R(x)$ and $C(x)$ in Figure 3.11 appears to occur where the slopes of the two curves are equal. Use the derivative of profit given in equation (3) to show that this is true.

2. Suppose a manufacturer can sell x items a week for a revenue of $r(x) = 200x - 0.01x^2$ cents, and the total cost for x items is $c(x) = 50x + 20000$ cents. What is the most profitable number of items to make each week?

3. Suppose the revenue and cost functions (in thousands of dollars) for a manufacturer are $r(x) = 9x$ and $c(x) = x^3 - 6x^2 + 15x$, where x is in thousands of units. What amount should be produced to generate the most profit?

4. For the cost function in problem 3, state a function that gives the average cost per item. What production level minimizes average cost?

5. Someone makes the statement: "Test scores are down again this year, but not as much as in previous years." Would you take this to mean that things are getting better? Justify your reasoning.

6. The President of the United States might claim that the economy is recovering because, during the last quarter, (a) the rate of increase in inflation declined; (b) the climb of unemployment slowed; or (c) the increase in the national debt

declined. How do these statements relate to the first derivative? How do the statements relate to the second derivative? By the reasoning used in these statements, what special point on each curve appears to signal the beginning of economic recovery? Is this reasoning always correct?

7. Being a point of inflection, the point of diminishing returns for the curve in Figure 3.13 is the location of the maximum value of $Q'(x)$, where $Q(x)$ is the quantity of output for x units of input. If the selling price of each unit of output is p and the cost per unit of input is c, the profit function P is given by the equation

$$P(x) = pQ(x) - cx - k,$$

where k includes costs that do not depend on the amount of input (such as marketing, general, and administrative costs). What is true of $P'(x)$ at the point of diminishing returns, *i.e.* the point where $Q'(x)$ is a maximum? How is profit changing at the point of diminishing returns?

3.4 Optimization and Developing Models: The Rest of the Story

The optimization problems that we have studied up to this point have had one thing in common: in each case we had a function for which we wanted to find minimum or maximum value over a particular domain. We started with a model and proceeded to use calculus to determine properties of that model. In contrast, sometimes mathematicians and scientists notice that a phenomenon has certain properties, including maximum and minimum values, and then use these properties to create a model for the phenomenon. In the following example, we start with an energy model and determine a variable value that minimizes energy. Following the example, we investigate how a model with the given properties could have been derived.

Example

Researchers have studied the way salmon swim upstream to spawn. In theory, the energy that the salmon expend in swimming against the current depends on the speed at which they swim as well as on the amount of time they spend swimming. Researchers theorize that the amount of energy E expended is given by

$$E(v,t) = cv^3t,$$

where v is the velocity of the salmon in still water (measured in miles per hour), t is the time the salmon spends swimming (measured in hours), and c is a positive constant of proportionality. Suppose a particular salmon needs to swim to a spawning ground that is 200 miles upstream in a river whose current is flowing at a speed of 4 miles per hour. At what speed should the salmon swim in order to minimize the energy expended?

Solution Assuming that the salmon's speed in still water is v, then the effective speed against the current is $v - 4$ miles per hour. If the salmon is to make any progress upstream, the speed at which it swims must exceed 4 miles per hour. The time t required to swim 200 miles is related to the salmon's effective speed by the equation

$$t = \frac{200}{v - 4},$$

since time is equal to distance divided by rate. This allows us to express the energy function in terms of v alone, which is

$$E(v) = cv^3 \frac{200}{v - 4}. \tag{6}$$

Differentiating the energy function $E(v)$ in equation (6) will allow us to find its minimum value. The derivative of $E(v)$ with respect to v is given by

$$E'(v) = cv^3 \cdot \frac{(-1)200}{(v - 4)^2} + 3cv^2 \cdot \frac{200}{(v - 4)},$$

which can be factored as

$$E'(v) = \frac{cv^2 \cdot 200}{v - 4} \cdot \left(\frac{-v}{v - 4} + 3 \right).$$

Since $v > 4$, the factor $\frac{cv^2 \cdot 200}{v - 4}$ is greater than zero. For $E'(v)$ to equal zero, the factor $\left(\frac{-v}{v-4} + 3 \right)$ must be zero. The equation $\frac{-v}{v-4} + 3 = 0$ has the solution $v = 6$. Since $E'(6) = 0$ we know that the graph of $E(v)$ has a horizontal tangent line at $v = 6$. To check if this corresponds to a minimum value, we can check the sign of the derivative to the left and to the right of $v = 6$. E' is continuous over its domain and $v = 6$ is the only value for which $E'(v)$ is zero. Because of this, the values for $E'(v)$ where $v < 6$ all have the same sign; similarly, the values for $E'(v)$ where $v > 6$ all have the same sign. Since $E'(5.5)$ is a negative number and $E'(6.5)$ is a positive number, the graph of $E(v)$ is decreasing to the left of $v = 6$ and is increasing to the right of $v = 6$. (We could have evaluated E' at any point in the intervals $4 < v < 6$ and $v > 6$, but both intervals must be checked.) By the first derivative test, this guarantees that E has a minimum value when $v = 6$. ∎

Generalizing our work in the preceding example, we observe that if the salmon swim a distance D miles upstream and the current flows at a rate of r miles per hour, then the energy function is given by

$$E(v) = cv^3 \cdot \frac{D}{v - r}.$$

We will show in the exercises that follow that the rate at which the salmon should swim to use the least energy is $v = 1.5r$, which means that the energy-minimizing speed for the salmon is 1.5 times the speed of the current. A graph of energy versus speed for the general function E over the domain $v > r$ is shown in Figure 3.14.

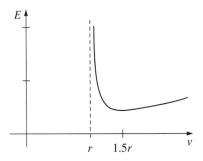

Figure 3.14 Graph of the energy function $E(v) = cv^3 \cdot \dfrac{D}{v - r}$

We can be sure that the salmon do not have the function E to guide them in determining how fast to swim. They instinctively know how fast to swim, and researchers assume that these fish instinctively use the minimum energy possible to reach their spawning waters. The researchers inferred the equation for $E(v)$ by observing the habits of the salmon. So, in a sense, we have done the problem backwards. The most energy-efficient speed to swim is not found by knowing the energy expenditure function E. The function E is actually found by knowing the most efficient speed to swim. By observing many fish in many streams, the researchers noticed that most fish swam approximately 1.5 times the rate of the current as they headed for their spawning waters. They then wrote a function to model the energy spent as a function of the salmon's speed. We now analyze how this model was created.

The model must be consistent with the observation that for v values close to r, the value of $E(v)$ is very large, since the salmon would take a long time to travel the given distance. In addition, the model should match the observation that $E(v)$ is minimized when $v = 1.5r$. The model was then designed so that the graph of $E(v)$ has two essential characteristics: a vertical asymptote at $v = r$, and a minimum value at $v = 1.5r$.

A simple form for the function E is a rational function whose equation is

$$E(v) = c \cdot D \cdot \frac{v^p}{v - r} \tag{7}$$

where the exponent p and the constant c are both positive. Equation (7) reflects the fact that the energy expended is dependent on the distance D the salmon must swim. The $v - r$ in the denominator corresponds to the vertical asymptote at $v = r$. The exponent p influences the location of the minimum of $E(v)$. If we take the derivative of E as given here we find that the minimum energy expenditure occurs at the velocity $v = 1.5r$ only when $p = 3$, which will be verified in the exercises.

The model for salmon energy expenditure was created to match an observed minimum value, which is a method for creating a model often used in the real world. Observing and using extreme points and inflection points in the process of building models is a useful tool for mathematicians and scientists.

Exercise Set 3.4

1. Consider the salmon energy expenditure equation

$$E(v) = c \cdot D \cdot \frac{v^p}{v-r}.$$

   a. In the example in the text we assumed that $p = 3$. Using $p = 3$, show that $v = 1.5r$ is the rate at which the salmon should swim to use the least amount of energy.

   b. Let $p = 1$. Find the velocity that results in the minimum expenditure of energy. Do you think using $p = 1$ results in a realistic model? Why or why not?

   c. Let $p = 2$. Find the velocity that results in the minimum expenditure of energy. Do you think using $p = 2$ results in a realistic model? Why or why not?

   d. Take the derivative of E with respect to v (assuming p is a constant) and show that $p = 3$ is the only value for which the energy is minimized when $v = 1.5r$.

2. Since tooth decay occurs when the pH level in the mouth is lowest, a candy company designs their candy so that the pH level reaches a minimum of 5.0 three minutes after a person eats candy. Assume that the pH level in the mouth is normally 6.5. A function of the form $A(t) = a - \frac{bt}{t^2+c}$ can be used to model this situation. Determine values of a, b, and c that allow the model to fit the company's designs.

3. Suppose that researchers notice that one type of salmon swims upstream slower than others. Instead of swimming at a rate 1.5 times the current, these fish swim at a rate 1.25 times the current. Making the assumption that these fish are still using a minimum amount of energy to get to their spawning grounds, revise the energy model for this type of salmon.

3.5   Finding Zeros Using Newton's Method

zero

root

For a function f, an x value at which the value $f(x)$ is zero is called a ***zero*** of the function. This x value is also called a ***root*** of the equation $f(x) = 0$. We know that we can learn a lot about a function f by finding the zeros of $f(x)$, $f'(x)$, or $f''(x)$. In particular, we know that:

–   The zeros of $f(x)$ tell us where the graph of f *might* change from positive to negative or vice versa, meaning the graph might cross the horizontal axis.

–   The zeros of $f'(x)$ tell us where the graph of $f(x)$ *might* change from increasing to decreasing or vice versa, meaning the graph might have a turning point.

–   The zeros of $f''(x)$ tell us where the graph of $f(x)$ *might* change from concave up to concave down, or vice versa, meaning the graph might have an inflection point.

The wealth of information that can be obtained by finding the zeros of a function as well as its first and second derivatives gives special importance to the general problem of finding zeros of functions. In algebra courses, zeros are often found using methods such as factoring polynomials. Finding zeros of linear functions is especially easy, and the quadratic formula provides a simple method to solve quadratic equations. In the real world, we often encounter functions which are not solved easily using algebra. In these cases a numerical method often is used. We now describe three frequently used numerical methods.

1.   Computer software or a graphing calculator can be used to zoom in on a small portion of the graph of a function containing the point where the graph crosses the horizontal axis. Listing a table of values over a small interval that contains the zero can be used in place of or in conjunction with the zooming strategy.

bisection algorithm

2.   The ***bisection algorithm*** is another method for locating zeros. We begin the algorithm by selecting points x_1 and x_2 on the x-axis, where one of the x values, say x_1, has a negative function value and the other, say x_2, has a positive function value. Assuming the function is continuous, this assures us the function has a zero between the two initial x values. We then find the value midway between these two x values. We replace one of the original x values (in Figure 3.15, we replace x_1) with the midpoint x value because the function value of the replaced x value has the same sign as the function value at the midpoint; call the new x value x_3. We still have two x values, x_2 and x_3, whose function values are opposite in sign, and thus a zero of the function must be between these two x values. We repeat the process of creating a new interval between x values that is half as long as the previous interval, always keeping the zero "trapped" within the new interval. We continue repeating the process until the length of the interval shrinks to a size that allows us to specify the zero to within the desired accuracy.

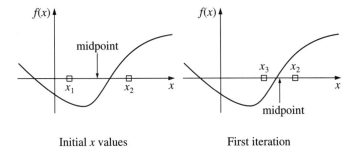

Figure 3.15 First iteration of the bisection algorithm

3. The *secant algorithm* is based on the concept that the algebraic solution of linear   *secant algorithm*
equations is easy, so we approximate the function near the zero with a secant line.
We expect the secant line to have a zero at approximately the same location as the
function. To begin this algorithm, we choose any two points p_1 and p_2 on the
graph of the function f and find an equation for the secant line through these two
points. We solve for the x-intercept of this secant line, which we call x^*. We
replace p_1 with a new point with coordinates $(x^*, f(x^*))$ and label the point p_2.
We change the label on the other point from p_2 to p_1 (see Figure 3.16). We now
draw another secant through the new points p_1 and p_2. We find the x-intercept of
the secant line and again repeat the replacement process for p_1 and p_2. This method
will take us toward a zero of the function provided we are on a part of the curve
that slopes toward the zero. The secant algorithm has several advantages over the
bisection algorithm. First, the two starting points do not have to have function
values that are opposite in sign. Second, in most cases, the secant algorithm
locates a zero much faster than the bisection algorithm.

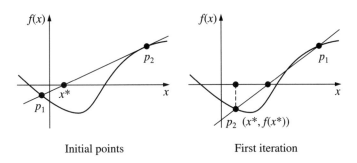

Figure 3.16 Secant algorithm

Newton's method

Since calculus allows us to find the equations of tangent lines to curves, we can refine the secant algorithm by using tangent lines instead of secant lines. The refined algorithm is called *Newton's method*. We begin our explanation of Newton's method with the following example.

Example

Use Newton's method to find a zero of the function $f(x) = x^2 - 7$.

Solution   Although we know the zeros of this function are $\pm\sqrt{7}$, we will examine how Newton's method finds one of these zeros. We begin by choosing some x value at which we find our initial tangent line. We choose $x = 1$ as our initial guess, which gives us a function value $f(1) = -6$ and a slope $f'(1) = 2$. The equation of the line tangent to the graph of f at the point $(1, -6)$ is

$$y = 2(x - 1) - 6.$$

The x-intercept of this tangent line is found by setting y equal to 0 and solving $0 = 2(x - 1) - 6$, which gives $x = 4$. Figure 3.17 shows the tangent line at $(1, -6)$ and its x-intercept $(4, 0)$. We now replace our initial x value of 1 with the x-intercept $x = 4$, and repeat the process of finding the x-intercept of the tangent line. Since $f(4) = 9$ and $f'(4) = 8$, the equation of the tangent line at $x = 4$ is $y = 8(x - 4) + 9$. This tangent line has an x-intercept of 2.875. As we observe from the graphs in Figure 3.17, the x-intercept of the second tangent is a better approximation for $\sqrt{7}$ than the x-intercept of the initial tangent.

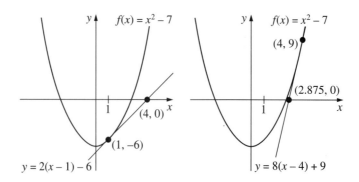

Figure 3.17 Graphs of Newton's method for $f(x) = x^2 - 7$

We continue the process of using the x-intercept of a tangent line to generate a new tangent line. Each iteration yields a new x-intercept that is closer to the zero than the previous x-intercept. Table 3.18 shows the results (rounded to three decimal places) for five iterations of Newton's method. Notice that the final two x-intercepts are the same rounded to three decimal places. This indicates that $x = 2.646$ is a good approximation for a zero of $f(x)$. We verify by noting that $f(2.646) \approx 0.001$. ∎

x value	tangent line	x-intercept
1	$y = 2(x - 1) - 6$	4
4	$y = 8(x - 4) + 9$	2.875
2.875	$y = 5.75(x - 2.875) + 1.266$	2.655
2.655	$y = 5.31(x - 2.655) + 0.049$	2.646
2.646	$y = 5.292(x - 2.646) + 0.001$	2.646

Table 3.18 Tangent lines and x-intercepts

Newton's method is based on the concept of linear approximation. So far in this course we have seen that a tangent line gives a reasonable local approximation for a function f. As a result, the zero of the tangent line is often a good approximation for a zero of the function. In each iteration of Newton's method, we essentially replace the given function with a tangent line, find the zero of the tangent line, and use this value as our guess for a zero of the function. The zeros of the succession of tangent lines provide a sequence of approximations for a zero of the function. If the difference between these approximations and the zero becomes arbitrarily small as we continue to iterate, then we say that Newton's method **converges** to the zero.

converges

We now generalize the procedure for Newton's method used in the preceding example. Let f be the function for which we are finding a zero, and let x_0 be the initial x value. The initial tangent line has the equation

$$y = f'(x_0)(x - x_0) + f(x_0).$$

The x-intercept of the tangent line occurs at the point with $y = 0$, so the x-intercept is the solution to the equation

$$f'(x_0)(x - x_0) + f(x_0) = 0.$$

Subtracting $f(x_0)$ from both sides yields

$$f'(x_0)(x - x_0) = -f(x_0).$$

Solving for x gives

$$x = x_0 - \frac{f(x_0)}{f'(x_0)}.$$

We use x_1 to represent this x-intercept, which is then used to generate the next tangent line. The tangent line through $(x_1, f(x_1))$ has equation

$$y = f'(x_1)(x - x_1) + f(x_1).$$

In similar fashion, we find the x-intercept of this tangent line and call it x_2. Therefore x_2 is given by the equation

$$x_2 = x_1 - \frac{f(x_1)}{f'(x_1)}.$$

In general, the next x value is determined from the previous x value by the equation

$$x_n = x_{n-1} - \frac{f(x_{n-1})}{f'(x_{n-1})}, \tag{8}$$

which generates a sequence of x values that approach a zero of $f(x)$. Eventually, each iteration of equation (8) generates a better approximation for a zero of the function. The process can be halted when the required number of digits in the approximation does not change from one iteration to the next. In this text, we will use this rule of thumb: any digit that remains unchanged in three consecutive iterations is considered accurate.

Exercise Set 3.5.A

1. Use Newton's method to find one zero of each function to the nearest thousandth.

    a. $y = x \cos x$ b. $y = t^5 - 12$ c. $y = \ln x - 3$

2. What problems could arise by having the derivative in the denominator of equation (8)?

3. In what situations will Newton's method fail to converge to a zero of the function?

4. Two different ways to specify a stopping criterion in Newton's method are: bounds on the y values, $|f(x_n)| < \varepsilon$, and bounds on the x values, $|x_n - x_{n-1}| < \varepsilon$, for some small number ε. In what situations would one criterion be preferred over the other? Which method is recommended in the text (the rule of thumb)?

5. A student was thinking about Newton's method and decided to try a new iterative formula using the equation

$$x_n = x_{n-1} - \frac{f'(x_{n-1})}{f''(x_{n-1})}$$

to generate a sequence of numbers. What important feature of the function f occurs at the point found with this iteration scheme?

6. On most computers, the computation of square roots is done with an iterative scheme. To find $\sqrt{k}$, the computer iterates the equation

$$x_n = \frac{1}{2}\left(x_{n-1} + \frac{k}{x_{n-1}}\right).$$

Show that Newton's method applied to $x^2 - k = 0$ can be rewritten in this form.

7. Find an iterative equation similar to the one in problem 6 for finding cube roots.

8. Use Newton's method to find the zero of $f(x) = \frac{1}{x} - c$. Use your result to explain how one can use technology to evaluate reciprocals without any division operations.

Basins of Attraction

One of the interesting aspects of Newton's method is how the choice of the initial value influences whether Newton's method converges and, if so, to which zero the algorithm converges. Let us investigate the behavior of Newton's method applied to $f(x) = x(x-2)(x+4)$. We know the zeros of f are 0, 2, and –4. Which starting values cause Newton's method to converge to each of these zeros? What choices for an initial value cause Newton's method to fail?

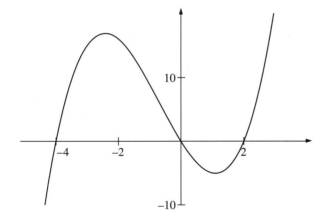

Figure 3.19 Graph of $f(x) = x(x-2)(x+4)$

The graph shown in Figure 3.19 suggests that two choices for initial x values will cause Newton's method to fail. These are the x values where the curve has a horizontal tangent, which is where the derivative of $f(x)$ equals zero. We can expand $f(x)$ so that

$$f(x) = x^3 + 2x^2 - 8x,$$

and its derivative is given by

$$f'(x) = 3x^2 + 4x - 8.$$

Using the quadratic formula, we find that the solutions to $f'(x) = 0$ are

$$x = \frac{-2 \pm 2\sqrt{7}}{3}.$$

If either of these two values is used as the initial x value, then Newton's method will fail on the first iteration. By extension, any initial x value that yields either $x = \frac{-2+2\sqrt{7}}{3}$ or $x = \frac{-2-2\sqrt{7}}{3}$ after one iteration will cause Newton's method to fail on the second iteration. One such value is shown graphically in Figure 3.20. Similarly, any initial x value that yields these x values after two iterations will cause Newton's method to fail on the third iteration. Extending this to three iterations, four iterations, and so on, tells us that there are many choices for an initial value that will cause Newton's method to fail eventually.

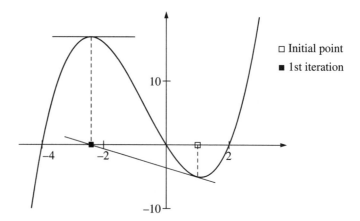

Figure 3.20 Location where Newton's method fails on the second iteration

How likely is it that Newton's method will fail at some iteration? By substitution into equation (8), the recursive equation for the algorithm applied to $f(x) = x(x-2)(x+4)$ is

$$x_n = x_{n-1} - \frac{x_{n-1}(x_{n-1}-2)(x_{n-1}+4)}{3x_{n-1}^2 + 4x_{n-1} - 8}. \tag{9}$$

If Newton's method fails on the nth iteration, then x_{n-1} must equal $\frac{-2+2\sqrt{7}}{3}$ or $\frac{-2-2\sqrt{7}}{3}$. If $x_{n-1} = \frac{-2+2\sqrt{7}}{3}$, then a previous value x_{n-2} that leads to x_{n-1} in one iteration is a solution to the equation

$$\frac{-2+2\sqrt{7}}{3} = x_{n-2} - \frac{x_{n-2}(x_{n-2}-2)(x_{n-2}+4)}{3x_{n-2}^2 + 4x_{n-2} - 8}.$$

Because the left side of this equation is an irrational number and all the coefficients and constants on the right side are rational, the solutions must be irrational. Just as the values that reach $\frac{-2+2\sqrt{7}}{3}$ in one iteration are irrational, the values that reach $\frac{-2+2\sqrt{7}}{3}$ in two iterations are likewise irrational. The same is true for values that iterate to $\frac{-2-2\sqrt{7}}{3}$.

This process can be extended to values that iterate to $\frac{-2\pm2\sqrt{7}}{3}$ in 3 iterations, 4 iterations, and so on. All of these values are irrational numbers. We cannot represent irrational numbers exactly on a computer or a calculator; therefore, Newton's method implemented with technology will never yield one of these numbers. So although in theory there are infinitely many such points, in practice Newton's method will not iterate to a point where the derivative is exactly zero.

To which zero of $f(x) = x(x-2)(x+4)$ will Newton's method converge if we choose an initial value that does not cause the algorithm to fail? If we choose an initial value of $x = 2$, the sequence of values generated by $x_n = x_{n-1} - \frac{f(x_{n-1})}{f'(x_{n-1})}$ will stay at $x = 2$ because $f(2) = 0$. Similar behavior will result if we choose an initial value of $x_0 = 0$ or $x_0 = -4$. What happens if we choose an initial value of $x_0 = -3$? Figure 3.21 shows a graph with a sequence of tangent lines, t_1, t_2, and t_3, created by Newton's method and a table with the beginning of the sequence of values generated by Newton's method. An initial value of $x_0 = -3$ causes the algorithm to converge to the zero at $x = -4$.

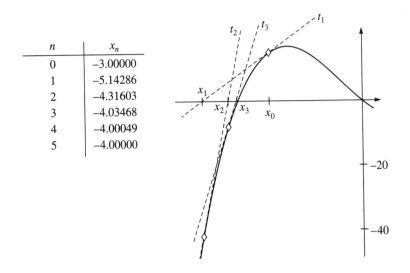

n	x_n
0	−3.00000
1	−5.14286
2	−4.31603
3	−4.03468
4	−4.00049
5	−4.00000

Figure 3.21 Graph and table of values for Newton's method with $x_0 = -3$

Suppose we choose an initial value of $x_0 = 1.5$. Table 3.22 shows the sequence of values generated by recursive equation (9). An initial value of $x_0 = 1.5$ causes Newton's method to converge to the zero at $x = 2$.

n	x_n
0	1.5
1	2.36842
2	2.06480
3	2.00262
4	2.00000

Table 3.22 Table of values for Newton's method with $x_0 = 1.5$

Suppose we choose an initial value of $x_0 = 0.9$. Table 3.23 shows the sequence of values generated by recursive equation (9). The table shows that, using an initial value of $x_0 = 0.9$, Newton's method converges to the zero at $x = 0$.

n	x_n
0	0.9
1	-1.56244
2	0.39648
3	-0.07388
4	-0.00122
5	0.00000

Table 3.23 Table of values for Newton's method with $x_0 = 0.9$

What will happen if we use an initial value of $x_0 = 3$, that is, to what zero of $f(x)$ will the sequence of values converge? Based on the shape of the graph and first two tangent lines of Newton's method shown in Figure 3.24, we predict that using $x_0 = 3$ causes Newton's method to converge to the zero at $x = 2$. Similarly, we can predict that an initial value of $x_0 = -5$ causes Newton's method to converge to the zero at $x = -4$. Table 3.25 shows some other initial values and the zeros to which Newton's method converges.

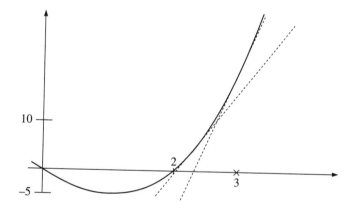

Figure 3.24 First approximations using $x_0 = 3$

Initial value x_0	Zero found	Initial value x_0	Zero found
−5.5	−4	0.92	−4
−2.6	−4	0.95	2
−2.45	−4	0.96	−4
−2.4305	2	1.09	−4
−1.9	2	1.0975	2
−1.5	0	1.1	2
0.5	0	1.8	2
0.8	0	3.6	2

Table 3.25 Zeros found by Newton's method for different initial values

basin of attraction

Associated with each zero of $f(x)$ is a set of x values called the **basin of attraction** for that zero. If used as the initial value for Newton's method, each point in the basin of attraction of a zero will generate a sequence of values that converges to that zero. Table 3.25, together with the graph of $f(x) = x(x-2)(x+4)$, indicates that x values less than the x value of the left-most turning point, which is $\frac{-2-2\sqrt{7}}{3}$ or approximately −2.431, belong to the basin of attraction of the zero at $x = -4$. Likewise, x values greater than the x value of the right-most turning point, which is $\frac{-2+2\sqrt{7}}{3}$ or approximately 1.097, belong to the basin of attraction of the zero at $x = 2$. An interval of x values between −2.431 and 1.097 that contains 0 belongs to the basin of attraction of the zero at $x = 0$. Near the turning points, however, the boundaries of the basins of attraction are not easily determined. For example, an initial value of $x_0 = -2$ converges to the right-most zero at $x = 2$, whereas an initial value of $x_0 = -1.7$ converges to the middle zero at $x = 0$. An initial value of $x_0 = -1.82$, which lies between these two values, converges to the left-most zero at $x = -4$. The boundary between the basins of attraction appears not to be simple; rather, the outcome of using Newton's method with initial values that are near the turning points on the graph is erratic and sensitive to small changes in the initial value.

In order to see the basins more clearly, it may help if we color each point on the real number line to indicate in which basin of attraction it belongs. (Technology that will allow you to assign color to points on a graph would be best.) Assign the color red to each point in the basin of attraction of the zero at $x = -4$, assign the color blue to each point in the basin of attraction of the zero at $x = 0$, and assign the color green to each point in the basin of attraction of the zero at $x = 2$. A picture of the basins of attraction is red to the left of the left-most turning point, blue in an interval containing $x = 0$, and green to the right of the right-most turning point. The coloring of the number line around the turning points is somewhat chaotic, as Newton's method does not behave in a simple way for these starting values. Between the band of red on the left and the blue in the middle are smaller bands of all three colors. Likewise, between the band of green on the right and blue in the middle are smaller bands of all three colors. Many points on the number line are not colored. They correspond to x values for which Newton's method fails to converge. When we color the number line to indicate basins of attraction, we are referring to all the points that actually converge with Newton's method. We ignore the non-convergent initial values that lie in the colored intervals.

Exercise Set 3.5.B

1. For the function $f(x) = x^2 - 9$, what are the zeros of f? What initial value causes Newton's method to fail on the first iteration? What are the basins of attraction of the zeros of f using Newton's method?

2. For the function $f(x) = (x+2)(x-2)(x+5)$, what initial values cause Newton's method to fail on the first iteration? Investigate what happens to starting values that are near the turning points of f. What are the basins of attraction of the zeros of f using Newton's method?

3.6 Related Rates

In many phenomena that we investigate, several variables change at the same time. For example, when we inflate a spherical balloon, the volume of the balloon, the radius of the balloon, and the surface area of the balloon all change over time. Since the volume, radius, and surface area are all related to each other, the rates of change of each of these variables are also related to each other. To study the relationships between these three rates of change, we begin with the relationships between the radius r, the volume V, and the surface area S, which are given by the following geometric formulas for a sphere.

$$V = \tfrac{4}{3}\pi r^3 \tag{10}$$

$$S = 4\pi r^2 \tag{11}$$

Since V, r, and S are all changing with respect to time, assume that each of these quantities can be expressed as a function of time. That is, $r(t)$ gives the radius in terms of time, $V(t)$ gives the volume in terms of time, and $S(t)$ gives the surface area in terms of time. We do not actually need to find equations for $r(t)$, $V(t)$, or $S(t)$ to investigate the relationships between their rates of change. The assumption that such functions exist allows us to find the relationship between the rates of change with respect to time of r, V, and S. The rates of change $\frac{dr}{dt}$, $\frac{dV}{dt}$, and $\frac{dS}{dt}$ are examples of *related rates*.

related rates

We begin the process of investigating these related rates by specifying relationships between the functions V, r, and S, which are given by the geometric formulas in (10) and (11). We next differentiate these functional relationships with respect to time, which involves applications of the chain rule. V is a function of r in equation (10), and r is a function of t, so when we use the chain rule to differentiate V with respect to t, we have $\frac{dV}{dt} = \frac{dV}{dr} \cdot \frac{dr}{dt}$. Since $\frac{dV}{dr} = 4\pi r^2$, by substitution we can write

$$\frac{dV}{dt} = 4\pi r^2 \frac{dr}{dt}. \tag{12}$$

Similarly, we know that S is a function of r and r is a function of t. When we differentiate S with respect to t, we have $\frac{dS}{dt} = \frac{dS}{dr} \cdot \frac{dr}{dt}$. Since $\frac{dS}{dr} = 8\pi r$, we know that

$$\frac{dS}{dt} = 8\pi r \frac{dr}{dt}. \tag{13}$$

Equation (12) specifies a relationship between the three quantities $\frac{dV}{dt}$, r, and $\frac{dr}{dt}$. If we know numerical values for any two of these quantities, equation (12) allows us to find a numerical value for the third quantity. Suppose we know that the balloon is inflated in such a way that 5 cubic inches of air are added to the balloon each second, which means that $\frac{dV}{dt}$ has a constant value of 5 cubic inches per second. Substituting $\frac{dV}{dt} = 5$ in equation (12) yields

$$5 = 4\pi r^2 \frac{dr}{dt}.$$

This equation allows us to determine the rate of change of the radius $\frac{dr}{dt}$ for any particular value of r. For example, to find the rate at which the radius is changing when the radius is 3 inches, we can substitute 3 for r and determine that

$$5 = 4\pi \cdot 3^2 \cdot \frac{dr}{dt},$$

or

$$\frac{dr}{dt} = \frac{5}{36\pi} \approx 0.044 \text{ in/sec.}$$

In addition, we can use equation (13) to find the rate at which the surface area is changing when the radius is 3 inches. Assuming that $\frac{dV}{dt} = 5$, $r = 3$, and $\frac{dr}{dt} = \frac{5}{36\pi}$, we substitute these numerical values in equation (13) to determine that

$$\frac{dS}{dt} = 8\pi \cdot 3 \cdot \frac{5}{36\pi}$$

which simplifies to

$$\frac{dS}{dt} = \frac{10}{3} \approx 3.333 \text{ in}^2/\text{sec.}$$

Therefore, if the volume of the balloon is changing at the constant rate of 5 cubic inches per second, at the time when the radius is 3 inches the radius is increasing at the rate of 0.044 inches per second and the surface area is increasing at the rate of 3.333 square inches per second.

Example 1

A truck driver is driving on a straight road toward a perpendicular intersection with a straight railroad track. The truck is moving at 55 mph, and a 1-mile long train is traveling toward the intersection at 80 mph. When the truck is 3 miles from the intersection and the train is 4 miles from the intersection, at what rate is the distance between the front of the truck and the front of the train changing? At that same moment, at what rate is the distance changing between the front of the truck and the end of the train?

Solution   First, we draw a diagram to illustrate the problem posed and identify our variables, as shown in Figure 3.26. Let x be the distance between the front of the train and the intersection, and let y be the distance between the front of the truck and the intersection. Let the distance between the front of the truck and the front of the train be denoted by h. The relationship between x, y, and h, given by the Pythagorean theorem, is

$$h^2 = x^2 + y^2.$$

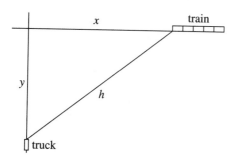

Figure 3.26 Train and truck approaching an intersection

Since we are interested in the rate of change of h, we need to differentiate both sides of this equation with respect to time. If two quantities are equal, then their derivatives are equal, and since x, y, and h are actually functions of time, we have

$$\frac{d}{dt}(h^2) = \frac{d}{dt}(x^2) + \frac{d}{dt}(y^2). \tag{14}$$

To illustrate the next step more clearly, let $S = h^2$, then

$$\frac{dS}{dt} = \frac{d}{dt}(h^2).$$

By the chain rule, since S is a function of h and h is a function of t, we know that

$$\frac{dS}{dt} = \frac{dS}{dh} \cdot \frac{dh}{dt}.$$

Substituting for $\frac{dS}{dh}$ gives

$$\frac{dS}{dt} = 2h \cdot \frac{dh}{dt},$$

so that

$$\frac{d}{dt}(h^2) = 2h \cdot \frac{dh}{dt}.$$

Similar relationships hold for the derivatives of x^2 and y^2, and equation (14) becomes

$$2h\frac{dh}{dt} = 2x\frac{dx}{dt} + 2y\frac{dy}{dt},$$

which simplifies to

$$h\frac{dh}{dt} = x\frac{dx}{dt} + y\frac{dy}{dt}. \tag{15}$$

Equation (15) tells us how the six quantities h, $\frac{dh}{dt}$, x, $\frac{dx}{dt}$, y, and $\frac{dy}{dt}$ are related to each other. The fact that the truck is driving toward the intersection at 55 mph means that $\frac{dy}{dt} = -55$. The value of $\frac{dy}{dt}$ is negative because the distance from the truck to the intersection is decreasing. Since the train is approaching the intersection at 80 mph we also know that $\frac{dx}{dt} = -80$. We are interested in the moment when $y = 3$ miles and $x = 4$ miles, at which time the truck and the train are $h = 5$ miles apart. Substituting all of these values into equation (15) gives

$$5\frac{dh}{dt} = 4(-80) + 3(-55). \tag{16}$$

We want to find the rate at which the distance between the truck and the front of the train is changing, which means that we want to find a value for $\frac{dh}{dt}$. Dividing both sides of equation (16) by 5 and simplifying yields

$$\frac{dh}{dt} = -97.$$

The distance between the front of the truck and the front of the train is decreasing at a rate of 97 miles per hour at the instant when $x = 4$ and $y = 3$.

To find how the distance between the front of the truck and the end of the train is changing, we use a new variable X to represent the distance from the intersection to the end of the train and a new variable H to represent the distance from the front of the truck to the end of the train, as shown in Figure 3.27. We know that $H^2 = X^2 + y^2$, so we can differentiate both sides of this equation with respect to time to get

$$2H\frac{dH}{dt} = 2X\frac{dX}{dt} + 2y\frac{dy}{dt},$$

or

$$H\frac{dH}{dt} = X\frac{dX}{dt} + y\frac{dy}{dt}. \tag{17}$$

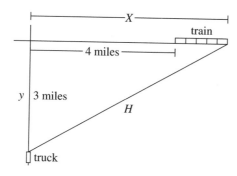

Figure 3.27 Distance to the end of the train

Since the train is 1 mile long, we know that $X = 5$ and $H = \sqrt{34}$ at the moment when $x = 4$ and $y = 3$. Since the train has a constant length, the end of the train approaches the intersection at the same rate as does the front of the train, so $\frac{dX}{dt} = -80$. Substituting these values along with $y = 3$ and $\frac{dy}{dt} = -55$ into equation (17) gives us

$$\sqrt{34} \cdot \frac{dH}{dt} = 5 \cdot (-80) + 3 \cdot (-55),$$

which simplifies to

$$\frac{dH}{dt} = \frac{5(-80) + 3(-55)}{\sqrt{34}} \approx -96.9.$$

The distance between the truck and the end of the train is decreasing at a rate of about 96.9 miles per hour. Note that at this instant of time the end of the train is approaching the truck at a slightly slower rate than the front of the train is approaching the truck. ∎

Example 2

In 1993 the United States population was approximately 258 million and increasing at a rate of about 2.79 million people per year. The per capita energy consumption in the United States was approximately 326 million BTU's (British thermal units) in 1993. In the same year the total energy use in the United States was increasing at a rate of about $1.9 \cdot 10^{15}$ BTU's per year. At what rate was the per capita energy consumption increasing in 1993?

Reference
U.S. Bureau of the Census, *Statistical Abstract of the United States*: 1994 (114th edition). Washington, DC, 1994.

Solution   We first need to identify our variables and specify a relationship between the variables. Let P represent the population, let E represent the per capita energy consumption, and let T represent the total amount of energy consumed. The relationship between these variables is

$$T = P \cdot E. \qquad (18)$$

All three of these variables are functions of time, so if we differentiate equation (18) with respect to time, we get

$$\frac{dT}{dt} = P\frac{dE}{dt} + E\frac{dP}{dt}. \qquad (19)$$

The given information tells us that in 1993, $P = 2.58 \cdot 10^8$, $\frac{dP}{dt} = 2.79 \cdot 10^6$, $E = 3.26 \cdot 10^8$, and $\frac{dT}{dt} = 1.9 \cdot 10^{15}$. Substituting these values into equation (19) yields

$$1.9 \cdot 10^{15} = (2.58 \cdot 10^8)\frac{dE}{dt} + (3.26 \cdot 10^8)(2.79 \cdot 10^6).$$

We can now solve for $\frac{dE}{dt}$, so that

$$\frac{dE}{dt} \approx 3.84 \cdot 10^6.$$

This tells us that in 1993 the per capita energy consumption in the United States was increasing at a rate of 3.84 million BTU's per year. ∎

Exercise Set 3.6

1. A balloon in the shape of a sphere is inflated at a rate of 100 cubic meters per minute. At what rate is the radius increasing when the radius is 10 meters? At what rate is the circumference increasing at this instant?

2. The length of each edge of a cube is increasing at a rate of 2 cm per second. When the length of each edge is 20 cm, at what rate is the volume of the cube changing? At what rate is the surface area changing?

3. When air expands without any temperature change, the pressure P and the volume V satisfy the relationship $PV^{1.4} = C$, with C a constant. At a certain instant the pressure is 50 lb/in^2, the volume is 32 in^3, and the volume is decreasing at a rate of 4 in^3/sec. How rapidly is the pressure changing at this instant?

4. Two students are traveling along a highway, trying to get to class on time. A police officer is hovering in a helicopter 1000 feet directly above the highway using a radar gun to determine the speed at which they are traveling. When the students are 2000 feet from the helicopter and moving away, the radar gun registers 85 feet per second. The speed limit on the highway is 95 feet per second (65 mph). Were they speeding? Assume that the highway is straight and level.

5. A coffee maker has the shape of a cone 10 cm high with a radius of 6 cm at the top. The coffee maker is filled with water, and the water is allowed to drain out

the bottom of the cone. Water is flowing out of the filter at a rate of 5 cm³/sec. At what rate is the water level in the cone falling when it is 5 cm from the bottom of the cone?

6. A point moves along the graph of $y = x^{2.5}$ so that its x-coordinate increases at a constant rate of 2 units/second. Find the rate at which the y-coordinate is increasing at the moment when the coordinates of the point are (4, 32).

7. Determine how the gravitational force between two bodies changes with respect to time if they are moving apart at a constant rate s. Newton's law for the gravitational force between two bodies is given by

$$F = G \cdot \frac{m_1 m_2}{R^2},$$

where G is the universal gravitational constant, m_1 and m_2 are the masses of the two bodies, and R is the distance between the bodies.

8. Kim stands 35 meters from the launch point of a hot air balloon.

   a. If the balloon rises straight up at 8 m/sec, how fast must Kim's eyes raise to keep the balloon in sight when it is 50 meters high?

   b. If the balloon rises along a path toward Kim at an angle of 70 degrees with the ground and at a rate of 8 m/sec along the path, how fast must Kim's eyes raise when the balloon is 50 meters high?

9. Fat deposits are building up on the wall of an artery with a cross section that is a circle of radius r. (For our purposes, assume the buildup is uniformly covering the wall.) At what rate is the cross-sectional area changing in terms of the thickness of the fat deposit?

10. A Ferris wheel with radius 10 meters turns at a uniform rate of 2 revolutions per minute.

    a. Find the rate at which a passenger's height above ground level changes if the seats clear the ground by one meter.

    b. Graph the rate at which the height is changing at any time .

    c. Use the fact that arc length is defined as radius times radian measure to relate the speed (change in arc length) to the angular velocity. If a law requires the speed to be less than 5 meters per second, how fast can the Ferris wheel turn?

11. A tanker accident has spilled oil in Pristine Bay. The oil slick has the form of a circular cylinder 500 feet in radius and 0.01 feet thick, but no new oil is added to the slick. As the slick spreads, the thickness of the oil is decreasing at a rate of 0.001 feet per hour. How fast is the area of the top of the slick changing?

12. Using the same coffee maker as in problem 5, suppose that water is flowing into the maker at a rate of 7 cm³/sec and coffee is flowing out the bottom at a rate of 6 cm³/sec.

a. If a total of 1000 cm^3 flows through the coffee maker, will the coffee maker overflow?

b. Find a function for the rate of change in the level of water in the coffee maker for any time t. Graph this function.

c. Modify your answer in part b to include the time after you stop pouring in the water until the time that the coffee maker is empty.

13. A light is on top of a pole h feet high. A ball is dropped from a point at the same height as the light but k feet horizontally away from it. How fast is the shadow of the ball moving along the ground $\frac{1}{2}$ second later? Assume that the ball falls a distance of $16t^2$ feet in t seconds.

14. The demand for certain goods is often determined by the price at which it is being sold. Suppose the number of one such item (n in thousands) that will be purchased per month is related to the price (p in dollars) by the equation $10n + 30p = 320$.

a. If p is currently $6.00, and p is increasing by $0.25 per month, at what rate is the demand changing?

b. If n is currently 3 (thousand items) and n is increasing at a rate of 0.25 (thousand) per month, at what rate is p changing?

15. The relation between price (p in dollars) and demand (d in thousands) for a certain product is given by $5p^2d + pd = 500,000$.

a. If the present price is $100 and the price is increasing at a rate of $15 per year, at what rate is the demand changing?

b. If the present demand is 19 (thousand) and is decreasing at a rate of 1 (thousand) per year, what is happening to the price?

3.7   Implicit Differentiation

On occasion we will want to study relationships that do not fit the mathematical definition of a function. In these relations, some values of the independent variable are paired with more than one value of the dependent variable. For example, suppose a man-made satellite follows an elliptical path around the earth with equation

$$\frac{x^2}{4620^2} + \frac{y^2}{4600^2} = 1 \tag{20}$$

where the units are in miles and the earth is located at the point $(-430, 0)$. A graph of this ellipse is shown in Figure 3.28. To find $\frac{dy}{dx}$, we could solve equation (20) for y in terms of x and then find the derivative using the techniques developed in Chapter 2.

This method is cumbersome, and in some problems solving for y in terms of x is not even possible, so we will not find $\frac{dy}{dx}$ by solving equation (20) for y in terms of x.

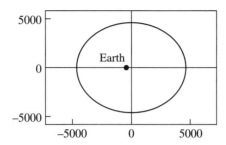

Figure 3.28 Graph of $\dfrac{x^2}{4620^2} + \dfrac{y^2}{4600^2} = 1$

As Figure 3.29 suggests, each point on the ellipse has a unique tangent line, although the tangents at (4620, 0) and (–4620, 0) are vertical lines and have undefined slopes. Even though this relation taken as a whole is not a function, the curve can be represented by a function around any point on the ellipse other than the two points (4620, 0) and (–4620, 0). Different sections of the curve may not be represented by the same functional formula; however, we do not have to be able to express the function with an explicit formula to find the derivative.

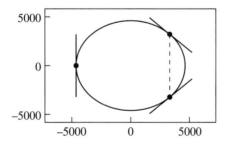

Figure 3.29 Graph of $\dfrac{x^2}{4620^2} + \dfrac{y^2}{4600^2} = 1$ with tangent lines

To find how the slopes of the tangent lines to the ellipse depend on x and y, we can differentiate both sides of equation (20) with respect to x. (Consider y to be a function of x.) This process relies on the fact that if two expressions are equal, then their derivatives are equal. Differentiating both sides of equation (20) with respect to x gives

$$\frac{d}{dx}\left(\frac{x^2}{4620^2}+\frac{y^2}{4600^2}\right)=\frac{d}{dx}(1),$$

which can be simplified to

$$\frac{1}{4620^2}\cdot\frac{d}{dx}(x^2)+\frac{1}{4600^2}\cdot\frac{d}{dx}(y^2)=0. \tag{21}$$

In Equation (21) the derivative expression $\frac{d}{dx}(x^2)$ equals $2x$. The derivative expression $\frac{d}{dx}(y^2)$ requires application of the chain rule, so that

$$\frac{d}{dx}y^2 = \frac{d}{dy}(y^2)\cdot\frac{dy}{dx}$$
$$= 2y\frac{dy}{dx}.$$

Substitution into equation (21) yields

$$\frac{1}{4620^2}(2x)+\frac{1}{4600^2}\left(2y\frac{dy}{dx}\right)=0. \tag{22}$$

Equation (22) can be solved for $\frac{dy}{dx}$ by isolating $\frac{dy}{dx}$ on one side of the equation, which gives

$$\frac{1}{4600^2}\left(2y\frac{dy}{dx}\right)=-\frac{1}{4620^2}(2x),$$

so that

$$\frac{dy}{dx}=-\frac{4600^2}{4620^2}\left(\frac{2x}{2y}\right)$$
$$=-\frac{52,900}{53,361}\cdot\frac{x}{y}. \tag{23}$$

Notice that the expression for $\frac{dy}{dx}$ given in (23) implies that the slope of the tangent line will be positive at any point on the curve in the 2nd or 4th quadrants and negative at any point on the curve in the 1st or 3rd quadrants. Equation (23) further shows that the slope of the tangent line is 0 at (0, 4600) and (0, −4600), and is undefined at (4620, 0) and (−4620, 0).

implicit differentiation   The process we used to find $\frac{dy}{dx}$ is called ***implicit differentiation***. This technique is used when the dependent variable is not written explicitly in terms of the independent variable. Even when solving for the dependent variable is possible, implicit differentiation may be easier than first obtaining an explicit formula for the dependent variable in terms of the independent variable and then differentiating.

Example 1

If $xy = 1$, find $\dfrac{dy}{dx}$ by implicit differentiation.

Solution   We take the derivative of both sides of the equation:

$$\frac{d}{dx}(xy) = \frac{d}{dx}(1).$$

Taking the derivative of xy with respect to x requires the product rule, which yields

$$x \cdot \frac{dy}{dx} + 1 \cdot y = 0.$$

Solving for $\dfrac{dy}{dx}$ gives

$$\frac{dy}{dx} = -\frac{y}{x}.$$

$\blacksquare$

This equation for the derivative gives the slope of a tangent line at the point (x, y) in terms of both the x- and y-coordinates. For instance, when $x = 2$ and $y = \frac{1}{2}$, then $\frac{dy}{dx} = -\frac{1}{4}$. This result can be verified by going back to the original equation $xy = 1$, solving for y,

$$y = \frac{1}{x},$$

and then taking the derivative so that

$$\frac{dy}{dx} = -\frac{1}{x^2}.$$

This equation for the derivative gives the slope of a tangent line at the point (x, y) in terms of the x-coordinate alone. When $x = 2$, the slope of the tangent line is $-\frac{1}{4}$, which agrees with the result found by implicit differentiation. Also, since $y = \frac{1}{x}$ we know $-\frac{y}{x} = -\frac{1}{x^2}$, so that the two methods of differentiation provide the same result.

Example 2

Find $\dfrac{dy}{dx}$ if $y = \tan^{-1} x$.

Solution   Knowing how to differentiate $\tan x$ enables us to differentiate its inverse $\tan^{-1} x$. Rewriting $y = \tan^{-1} x$, we have

$$\tan y = x.$$

We differentiate both sides of this equation with respect to x and use the chain rule on the left side to obtain

$$(\sec^2 y)\frac{dy}{dx} = 1.$$

Solving for $\frac{dy}{dx}$, we find

$$\frac{dy}{dx} = \frac{1}{\sec^2 y}$$
$$= \cos^2 y.$$

By substitution,

$$\frac{dy}{dx} = \cos^2(\tan^{-1} x),$$

which gives an expression for the derivative in terms of x only. We can write $\cos^2(\tan^{-1} x)$ as an algebraic expression without trigonometric functions by referring to the right triangle shown in Figure 3.30. If we let $\theta = \tan^{-1} x$, so that $x = \tan\theta$, then the ratio of the length of the side opposite angle θ to the length of the side adjacent to angle θ must equal x. We can achieve this relationship in many ways, but the most simple is to let the opposite side be x and the adjacent side be 1. By definition, $\cos\theta$ equals the length of the adjacent side divided by the length of the hypotenuse, which means

$$\cos\theta = \frac{1}{\sqrt{1 + x^2}}.$$

Since $\theta = \tan^{-1} x$, we substitute into $\frac{dy}{dx} = \cos^2(\tan^{-1} x)$ to get $\frac{dy}{dx} = \cos^2\theta$, or

$$\frac{dy}{dx} = \frac{1}{1 + x^2}. \qquad\qquad \blacksquare$$

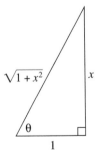

Figure 3.30 Right triangle with $\theta = \tan^{-1} x$

Example 3

Find $\frac{dy}{dx}$ if $y = \ln x$.

Solution   We estimated from the graphs in Lab 1 that the derivative of $\ln x$ is $\frac{1}{x}$. The following analytical argument is based on the method of implicit differentiation. Since $y = \ln x$, we can use the concept of function inverses and rewrite the equation:

$$e^y = x.$$

Differentiating both sides of the previous equation with respect to x gives

$$e^y \frac{dy}{dx} = 1,$$

so that

$$\frac{dy}{dx} = \frac{1}{e^y}.$$

Since $y = \ln x$, we substitute to get

$$\frac{dy}{dx} = \frac{1}{e^{\ln x}},$$

which simplifies to

$$\frac{dy}{dx} = \frac{1}{x}. \qquad\qquad \blacksquare$$

Examples 2 and 3 use implicit differentiation to find the derivatives of specific inverse functions. Suppose we are interested in knowing the derivative of a general inverse function $y = f^{-1}(x)$. If $y = f^{-1}(x)$, then

$$f(y) = x.$$

Differentiating both sides of the previous equation with respect to x (and using the chain rule on the left side) gives

$$f'(y) \cdot \frac{dy}{dx} = 1.$$

Solving for $\frac{dy}{dx}$ gives

$$\frac{dy}{dx} = \frac{1}{f'(y)},$$

or, in terms of x,

$$\frac{d}{dx}\left[f^{-1}(x)\right] = \frac{1}{f'\left[f^{-1}(x)\right]}.$$

This result implies that the slope of the tangent line to $y = f^{-1}(x)$ at (b, a) is the reciprocal of the slope of the tangent line to $f(x)$ at (a, b). Figure 3.31 gives a graphical representation of this concept. Because f and f^{-1} are inverses, the graphs are reflections across the line $y = x$, and the point (a, b) on the graph of f becomes (b, a) on the graph of f^{-1}. If a line tangent to the curve $y = f(x)$ is likewise reflected so that it becomes a line tangent to the curve $y = f^{-1}(x)$, then the slopes of these tangents are reciprocals. We confirm this by determining the slope of each tangent using the point (c, c) common to both tangents. The slope of the tangent to f is $\frac{b-c}{a-c}$, and the slope of the tangent to f^{-1} is $\frac{a-c}{b-c}$; thus, the slopes are reciprocals.

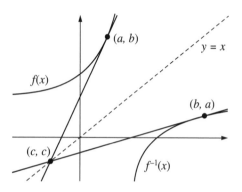

Figure 3.31 Tangent lines of inverse functions

Logarithmic Differentiation

The function x^x is neither a power function nor an exponential function; however, the similarity of this function to those types of functions suggests that we can take the derivative using a rule for one of those forms. If the derivative of this function followed the pattern of a power function, it would be $x \cdot x^{x-1}$. If the derivative followed the pattern of an exponential function, it would be $x^x \ln x$. Which, if either, of these is correct?

Class Exercises

1. Graph difference quotients for the function with equation $y = x^x$ for $0 < x < 5$. Compare the graph of these difference quotients to the graphs of $y = x^{x-1}$ and $y = x^x \ln x$, two candidates for the derivative. Does either curve fit the difference quotients?

2. Write the limit definition of the derivative for the function x^x. Does this appear to be an efficient method for finding the derivative of this function?

As you have seen in the preceding class exercises, the rules for power and exponential functions do not work for the function x^x, and the limit definition of the derivative appears to be a very difficult way to find the derivative. This function creates a problem because a variable is in both the base and the exponent. Taking logarithms of both sides of the equation $y = x^x$ helps because

$$\ln y = \ln x^x$$
$$= x \ln x. \qquad (24)$$

We now have y defined implicitly in terms of x, and we have a simple product instead of a more complicated power.

Using the chain rule on the left side and the product rule on the right side to differentiate both sides of equation (24) with respect to x gives

$$\frac{1}{y} \cdot \frac{dy}{dx} = 1 \cdot \ln x + x\left(\frac{1}{x}\right).$$

Simplifying and solving for $\frac{dy}{dx}$ yields

$$\frac{dy}{dx} = y(\ln x + 1),$$

which by substitution of x^x for y results in

$$\frac{dy}{dx} = x^x(\ln x + 1). \qquad (25)$$

The technique in which we first take the logarithm of a function and then differentiate implicitly is called *logarithmic differentiation.*

logarithmic differentiation

Exercise Set 3.7

---

1. Write the definition of the derivative for $\tan^{-1} x$ and $\ln x$, and discuss why finding these derivatives directly from the limit definition is difficult or impossible.

2. Graph the derivative of $y = x^x$ given in equation (25) along with difference quotients that approximate the derivative. Verify that the derivative seems to agree with the difference quotients.

3. Find $\frac{dy}{dx}$ if:

    a. $x^2 + xy - \sqrt{y} = 3$ b. $x^2 y^2 = 1$ c. $x = e^y$

    d. $y = x^{1/x}$ e. $y = \sin^{-1} x$ f. $y = \cos^{-1} x$

g. $y = \cos^{-1}(\ln x)$ h. $y = \tan^{-1}(x^2)$ i. $y = (\sin x)^x$

j. $x \sin y = y \sin x$

4.   Use the limit definition of the derivative to find $f'(x)$ for $f(x) = \sqrt[3]{x}$. Compute the same derivative by considering f as the inverse of $g(x) = x^3$.

5.   The equation $y = x^{2/3}$ can be written as $y^3 = x^2$. Use implicit differentiation to find y'. Generalize your result for functions of the form $y = x^{p/q}$. How does your result compare with the formula for the derivative of $y = x^n$, where n is an integer?

6.   A picture with height a cm hangs on a wall such that the bottom of the picture is b cm above eye level. The best viewing angle for the picture is the one where the angle between your eye and the top and bottom of the picture has the largest measure. How far away from the picture should you stand to obtain the best viewing angle?

7.   In rugby, after a try is scored (similar to a touchdown) the team that scores attempts to kick the ball through the goal posts. The kick may be taken anywhere along a line perpendicular to the goal line that emanates from the point on the goal line where the ball was touched down. If the goal posts are 5.6 meters apart and a try is scored on the goal line a meters from the goal post, how far from the goal line should the kick be taken to maximize the chance of success? This chance is maximized by maximizing the angle α shown in Figure 3.32. Three choices for the angle α are shown.

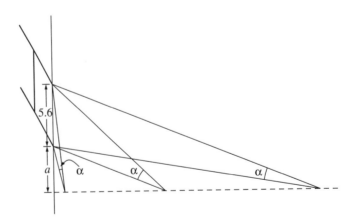

Figure 3.32 The rugby problem

3.8   Investigations

We now investigate extended problems involving the first and second derivatives. These problems link the ideas of this chapter with other areas of study, including statistics, physics, history, and economics. The investigations are designed for group work, and each group should choose one investigation from the set. Questions are provided with each investigation; however, these questions are not meant to limit the investigations. Additional ideas and questions can be generated that build upon this base, and you should feel free to explore your ideas and questions.

Investigation 1:   Fitting a Least Squares Line to Data

A widely used method of finding a linear model for a set of ordered pairs is the method of least squares. Although the equation of the least squares line can be found without using calculus, calculus is an efficient way to find the slope m and y-intercept b of the least squares line. The principle of the least squares line is that the sum of the squares of the residuals should be minimized. If we write an expression for the sum of the squares of the residuals for a given set of data, we can then use the derivative to help us find the minimum value of this expression. Since the equation of the least squares line contains two unknown quantities, slope and intercept, whose values we must determine, our expression for the sum of the squares of the residuals will include two unknowns.

We will develop the method used to determine the slope and intercept of a least squares line using sample data gathered in an experiment. In the experiment, weights were suspended from a spring, causing the spring to stretch. Table 3.33 gives the weight suspended from the spring (in decagrams) and the resultant length of the spring (in centimeters).

Weight (in dg)	1	2	3	4	5
Length (in cm)	6	7	9	10	11

Table 3.33 Weights and lengths of spring

The scatter plot of length versus weight in Figure 3.34 shows that a line is an appropriate model for this data set. The residuals are represented graphically by the vertical distances between the data points and the linear model. Residual values can be calculated by subtracting the y-coordinate of a point on the linear model from the y-coordinate of a corresponding data point. The points on the linear model $y = mx + b$ have coordinates $(1, 1m + b)$, $(2, 2m + b)$, $(3, 3m + b)$, $(4, 4m + b)$, and $(5, 5m + b)$. The residuals associated with each of the five data points are $6 - (m + b)$, $7 - (2m + b)$,

$9 - (3m + b)$, $10 - (4m + b)$, and $11 - (5m + b)$.   Therefore, an expression for S, the sum of the squares of the residuals, is given by

$$S = [6 - (m + b)]^2 + [7 - (2m + b)]^2 + [9 - (3m + b)]^2 +$$
$$[10 - (4m + b)]^2 + [11 - (5m + b)]^2.$$

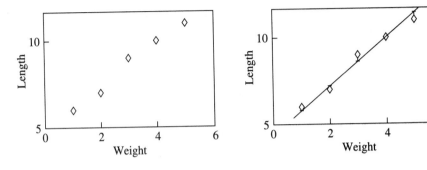

Figure 3.34 Scatter plot of length versus weight and a linear model

Exercises for Investigation

1.   Treat m as a constant and find the derivative of S with respect to b.

2.   Treat b as a constant and find the derivative of S with respect to m.

3.   If S is to be minimized, both of the derivatives from problems 1 and 2 must be equal to zero.   Find the values of m and b so that both of these derivatives are equal to zero.

partial derivatives

*Taking the derivative of a function with respect to one independent variable while treating the other variables as if they are constants produces derivatives that are called **partial derivatives**.   The partial derivative of S with respect to b is written $\frac{\partial S}{\partial b}$, which means m is treated as a constant.   (The symbol ∂ is the lowercase Greek letter delta; Δ is the uppercase delta.)   The partial derivative of S with respect to m is written $\frac{\partial S}{\partial m}$, which means b is treated as a constant.   In general, the concept of partial derivatives is more complicated than we have made it seem here and is usually covered in a multivariable calculus course.*

summation notation

4.   Suppose we replace the specific data points for the spring with the general data points (x_1, y_1), (x_2, y_2), (x_3, y_3), (x_4, y_4), and (x_5, y_5).   Repeat the process used in problems 1–3 to find formulas for m and b in terms of the x_i's and the y_i's, where i is an integer that ranges in value from 1 to 5.   **Summation notation** is useful in these expressions.   For example, in summation notation, the sum of the x_i's is written as

$$\sum_{i=1}^{5} x_i,$$

and the sum of the y_i's is written as

$$\sum_{i=1}^{5} y_i.$$

(The symbol Σ is the uppercase Greek letter *sigma*, which in this case represents *sum*.) Here, i is the index and the numbers 1 and 5 indicate that integer values from 1 to 5, inclusive, are to be used in place of i in the summand. That is,

$$\sum_{i=1}^{5} x_i = x_1 + x_2 + x_3 + x_4 + x_5.$$

5. Now that we have found expressions for the slope and intercept of the least squares line for any data set of 5 points, we can generalize the results for any size set of data. Assume we have N points with coordinates (x_i, y_i), where i is an integer from 1 to N. Solve for m and b in terms of the x_i's and the y_i's, and write the expressions using summation notation.

6. The data set shown in Table 3.35 gives the data that students gathered to test the hypothesis that a person's arm span and height are equal. Use the formulas found for m and b in problem 5 to determine the least squares linear model for this data. Compare the results with the least squares line determined by a calculator or computer.

Arm span (in cm)	Height (in cm)
157	171
164	168
177	178
162	167
193	181
183	186
158	164
175	175
156	159

Table 3.35 Arm span and height data

7. The method for finding the equation of the least squares line can be extended to finding a least squares quadratic model. Suppose we want to find a model of the form $y = ax^2 + bx + c$ for a set of data. Write an expression for the sum of the squared residuals S in terms of the constants a, b, and c. Use partial derivatives $\frac{\partial S}{\partial a}$, $\frac{\partial S}{\partial b}$, and $\frac{\partial S}{\partial c}$ to find values for a, b, and c that minimize S. Your answers should be in terms of the data points (x_i, y_i), where i is an integer from 1 to N. It may help to use the fact that

$$\frac{d}{dx}\left\{\sum_{i=1}^{N}[f(x_i)]^2\right\} = \sum_{i=1}^{N}\frac{d}{dx}[f(x_i)]^2.$$

(Hint: Is it necessary to expand $[f(x_i)]^2$ before taking the derivative?)

Investigation 2: Interatomic Forces

The attractive and repulsive forces that exist between two atoms vary as the distance between the atoms varies. Attractive forces exist between the negative charges in one atom and the positive charges in the other atom. Repulsive forces exist between the positive charges in the nuclei of the atoms. A typical graph of interatomic force as a function of separation distance is shown in Figure 3.36, where attraction is a positive force and repulsion is a negative force. The greatest net attractive force typically occurs at a separation of about $1 \cdot 10^{-10}$ m. For simplicity, we assign a value of 1 to the force and a value of 1 to the separation distance at this point. As the distance between the atoms increases, the attractive force decreases, so that for separations greater than 2, the attractive force is negligible. An equation for a function whose graph behaves in this manner can be written by combining an inverse square function $\frac{c}{x^2}$ with an inverse cubed function $\frac{d}{x^3}$, such as

$$f(x) = \frac{c}{x^2} + \frac{d}{x^3}.$$

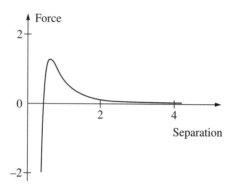

Figure 3.36 Interatomic force versus separation distance

Exercises for Investigation

1. Determine values for c and d which give a maximum attraction of 1 at $x = 1$. If the force at this point is k instead of 1, what are the values of c and d?

2. Describe how the force is changing at the inflection point on the graph in Figure 3.36. For the values of c and d found in problem 1, determine where the force function has an inflection point.

3. Another possible model for interatomic force combines an inverse square function $\frac{a}{x^2}$ with the reciprocal of the fifth power $\frac{b}{x^5}$, as in

$$g(x) = \frac{a}{x^2} + \frac{b}{x^5}.$$

Determine values for a and b in the equation for $g(x)$ so that a maximum force with value 1 occurs when the distance is approximately 1.

4. Compare the models in problems 1 and 3. How do their inflection points differ? Describe any other qualitative differences between the two functions.

5. What other types of functions might model the relationship between force and separation distance? Determine equations for some of these functions.

Investigation 3: The Nuclear Arms Race

A study of history has shown that almost all modern wars have been preceded by an arms race. Between 1816 and 1965, disputes preceded by an unstable arms race escalated to war 23 out of 28 times, while disputes not preceded by an arms race resulted in war only 3 out of 71 times. [Michael Wallace, "Arms Races and Escalation: Some New Evidence," in *Explaining War*, ed. J. David Singer (Beverly Hills, Calif.: Sage, 1979), pp. 240-252.] Rapid competitive military growth may appear closely associated with a willingness, possibly even an eagerness, to go to war. The ability to control an arms race, then, seems important to maintaining world peace.

For nearly 45 years, the former Soviet Union and the United States were engaged in a competitive arms race involving nuclear arms. To help us understand why the arms race between the Soviet Union and the United States continued for such a long time, we will analyze the strategy that may have been used by each country and develop a model of this situation.

Former chairman of the Joint Chiefs of Staff, General Maxwell D. Taylor, explains the American strategy as follows [M. D. Taylor, "How to Avoid a Nuclear Arms Race," *The Monterey Peninsula Herald*, January 24, 1982, p. 3C]:

The strategic forces, having the single capability of inflicting massive destruction, should have the single task of deterring the Soviet Union from resorting to any form of strategic warfare. To maximize their deterrent effectiveness they must be able to survive a massive first strike and still be able to destroy sufficient enemy targets....

General Taylor was saying that the United States was not planning a first strike; however, the United States believed that the Soviet Union was planning a first strike. One way to view what General Taylor said is that the United States thought of itself as the "good guys" who would not carry out a first strike and of the Soviet Union as the "bad guys". Of course, the Soviets viewed the situation exactly reversed, with themselves as the "good guys" and the United States as the "bad guys".

Suppose that two countries X and Y are engaged in a nuclear arms race and that each country adopts the strategy described by General Taylor. Let x represent the number of missiles possessed by country X, and let y represent the number of missiles possessed by country Y. We will consider this problem from the point of view of country Y and determine the number of missiles required by country Y when country X has a given number of missiles. That is, we will determine the characteristics of a function for y in terms of x where y represents the minimum number of missiles that are needed to accomplish the strategic objectives of country Y.

Since y is a function of x, we are assuming that the number of missiles country Y needs for successful deterrence depends upon the number of missiles country X has. In developing a model for the relationship between x and y, we will assume that the number of missiles possessed by country X is an integer multiple of the number of missiles possessed by country Y. In addition, when a missile is fired at a target, there is no guarantee that the target will be destroyed, so we will assume that a target will survive an attack with probability p. Suppose Y needs M missiles to destroy the intended targets. This means that after an initial attack by X, Y must have M missiles left in its arsenal.

How will Y respond if $x = 0$, that is, if X has no missiles? Then country Y simply needs $y = M$ missiles in its arsenal.

How will Y respond if $x = y$, that is, if X has the same number of missiles as Y? When X launches a first strike, each of Y's missiles is attacked by one of X's missiles. As a result, the number of Y's missiles that survive the attack is py. Since Y needs to have M missiles left after the attack, we know that $M = py$. This means that Y needs $y = \frac{M}{p}$ missiles to carry out its strategy of deterrence.

How will Y respond if $x = 2y$, that is, if X has twice as many missiles as Y? If we assume that X fires all of its missiles simultaneously in a first strike, each of Y's missiles is attacked by two of X's missiles. The probability of surviving the first missile is p, and the probability of surviving the second missile is also p. The probability of surviving both missiles is p^2, the product of these two probabilities. As a result, the number of Y's missiles that survive the attack is $p^2 y$. Since Y needs to have

M missiles left after the attack, we know that $M = p^2 y$. Thus, in this situation Y needs $y = \dfrac{M}{p^2}$ missiles to carry out its strategy of deterrence.

Continuing in this manner, we find that if $x = ny$, then $M = p^n y$ missiles are left in Y's arsenal after X carries out a first strike. In general, the number of missiles required to satisfy Y's strategic goal of deterrence is

$$ y = \frac{M}{p^n}, $$

where n is a non-negative integer. Since $x = ny$, our model for the number of missiles needed by country Y can be written as

$$ y = \frac{M}{p^{x/y}}. \tag{26} $$

If we assume that country X adopts the same strategy as country Y, then their function will be similarly defined as

$$ x = \frac{N}{q^{y/x}}, \tag{27} $$

where N is the number of missiles X needs to destroy the intended targets and q is the probability that a target in country X will survive an attack by one of Y's missiles. In general, since the specifics of the two countries differ, $N \neq M$, and since the accuracy of the missiles differs, $p \neq q$.

Recall that the purpose of equation (26) is to give us a function for the number of missiles y required by country Y that depends on the number of missiles x owned by country X. Each value of x should give us only one value of y, but we cannot solve for y explicitly. Equation (26) gives an implicit definition for the functional relationship between x and y, and this implicit form is the only way we can express the relationship between x and y. We will use calculus as a tool to investigate the relationship between x and y.

Reference

Based on *A First Course in Mathematical Modeling*, pp. 5–15, by F.R. Giordano and M.D. Weir. Copyright © 1985 by Brooks/Cole Publishing Company, a division of Thomson Publishing Inc., Pacific Grove, CA 93950. Reprinted by permission of the publisher.

Exercises for Investigation

1. Find $\dfrac{dy}{dx}$ for the function $y = f(x)$ defined implicitly by equation (26), assuming $M > 0$ and $0 < p < 1$.

   a. Over what intervals is f increasing? Over what intervals is f decreasing?

   b. Is the graph of the function f concave up or down? What information about the arms race does the concavity give you?

    c. Is there a point at which country Y can be sure it has enough missiles to accomplish its goal of deterrence? That is, is it possible for Country Y to have enough missiles no matter how many missiles Country X has? Use characteristics of f to justify your answer.

    d. What happens to the slope of the graph of f if the value of p decreases? In terms of the arms race, what phenomenon would a decrease in the value of p represent?

    e. What effect does changing the value of M have on the graph of f? In terms of the arms race, what event would a change in the value of M represent?

2. How does the graph of the function defined by equation (27) compare to the graph of the function in problem 1? In terms of the arms race, what is the significance of the point of intersection of these graphs?

3. Analyze the effect on the arms race of each of the following changes by investigating how the change influences the point of intersection of the graphs of f and g.

    a. Country X increases the accuracy of its missiles.

    b. Country Y puts its missiles on mobile launching pads to make them more difficult for country X to target.

4. What changes in the offensive or defensive capabilities of each country would reduce the number of missiles each country believes it needs? Is it possible for a country to alter its offensive and defensive capabilities so that the number of missiles it needs decreases but the number of missiles its opponent needs increases?

Investigation 4: Elasticity of Demand

Econometric studies typically encompass entire economic sectors, for instance the corn, wheat, or soybean markets, the gasoline market, *etc*. To evaluate related legal and fiscal policies, such studies often focus on the effect of a price change upon the volume of sales and total revenues. To perform such studies, economists need a measure of market responsiveness that is independent of units and market size to allow for comparisons between markets.

    We start by modeling the market for a product with a demand function $D(p)$ that gives the amount of the product that can be sold over a certain time if the product is priced at p dollars per unit. In general, a demand function is obtained by fitting a curve to data from test markets and surveys. For our study, we assume $D(p)$ is a continuous function whose graph is a decreasing curve. This assumption is justified by considering what may happen to the demand for a product as the price of the product increases. For some products the demand declines because consumers are willing to buy less at a higher price. (For other products, such as jeans and athletic shoes, many people would be willing to pay high prices because some brand names carry a certain level of prestige or quality, but we are not considering such a product here.)

What else happens when a manufacturer raises the price of a product? For our example, the demand goes down, but what happens to the manufacturer's revenue? Assuming that all of the product in demand is actually sold, revenue is given by

$$R(p) = p \cdot D(p).$$

A price increase can lead to an increase in revenue; however, a price increase can also cause revenue to decrease. For instance, suppose the demand for a particular product is 1000 units when the price is $10 per unit. Since revenue is equal to price times demand, the revenue associated with a price of $10 is $10,000. What will happen if the price increases to $11 per unit? Again, in this case, the price increase will lead to a decrease in demand. If demand falls to 990 units, then revenue will be given by $11 per unit times 990 units, so revenue increases to $10,890. If the demand falls to 900 units, then revenue will be given by $11 per unit times 900 units, so revenue decreases to $9,900. Thus, an increase in price to $11 may lead to either an increase or a decrease in revenue. Whether revenue rises or falls when prices go up depends on the extent to which demand falls in response to the price increase.

Economists have quantified the responsiveness of the demand function to increases in price by comparing changes in the demand to changes in the price. Changes in demand and price are more meaningful when we also know the current demand and price level; for example, a change of $0.10 in the price for an item means one thing if the current price is $10.00 and means something very different if the current price is $0.20. For this reason, economists compare relative changes rather than absolute changes. The relative change in demand is the ratio $\frac{\Delta D}{D}$, and the relative change in price is $\frac{\Delta p}{p}$. If we now look at the ratio of these two relative changes,

$$\frac{\text{relative change in demand } D(p)}{\text{relative change in price } p},$$

we get the expression

$$\frac{\Delta D}{D(p)} \bigg/ \frac{\Delta p}{p}. \tag{28}$$

The expression given in (28) is equivalent to $\frac{\Delta D \cdot p}{\Delta p \cdot D(p)}$, which we can separate into a product of two ratios as

$$\frac{\Delta D}{\Delta p} \cdot \frac{p}{D(p)}.$$

This quantity gives the **_elasticity of demand_** over a price interval from p to $p + \Delta p$.   _elasticity of demand_

The function that gives the elasticity of demand for each point on the demand curve is defined as the limit of the elasticity of demand over an interval as the length of the interval Δp goes to zero. In symbols, elasticity of demand is given by

$$E(p) = \lim_{\Delta p \to 0} \frac{\Delta D}{\Delta p} \cdot \frac{p}{D(p)},$$

which, by the limit definition of the derivative, is equivalent to

$$E(p) = \frac{dD}{dp} \cdot \frac{p}{D(p)}.$$

Although it is not always so, we usually assume that an increase in price will result in a decrease in demand, and a decreasing demand curve implies that $\frac{dD}{dp}$ is negative for all p values. Since p and $D(p)$ both represent quantities that are positive, this implies that $E(p)$ is also negative for all p values.

To see how the elasticity of demand relates to revenue we can look at the rate of change of the revenue function R. The rate of change of revenue with respect to price is given by the derivative $\frac{dR}{dp}$, which is obtained by applying the product rule to the equation $R(p) = p \cdot D(p)$ to give

$$\frac{dR}{dp} = p \cdot \frac{dD}{dp} + D(p). \tag{29}$$

Rewriting equation (29) by factoring out $D(p)$ gives us

$$\frac{dR}{dp} = D(p) \cdot \left(1 + \frac{p \cdot \frac{dD}{dp}}{D(p)}\right). \tag{30}$$

The quantity $\dfrac{p \cdot \frac{dD}{dp}}{D(p)}$ is the elasticity of demand, so equation (30) can be written as

$$\frac{dR}{dp} = D(p)[1 + E(p)]. \tag{31}$$

The way that revenue responds to price increases can be analyzed in terms of $\frac{dR}{dp}$, the rate of change of revenue with respect to price. When $\frac{dR}{dp} > 0$ then revenue is an increasing function of price and price increases will lead to revenue increases. Since $D(p)$ is positive or zero for all values of p, equation (31) tells us that $\frac{dR}{dp}$ will be greater than zero only if $1 + E(p)$ is greater than zero, which occurs only when $E(p) > -1$. When $\frac{dR}{dp} < 0$ revenue is a decreasing function of price so price increases will lead to revenue decreases. Again, since $D(p)$ is positive or zero for all values of p, we know that $\frac{dR}{dp}$ will be less than zero only if $1 + E(p)$ is less than zero, and this occurs only when $E(p) < -1$.

The three possibilities for the behavior of revenues at a price $p = p^*$ that follow from equation (31) are summarized below. In this case, we are assuming that $D(p)$ has a negative slope, so demand decreases as price increases.

1. $-1 < E(p^*) < 0$: $R'(p^*)$ is positive, so $R(p)$ is increasing near p^*. If the price increases, then revenues rise. This occurs because the price rise does not reduce demand enough to cause revenues to decrease. In this case, demand is said to be ***inelastic***.

inelastic

2. $E(p^*) < -1$: $R'(p^*)$ is negative, so $R(p)$ is decreasing near p^*. If the price increases, then revenues decline. This occurs because the price rise reduces demand sufficiently to decrease revenues. In this case, demand is said to be *elastic*.

elastic

3. $E(p^*) = -1$: $R'(p^*)$ is zero, so the graph of $R(p)$ is horizontal near p^*. An increase in price will result in no change in revenues. This occurs because the increased selling price is offset exactly by the decrease in demand. In this case, demand is said to be **unit elastic**.

unit elastic

The elasticity of demand for a particular product represents the responsiveness of the demand function to increases in price. It is a relative measure of responsiveness and is thus independent of market size, so we can compare the elasticity of different products. The demand for necessities like food and clothing is usually inelastic. Because consumers need to purchase these products no matter what the price, demand does not fall very much as prices increase, and thus price increases usually result in increases in revenue. The demand for discretionary items like diamonds and fur coats is often elastic. The willingness of many consumers to purchase these products can be very sensitive to their price. Demand falls a great deal when prices increase, and thus price increases usually result in decreases in revenue. However, if one could buy such goods at extremely low prices, the demand may drop as they would no longer be considered valuable or prestigious. The economics of supply and demand can be very complicated.

Exercises for Investigation

1. Suppose the demand function for a particular graphing calculator is

    $$D(p) = 1000 \cdot e^{-0.015p}.$$

    a. What is the elasticity of demand associated with a price of $100?

    b. What is the elasticity of demand associated with a price of $60?

    c. Investigate the change in revenue associated with each price to determine how the elasticity of demand is related to the way that revenue responds to price increases.

2. What demand function has an elasticity of -1 for all values of p?

3. Suppose the demand for orders of tea from a certain distributor is given by the equation $q = -0.34p + 52.6$, with p representing the retail price of the tea in dollars per case, and q representing the average number of cases (in thousands) of tea purchased annually by consumers.

   a.   What is the function for elasticity in terms of price?

   b.   What is the elasticity if tea costs $30.50 per case? What does this value mean?

4. Suppose the market for diet soft drinks at State University can be modeled with the demand equation $D(p) = 332.1p^{-1.1}$, where p is the price in cents of a 12 ounce diet soft drink and $D(p)$ is the corresponding average weekly demand per student.

   a.   What is the elasticity of demand?

   b.   How much would price have to increase for demand to decrease by 10%?

   c.   With a price increase, do sales revenues increase or decrease?

5. Suppose you want to open a restaurant and have determined that there is suffi-cient demand to make a profit. Preliminary research shows that the demand curve for your business can be modeled by $D(p) = 180 - 50 \ln p$, where demand is measured in customers per hour and price is in dollars.

   a.   What is the revenue function R?

   b.   What is the elasticity of demand E?

   c.   Graph $D(p)$, $R(p)$, $E(p)$, and $R'(p)$. Explain the relationships between the graphs of $D(p)$, $R(p)$, $E(p)$, and $R'(p)$.

Investigation 5: Function Iteration

To find a zero of the function f using Newton's method, we iterate the equation $x_n = x_{n-1} - \dfrac{f(x_{n-1})}{f'(x_{n-1})}$. Newton's method is a particular instance of the more general concept of function iteration, which we introduced in Chapter 1. The term iteration implies repeating a process. Function iteration produces a sequence of numbers by using a function evaluated at the current value to produce the next value. This new value then becomes the current value and the process is repeated. The values gener-*iterates* ated by iteration are known as the ***iterates*** of a function. A general rule for function iteration is represented by the iterative equation

$$x_n = f(x_{n-1}).$$

    We can display function iteration in several graphical forms. One type of display shows the values of the iterates versus the iteration number. The first ten iterates of $f(x) = \cos x$ with initial value $x_0 = 0.1$ are shown in this form in Figure 3.37. The accompanying table of values gives further insight into the behavior of the iterates.

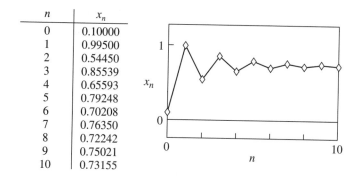

n	x_n
0	0.10000
1	0.99500
2	0.54450
3	0.85539
4	0.65593
5	0.79248
6	0.70208
7	0.76350
8	0.72242
9	0.75021
10	0.73155

Figure 3.37 Graph and table of values of iterates of $f(x) = \cos x$

Another method of displaying function iteration is a ***web diagram***, so named be-   ***web diagram***
cause the graphs resemble cobwebs. A web diagram gives a picture of how the iterates
move around the graph of the function we are iterating, which in this case is
$f(x) = \cos x$. We begin the web diagram by sketching the graph of $y = \cos x$. We
then locate the starting value for the iterations, x_0, on the x-axis. The first iterate $x_1 =$
$\cos x_0$ is located on the graph by moving vertically from x_0 on the x-axis to the curve
representing $y = \cos x$, as shown in Figure 3.38. The vertical distance to the curve is
the value of $\cos x_0$, which is x_1; therefore, the point we reach on the curve has coordinates (x_0, x_1). To find the second iterate $x_2 = \cos x_1$ on the graph, we first need to
locate x_1 on the x-axis. To do this, we move horizontally from the point (x_0, x_1) on the
curve to the line $y = x$. The point we arrive at on the line has coordinates (x_1, x_1).
Moving vertically from this point to the curve takes us to the point on the curve with x-
coordinate x_1. The vertical distance from the x-axis to this point is the value of $\cos x_1$,
which is x_2, so this point has coordinates (x_1, x_2). This point gives the location on the
graph of the second iterate. We continue this process of bouncing off the line $y = x$ to
obtain the web diagram shown in Figure 3.38.

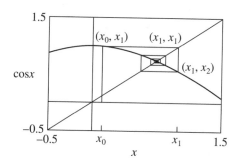

Figure 3.38 Web diagram for iterates of $f(x) = \cos x$

Notice that the web diagram converges to the point of intersection of $y = \cos x$ and $y = x$, which we can discern by extending the table in Figure 3.37 is where $x \approx 0.739$. A web diagram can have other types of appearances, such as a staircase, that are explored in this investigation. The displays of function iteration described above can be used to investigate iterations of various functions.

Exercises for Investigation

1. Iterate the function $f(x) = \frac{1}{2}x - 1$ with a starting value $x_0 = 1$. What is the long-term behavior of the iterates? Do the same with starting value $x_0 = -3$. Try other starting values. Does the starting value affect the long-term behavior of the iterates?

2. Iterate the function $f(x) = 2x + 1$ with a starting value $x_0 = 0$. What is the long-term behavior of the iterates? Do the same with starting value $x_0 = -2$. Try other starting values. Does the starting value affect the long-term behavior of the iterates?

fixed point

3. A point that does not change under iteration of a function is known as a *fixed point* of that function. For example, if $f(x) = x^2 - x$, $f(2) = 2$. The point $(2, 2)$ is a fixed point of f. The point $(0, 0)$ is also a fixed point of f. What are the fixed points of the functions in problems 1 and 2?

If a fixed point of a function is the limit of the iterates for starting values near the fixed

attracting

*point, then the fixed point is called an **attracting** fixed point. If the iterates move away from the fixed point no matter how close the starting values are to the fixed point, then*

repelling

neutral

*the fixed point is called a **repelling** fixed point. A fixed point that is neither attracting nor repelling is called a **neutral** fixed point.*

4. Why is the fixed point of $f(x) = -x + 2$ considered neutral?

5. Based on the introductory section of this investigation, what is the fixed point of $f(x) = \cos x$? Is it attracting or repelling?

6. Classify the fixed points found in problem 3 as attracting, repelling, or neutral.

7. Find the fixed point of the general linear function $f(x) = mx + b$. Experiment with various values of m to determine answers to the following questions.

   a. What value of m can cause f to have no fixed points?

   b. What combination of values for m and b give f an infinite number of fixed points?

   c. What values of m cause the fixed point to be attracting? What values of m cause the fixed point to be repelling? What values of m cause the fixed point to be neutral?

8. Iterating a function at a fixed point yields the same value over and over again. A fixed point of a function f is a solution to what algebraic equation? A fixed point is also a point of intersection of the graphs of $y = f(x)$ and what line?

9. Investigate iteration of $f(x) = x^2 - R$. Let the constant R take on values no more than 0.1 unit apart on the interval from 0 to 2. What different types of behavior do the iterates exhibit? Find the approximate R values at which you observe changes in the behavior of the iterates. Among others, be sure to include the following R values in your investigation: 0.1, 1.1, 1.31, 1.38, 1.55, 1.76, 2.

10. Using the results of problems 7 and 9, determine how the slope of the curve $y = f(x)$ at a fixed point is related to whether the fixed point is attracting or repelling.

11. a. What are the fixed points of $f(x) = x^2 - R$ in terms of R?

    b. Use the results of problem 10 to determine the exact interval of R values for which $f(x) = x^2 - R$ has an attracting fixed point.

Investigation 6: Newton's Method in the Complex Plane

In this investigation we extend Newton's method to the problem of finding the complex zeros of polynomials. In particular, we are interested in finding the basins of attraction of these zeros, which will be regions in the complex plane rather than intervals on the real number line. This investigation requires the use of technology that can operate on complex numbers and generate images in the complex plane.

We begin by using Newton's method to investigate the basins of attraction of the zeros of $f(z) = z^2 - 1$. The zeros of $f(z)$ are complex numbers z for which $z^2 = 1$, that is 1 and -1. These numbers, called the **square roots of unity**, are written as $1 + 0i$ *square roots of unity* and $-1 + 0i$ in complex number notation. Newton's method generates a sequence of estimates of the zeros of $f(z)$ using the iterative equation

$$z_n = z_{n-1} - \frac{f(z_{n-1})}{f'(z_{n-1})}.$$

Since $f(z) = z^2 - 1$ and $f'(z) = 2z$, Newton's method generates a sequence using the recursive equation

$$z_n = z_{n-1} - \frac{z_{n-1}^2 - 1}{2z_{n-1}},$$

which can be written as

$$z_n = \frac{z_{n-1}^2 + 1}{2z_{n-1}}. \tag{32}$$

Because the numbers $z = 1 + 0i$ and $z = -1 + 0i$ both have an imaginary part of $0i$, the zeros are both located on the real axis. This can be demonstrated by graphing the numbers in the complex plane. We can investigate the basins of attraction for these zeros by iterating Newton's method for various starting values z_0. For complex values of z_0 with positive real parts (such as $1 + i$, $0.5 - 0.75i$, and $0.1 + 2i$), Newton's method converges to 1. For complex values of z_0 with negative real parts, Newton's method converges to -1. Every starting value z_0 that converges to 1 is in the basin of attraction of 1. Likewise, every starting value z_0 that converges to -1 is in the basin of attraction of -1. The imaginary axis in the complex plane appears to form a boundary between the two basins of attraction for the zeros of $f(z)$.

To see the basins more easily, we can assign the color red to each point in the basin of attraction of 1 and assign the color green to each point in the basin of attraction of -1. (Technology that allows you to color points on a graph would be helpful.) Then a picture of the basins of attraction will be red to the right of the imaginary axis and green to the left of the imaginary axis. What happens to points located exactly on the imaginary axis?

The values generated by using equation (32) and initial value z_0 on the imaginary axis (that is, z_0 of the form $0 + ki$) do not belong to either basin of attraction. Some of these values remain on the imaginary axis and never converge to any value. Others lead to failure of Newton's method. To understand why this occurs, consider what happens if we choose 0 as the starting value. Newton's method fails due to division by 0 in equation (32). In addition, if we start with $\pm i$, one iteration will take us to 0, and then Newton's method fails on the second iteration. By extension, a starting value that yields $\pm i$ in one iteration will go to 0 in two iterations, and again the method fails. The starting values that go to 0 in two iterations are the solutions to

$$i = \frac{z^2 + 1}{2z} \quad \text{and} \quad -i = \frac{z^2 + 1}{2z}.$$

Each equation is quadratic and has two solutions of the form $0 + ki$, so a total of four starting values on the imaginary axis go to 0 in two iterations. Each of these four values can be reached in one iteration by two other values, so eight starting values go to 0 in three iterations. Putting all this together, observe that 2 starting values go to 0 in one iteration, 4 starting values go to 0 in 2 iterations, and 8 starting values go to 0 in 3 iterations. In general 2^n starting values go to 0 in n iterations. In fact, an infinite number of points on the imaginary axis will eventually iterate to 0, thus failing, under Newton's method. Other than 0, i, and $-i$, all of these numbers have irrational imaginary parts. As we noted in the situation with basins of attraction on the real number line, we cannot represent irrational numbers exactly with technology; we are able to work only with finite decimal approximations of these numbers. In practice, then, these numbers will not appear in the sequence of iterates.

Exercises for Investigation

These questions assume the use of technology to iterate Newton's method in the complex plane and to generate images of the basins of attraction.

1. Verify with technology that the basins of attraction for the zeros of $f(z) = z^2 - 1$ are the left and right half-planes divided by the imaginary axis, as described in the text.

2. For initial values on the imaginary axis, how do the iterates of Newton's method for $f(z) = z^2 - 1$ move around on the imaginary axis?

3. Let $f(z) = z^3 - 1$.

    a. Find the zeros of $f(z)$. These numbers are also the roots of $z^3 = 1$, and they are called the **cube roots of unity**. Graph these numbers in the complex plane.   *cube roots of unity*

    b. Write the iterative equation to use with Newton's method to find the zeros of $f(z)$.

    c. If the technology you are using allows it, generate a three-color image of the basins of attraction in the complex plane. What do you notice about the three colors near the boundaries of the basins?

*The image generated in problem 3c is an example of a **fractal**. Because of the infi-   fractal nitely complicated boundaries of the basins of attraction, the dimension of the boundaries is not 1, as it would be if the boundary were a smooth curve. On the other hand, the dimension is not 2, which would imply that the boundary fills a region of the plane. Instead, the dimension of each boundary is a fraction between 1 and 2, hence the name fractal. Fractals often possess the characteristic of self-similarity, meaning the image looks the same at varying levels of magnification. The Newton's method fractal in problem 3 is self-similar, as zooming in on the boundaries of the basins of attraction reveals.*

4. Let $f(z) = z^4 - 1$.

    a. Find the zeros of $f(z)$, which are the fourth roots of unity. Graph these numbers in the complex plane.

    b. Write the iterative equation to use with Newton's method to find the zeros of $f(z)$.

    c. If the technology you are using allows it, generate a four-color image of the basins of attraction in the complex plane. What do you notice about the four colors near the boundaries of the basins?

5. Repeat problem 4 for $f(z) = z^5 - 1$ (the fifth roots of unity). Use five colors in part c.

6. If n is an integer greater than 2, where do you think the nth roots of unity are located in the complex plane? The nth roots of unity are zeros of the function $f(z) = z^n - 1$. Describe the basins of attraction of these numbers.

7. Suppose a fifth-degree polynomial has complex zeros given by 2, $\pm 2i$, and $1 \pm i$. Find a formula for the polynomial and generate an image of the basins of attraction of the zeros.

END OF CHAPTER EXERCISES

Review and Extensions

1. Use the graph for $y = f'(x)$ to sketch a graph of $y = f(x)$. Label maximum, minimum, and inflection points.

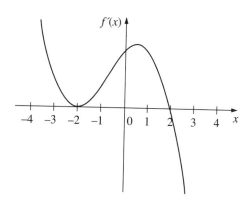

2. Knowing that $f''(a) = 0$ is not sufficient to guarantee that $f(x)$ has an inflection point at $x = a$.

    a. Give an example to support this statement.

    b. If $f''(a) = 0$, what additional information will guarantee that there is an inflection point at $x = a$?

3. Given $f(x) = \frac{\ln x}{x}$, use calculus ideas to find the coordinates of all maximum, minimum, and inflection points.

4. A cotton candy vendor at the state fair wants you to find the dimensions of a conical cup that will hold a given volume and have minimum surface area. Assume the volume is V and give the height and radius in terms of V.

5. The attracting force between two molecules can be modeled by a function of the form $f(x) = \frac{ax+b}{x^3}$. If the graph of this function has a maximum point at $(1, 1)$, find the values of a and b.

6. a. Use Newton's method to approximate a zero of the function $f(x) = x^4 + x - 3$ correct to three decimal places. What was your initial value?

    b. Examine a technology-generated graph of $y = f(x)$. You should see two zeros. Explain why your initial value resulted in the zero you found.

7. A softball game is being played in a field where the baselines are 60 feet long. A player throws the ball from second to third base, and it is estimated that the ball travels at a speed of 50 feet per second. At what rate is the distance from the ball to home plate changing when the ball is 20 feet from third base?

8. Find $\frac{dy}{dx}$ if :

    a. $y = \cos^{-1}(4x)$ b. $y = x^{\sqrt{3x}}$

9. Find the equation of the tangent line to the curve defined by $e^x + e^y = 2 + xy$ at the point where $x = 0$.

10. The following chart gives information about a function $f(x)$ and its derivative $f'(x)$. Use this information to write the equation of the tangent line to $f^{-1}(x)$ at $x = 6$.

x	$f(x)$	$f'(x)$
0	2	6
2	6	4
6	5	1

Mean Value Theorem

11. The ***Mean Value Theorem*** states that if a function f is continuous on the closed interval $[a,b]$ and differentiable on the open interval (a,b) then there is a number c in (a,b) such that $f'(c) = \frac{f(b)-f(a)}{b-a}$.

    a. Draw a graph to demonstrate what this theorem says about secant and tangent lines.

    b. Interpret this theorem based on ideas concerning average and instantaneous rates of change.

    c. Let $f(x) = x^3 - 2x + 1$, $a = 0$, $b = 3$. Find a number c that satisfies the results of this theorem.

    d. Let $f(x) = ax^2 + bx + d$. Find the number c that satisfies the Mean Value Theorem on the interval (x_1, x_2). Your answer will be in terms of x_1 and x_2.

    e. Let $f(x) = |x|$, $a = -2$, $b = 2$. Evaluate $\frac{f(b)-f(a)}{b-a}$. Explain why the Mean Value Theorem does not apply in this example.

CHAPTER **4**

Numerical Solutions to Differential Equations

4.1   Introduction to Differential Equations

In Chapter 1 we investigated quantities that change at a rate proportional to the current value of the quantities. Examples include savings account balances, in which the interest earned per period of time is proportional to the balance, and populations, in which the growth in the population is proportional to the existing population. In both of these situations the fact that the phenomenon experiences a proportional rate of change can be represented by the equation

$$\frac{dP}{dt} = k \cdot P, \tag{1}$$

where P is the dependent variable, t is the independent variable, and k is the constant of proportionality. In Chapter 1 we generated values of the dependent variable using recursive equations of the form

$$P_n = P_{n-1} + (k \cdot P_{n-1}) \cdot \Delta t$$

and found that P was an exponential function of t. If k is positive, P continues to increase with time, and never becomes constant or decreases. A phenomenon with a rate of change represented by equation (1) with positive k experiences **unconstrained exponential growth**. Equation (1) is also known as the Malthusian model of population, named for Thomas Malthus (1766–1834) who first studied population growth.

unconstrained exponential growth

   Not all population growth is unconstrained. There are many situations in which populations increase up to a point and then become constant, or "level off." This leveling off usually occurs because of limited food, space, or other resources. How can we modify equation (1) to reflect the fact that there is an upper limit, or ceiling, past which the population cannot grow?

Suppose a constant M is the maximum sustainable population. If P is near M, then the limited resources constrain the growth of the population. In terms of derivatives, this means that the closer P is to M, the closer $\frac{dP}{dt}$ is to 0. It may seem reasonable, then, that $\frac{dP}{dt}$ is proportional to $M - P$, or

$$\frac{dP}{dt} = c \cdot (M - P)$$

where c is a constant.

However, if the population P is far from M, then the existence of a ceiling has little influence on the rate at which P grows, and $\frac{dP}{dt}$ is approximately proportional to P. In the equation above, if P is small, $\frac{dP}{dt}$ is approximately cM. The derivative $\frac{dP}{dt}$ has to be proportional to both P and $M - P$, so

$$\frac{dP}{dt} = c \cdot P \cdot (M - P).$$

Now, if P is far from M, the term $M - P$ is approximately equal to M, so $\frac{dP}{dt}$ is proportional to PM. If P is close to M, $\frac{dP}{dt}$ is still close to 0. So this is almost what we want.

Since M is a constant, $\frac{dP}{dt}$ proportional to PM actually means $\frac{dP}{dt}$ is proportional to P. To see this more easily, define a constant k such that $c = \frac{k}{M}$. Now we have

$$\frac{dP}{dt} = \frac{k}{M} \cdot P \cdot (M - P)$$

$$= k \cdot P \cdot \left(\frac{M - P}{M}\right).$$

The factor $(\frac{M-P}{M})$ expresses the difference between P and M as a fraction of M. Notice that we now have $\frac{dP}{dt}$ is approximately 0 when P is near M and $\frac{dP}{dt}$ is approximately $k \cdot P$ when P is far from M. Rewriting the final factor gives us

$$\frac{dP}{dt} = k \cdot P \cdot \left(1 - \frac{P}{M}\right). \tag{2}$$

constrained exponential growth

logistic growth

differential equations

A phenomenon with a rate of change given by equation (2) experiences ***constrained exponential growth***, which is also known as ***logistic growth***.

Equations (1) and (2) are examples of ***differential equations*** because some of the terms include derivatives. Other examples of differential equations include the following.

$$\frac{dy}{dt} = 2t \qquad\qquad \frac{d^2y}{dx^2} + \frac{dy}{dx} = 2x - y$$

$$\frac{dP}{dt} = 0.2P - 500 \qquad\qquad x + y\frac{dy}{dx} = x^2$$

A function that satisfies a differential equation is called a *solution* of the differential equation. For example, $y = x^2$ is a solution to the differential equation

$$\frac{dy}{dx} = 2x \qquad (3)$$

because $2x$ is the derivative of x^2 with respect to x. The functions with equations $y = x^2 + 5$ and $y = x^2 - 72$ also satisfy equation (3), so this differential equation does not have a unique solution. In fact, any function of the form $y = x^2 + c$ is a solution of equation (3), so we call $y = x^2 + c$ the **general solution** of equation (3). The equation $y = x^2 + c$ represents a family of functions that are the solutions to equation (3). Each of the functions with equations $y = x^2$, $y = x^2 + 5$, and $y = x^2 - 72$ is called a **particular solution** of equation (3).

general solution

particular solution

We can solve many differential equations based on our knowledge of derivatives. For instance, we recognize that the general solution of $\frac{dy}{dx} = \cos x$ is $y = \sin x + c$ because we know that $\sin x + c$ is a function whose derivative with respect to x is $\cos x$. By similar reasoning, we see that the general solution to $\frac{dy}{dx} = 3x$ is $y = \frac{3}{2}x^2 + c$. If, in addition, we are given that the solution function contains the ordered pair $(2, 5)$, then we can substitute $(2, 5)$ into $y = \frac{3}{2}x^2 + c$ and find the value of c. This yields the equation $5 = \frac{3}{2}(2^2) + c$, which can be solved to find $c = -1$. Therefore, the particular solution of $\frac{dy}{dx} = 3x$ containing $(2, 5)$ is $y = \frac{3}{2}x^2 - 1$.

Exercise Set 4.1

Use your knowledge of the derivatives of elementary functions (see Table 2.24 in Section 2.5) to find the general solution of each differential equation. Then find the particular solution that includes the given ordered pair.

1. $\dfrac{dy}{dx} = 2x + 3$, $(-1, 2)$

2. $\dfrac{dy}{dx} = \sec^2 x$, $\left(\dfrac{\pi}{4}, 2\right)$

3. $\dfrac{dy}{dx} = \dfrac{1}{\sqrt{x}}$, $(4, 2)$

4. $\dfrac{dy}{dx} = -\dfrac{1}{x^2}$, $(4, 3)$

5. $\dfrac{dy}{dx} = x^2 - 3x$, $(0, 0)$

6. $\dfrac{dy}{dx} = \pi$, $(1, 1)$

4.2 Slope Fields

Suppose we are investigating a phenomenon and we write a differential equation that describes how the phenomenon changes. Even if we cannot solve this differential equation, we can visualize solutions of the differential equation. Because any differentiable function is locally linear, we can approximate its behavior about any point by using a line tangent to the function at that point. If we graph many of these approximating tangent lines, we can get a picture of the general solution of the differential equation. This picture is called a **slope field**. We can create a slope field using the following procedure:

slope field

1. select regularly spaced points in the coordinate plane;

2. calculate the slope of the line tangent to the solution function at each of the points by evaluating the derivative at that point; and,

3. draw a short line segment through each point using the slope from step (2).

We draw our first slope field for the differential equation $\frac{dy}{dx} = 2x$, which we know has the general solution $y = x^2 + c$. This differential equation implies that the slope of a solution function at any point is equal to 2 times the x-coordinate at that point. At the point (0, 1), the slope is $2 \cdot 0 = 0$, so the line segment through (0, 1) in the slope field has slope 0. At (0, −1) and at (0, 0) the slope is also zero. At the point (1, 0) the slope is $2 \cdot 1 = 2$, so the line segment through (1, 0) in the slope field has slope 2. In general, the slope is 2 at any point whose x-coordinate is 1. At (−1, 1) the slope is $2 \cdot (-1) = -2$, and any point with x-coordinate −1 also has slope −2. Tangent segments at these and other points are shown in Figure 4.1.

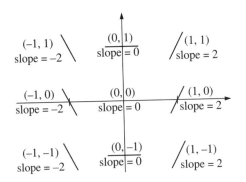

Figure 4.1 Slope field near the origin for $\frac{dy}{dx} = 2x$

Continuing to draw segments with slopes determined by $\frac{dy}{dx} = 2x$ produces the slope field in Figure 4.2. This configuration of slopes agrees with the shape of the general solution $y = x^2 + c$. We can graph the particular solution that includes the point (−1, 0) by starting at (−1, 0) in the slope field and following the direction indicated by the tangent lines, as shown in Figure 4.2. This curve agrees with the particular solution that includes the ordered pair (−1, 0), which is $y = x^2 - 1$.

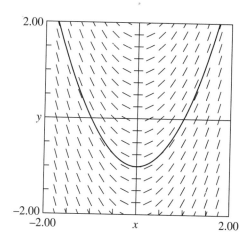

Figure 4.2 Slope field for $\dfrac{dy}{dx} = 2x$ with particular solution

Cooling Curves

Suppose that we place a cup of hot water in a cool room and measure its temperature over time. Newton's law of cooling states that the water cools at a rate that is proportional to the difference between the temperature of the water and the ambient temperature (the temperature of the surrounding environment). A differential equation for the temperature T of the water as a function of time t is

$$\frac{dT}{dt} = k(T - A), \qquad (4)$$

where A is the ambient temperature. The constant of proportionality k must be negative because $T - A > 0$ and cooling implies that T must be decreasing. This model is accurate in general only if the temperature of the cooling substance is relatively close to the ambient temperature. The slope field for equation (4) with $k = -0.01$ and $A = 25$ (in degrees Celsius) is shown in Figure 4.3.

The slope field in Figure 4.3 helps us visualize the general solution of $\frac{dT}{dt} = -0.01(T - 25)$. For an initial water temperature greater than 25 degrees, the slope field shows that the temperature of the water decreases over time and eventually levels off. In the differential equation, the factor $(T - 25)$ is positive when the temperature of the water is greater than 25 degrees. Therefore, $\frac{dT}{dt}$ is negative for $T > 25$, and the graph of T versus t is decreasing everywhere. Also notice that as T gets close to 25, the segments in the slope field become nearly horizontal. This is because as T approaches 25, the factor $(T - 25)$ approaches 0, so $\frac{dT}{dt}$ likewise approaches 0. These horizontal segments indicate that the temperature of the water eventually stabilizes at 25 degrees.

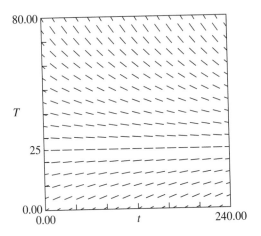

Figure 4.3 Slope field for $\dfrac{dT}{dt} = -0.01(T - 25)$

Unconstrained Exponential Growth

As we discussed in Section 4.1, a population experiencing unconstrained exponential growth can be modeled by the differential equation

$$\frac{dP}{dt} = k \cdot P, \tag{5}$$

where P is the population at time t and k is a positive constant of proportionality. Figure 4.4 shows a slope field for differential equation (5) using $k = 0.06$, a value chosen arbitrarily. At regularly spaced points within the viewing window, segments of tangent lines have been drawn. In this particular slope field, the regularly spaced points are 2.5 units apart both horizontally and vertically. To demonstrate how this slope field is drawn, consider the point $(15, 15)$. At this point the value of $\frac{dP}{dt}$ is $0.06 \cdot 15 = 0.90$, and this value is used as the slope of the line segment drawn in the slope field. This means that if $(15, 15)$ is on the graph of a solution of the differential equation, then the slope of the solution at $(15, 15)$ is 0.90.

The slope field in Figure 4.4 tells us a lot about the general solution to equation (5). The slopes of the tangents reveal that when P is positive and relatively small, P increases at a relatively slow rate. As P increases, the rate of change of P also increases. In Figure 4.4, the tangents with the largest slopes occur at the points with the largest values of P. If P is negative, the slopes of the tangents indicate that P decays. The slope field helps us visualize the family of functions that is the general solution to equation (5). Any particular solution can be determined by specifying an ordered pair (t, P) that represents the size of the population at a specified time and then following the direction of the tangent segments beginning at that point.

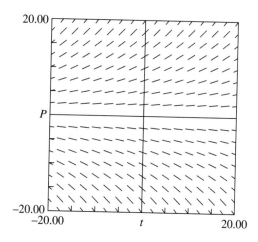

Figure 4.4 Slope field for $\frac{dP}{dt} = 0.06 \cdot P$

Constrained Exponential Growth

As we discussed in Section 4.1, a population experiencing constrained exponential growth, or logistic growth, can be modeled by the differential equation

$$\frac{dP}{dt} = k \cdot P \cdot \left(1 - \frac{P}{M}\right), \tag{6}$$

where P is the population at time t, M is the maximum sustainable population, and k is a positive constant of proportionality. Figure 4.5 shows a slope field for differential equation (6) using $M = 80$ and $k = 0.06$, chosen arbitrarily. At regularly spaced points within the viewing window, segments of tangent lines have been drawn. In this particular slope field, the regularly spaced points are 6.25 units apart both horizontally and vertically. As an example of how to interpret this slope field, we consider the point $(25, 25)$. The value of $\frac{dP}{dt}$ at $(25, 25)$ is $0.06 \cdot 25 \cdot (1 - \frac{25}{80}) \approx 1.03$, and this value is used as the slope of the line segment drawn in the slope field. This means that if $(25, 25)$ is on the graph of a solution of the differential equation, then the slope of the solution at $(25, 25)$ is approximately 1.03.

The slope field in Figure 4.5 tells us a lot about the general solution to equation (6). The slopes of the tangents reveal that when P is relatively small, P increases at a relatively slow rate. As P increases (but remains small compared to M), the rate of change of P also increases. Finally, as the value of P approaches M, the rate of change of P decreases and approaches zero. The slope field helps us visualize the family of functions that is the general solution of equation (6). Any particular solution can be determined by specifying an ordered pair (t, P) that represents the size of the population at a specified time and then following the direction of the tangent segments beginning at that point.

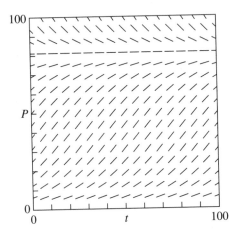

Figure 4.5 Slope field for $\frac{dP}{dt} = 0.06 \cdot P \cdot \left(1 - \frac{P}{80}\right)$

logistic curves

Figure 4.6 shows a slope field with three particular solutions of equation (6), which are called **logistic curves**. The overall shape of each logistic curve is sensitive to the initial size of the population. Each solution is determined by specifying the initial value P_0 of the population at time t_0. In solution A, we see that if P_0 is small compared to M, the graph of the population is concave up at first, changes concavity as P increases, and then levels off at M. In solution B, we see that if P_0 is close to M but less than M, the graph of the population is concave down, is increasing, and levels off at M. In solution C, we see that if P_0 is greater than M, the graph of the population is decreasing until it levels off at M. As you examine the diagram, you may notice that the slopes for $P = 81.25$ appear to be horizontal. As always, when using diagrams and technology you must be wary when such tools show potentially misleading information.

In Figure 4.6, notice that for P_0 between 0 and M, the population is always increasing. This agrees with equation (6), which shows that $\frac{dP}{dt}$ is positive for values of P such that $0 < P < M$, since the factors P and $(1 - \frac{P}{M})$ are positive when P is in this interval. On the other hand, if P is greater than M, then the $(1 - \frac{P}{M})$ factor in equation (6) is negative. Thus $\frac{dP}{dt}$ is also negative, which implies that the graph of P is decreasing, as seen in solution C.

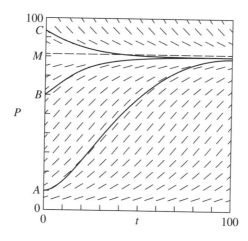

Figure 4.6 Slope field with three particular solutions

Exercise Set 4.2

1. i.   Make a pencil and paper sketch of the slope fields of two of the differential
         equations from each of groups I, II, and III.

    ii. Use the technology of your choice (calculator or computer) to create a slope
         field for each of the differential equations listed below.

         Group I

         a. $\dfrac{dy}{dx} = x - 1$ b. $\dfrac{dy}{dx} = 2x - 1$ c. $\dfrac{dy}{dx} = 2x + 1$

         d. $\dfrac{dy}{dx} = \cos x$ e. $\dfrac{dy}{dx} = e^x$

         Group II

         f. $\dfrac{dy}{dx} = y$ g. $\dfrac{dy}{dx} = 2y$ h. $\dfrac{dy}{dx} = y - 1$

         i. $\dfrac{dy}{dx} = 0.3(4 - y)$ j. $\dfrac{dy}{dx} = 0.1y(8 - y)$

         Group III

         k. $\dfrac{dy}{dx} = x + y$ l. $\dfrac{dy}{dx} = x - y$ m. $\dfrac{dy}{dx} = xy$

         n. $\dfrac{dy}{dx} = -xy$ o. $\dfrac{dy}{dx} = \dfrac{x}{y}$ p. $\dfrac{dy}{dx} = -\dfrac{x}{y}$

         q. $\dfrac{dy}{dx} = -\dfrac{2x}{5y}$ r. $\dfrac{dy}{dx} = -\dfrac{4y}{3x}$ s. $\dfrac{dy}{dx} = \dfrac{y - x}{y + x}$

2. Describe the characteristic shared by the slope fields from Group I, in which the derivative depends only on x. Explain why the slope fields have this characteristic.

3. Describe the characteristic shared by the slope fields from Group II, in which the derivative depends only on y. Explain why the slope fields have this characteristic.

4. Choose one differential equation from Group I, II, or III whose solution you can find analytically. Write a sentence or two comparing the graph of the solution with the slope field.

5. Choose one differential equation from Group I, II, or III whose solution you cannot find analytically. Describe what the slope field tells you about the graph of the solution.

6. Refer to differential equation i from Group II for the following.

    a. Complete the following statement: The rate of change of y with respect to x is proportional to….

    b. Describe the shape of the slope field and how it is related to the differential equation.

    c. Investigate the effect of the constants 0.3 and 4 on the slope field by changing the values of these constants. Describe the corresponding changes in the slope field.

    d. Describe a phenomenon for which this differential equation is a good model.

7. Refer to differential equation j from Group II for the following.

    a. Complete the following statement: The rate of change of y with respect to x is proportional to….

    b. Describe the shape of the slope field and how it is related to the differential equation.

    c. Investigate the effect of the constants 0.1 and 8 on the slope field by changing the values of these constants. Describe the corresponding changes in the slope field.

    d. Describe a phenomenon for which this differential equation is a good model.

8. Figure 4.3 shows the slope field for the differential equation $\frac{dT}{dt} = -0.01(T - 25)$. What does the graph of a particular solution look like if the initial temperature of the water is less than 25 degrees? Explain how the behavior of this particular solution is related to the differential equation.

9. Learning theorists have hypothesized that the rate at which a person learns a body of facts or a particular skill is proportional to the difference between a maximum attainable level and the present level of mastery. This means that when a person's

skill level is far from that individual's maximum level, then the person will improve quickly. When a person's skill level is near that individual's maximum level, then the person will improve slowly. If y represents the amount that has been learned at time t, the relationship between y and t can be represented by the differential equation $\frac{dy}{dt} = k \cdot (M - y)$, where k is a positive constant of proportionality and M is the maximum attainable value of y. Some of the particular solutions of this differential equation are called **learning curves.** Create a slope field for this differential equation with $k = 0.2$ and $M = 100$. What does the graph of a particular solution look like with an initial value of y below 100? What if the initial value of y is above 100? Which of these graphs should be called a learning curve? Explain how the behavior of these particular solutions is described by the differential equation.

learning curves

10. The differential equation $\frac{dP}{dt} = k \cdot P \cdot (1 - \frac{P}{M})(\frac{P}{m} - 1)$ is called the **threshold model** for population growth. According to this model, the rate of growth of the population is proportional to the product of three factors: (1) the current population level, (2) the fraction of the maximum population M that the current population is below the maximum, and (3) the fraction of the minimum threshold value m that the current population is above the threshold. Create a slope field for this differential equation with $k = 0.06$, $M = 80$, and $m = 10$. What does the graph of a particular solution look like for an initial value of the population

threshold model

a.   greater than 80?   b.   between 10 and 80?

c.   less than 10?

Explain how the behavior of these particular solutions is described by the differential equation.

4.3   Euler's Method and Differential Equations

Suppose we know the following facts about the world population.

−   The world population in 1990 was approximately 5.333 billion.

−   The world population changes at a rate proportional to the current population.

−   On the average, the constant of proportionality is approximately 0.017 (based on historical population data).

How can we use this information to predict the world population in the years following 1990?

The second two pieces of information imply that, in general, population growth is governed by the differential equation

$$\frac{dP}{dt} = 0.017 \cdot P. \tag{7}$$

When we write this differential equation, we assume that P is a differentiable function of t, even though we may not know an equation for this function.

The first piece of information given above tells us that the point $(1990, 5.333)$ is on a graph of P versus t. Since differentiable functions are locally linear, P must be locally linear. We can use a tangent line to approximate the graph of P versus t over a reasonably small interval around the point $(1990, 5.333)$. Figure 4.7 shows this point and a segment of the line tangent to the graph at this point. Differential equation (7) allows us to determine that the slope of the tangent line at the point $(1990, 5.333)$ is $0.017 \cdot 5.333$, or approximately 0.091.

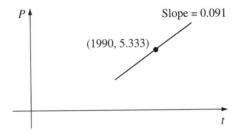

Figure 4.7 Tangent line segment at $(1990, 5.333)$

The equation of the tangent line in Figure 4.7 is

$$P = 0.091(t - 1990) + 5.333. \tag{8}$$

To estimate the population in 1994, we can let $t = 1994$ in equation (8) and calculate the corresponding value of P, which is about 5.696. Thus, 5.696 billion is an approximation for the population P when $t = 1994$. **Note:** In all of the calculations in this chapter, additional decimal places were retained to keep errors due to rounding at a minimum. However, only three places are shown in the text.

We could predict the world population in any year using the tangent line equation in (8). For example, substituting $t = 1998$ in equation (8) tells us that the population will be approximately 6.058 billion in 1998. However, as we move farther away from the point $(1990, 5.333)$, we have less confidence in our estimates because we know that the tangent line is less accurate as we move away from the point of tangency. We can improve our prediction for the population in 1998 if we base it on what we have already predicted about 1994, rather than basing it on our original information for 1990. We have estimated that the population in 1994 was about 5.696 billion. Thus, the point $(1994, 5.696)$ is close to a point on the graph of P versus t. We can now use differential equation (7) to find that the slope of the graph at a point with $P = 5.696$ is

0.017 · 5.696, or about 0.097. The equation of the line with slope 0.097 and containing the point (1994 , 5.696) is

$$P = 0.097(t - 1994) + 5.696. \tag{9}$$

To estimate the population in 1998, we can substitute $t = 1998$ into equation (9) and calculate the corresponding value of P, which is about 6.083. This tells us that the population is approximately 6.083 billion in 1998, a larger estimate than was obtained using $t = 1998$ in equation (8).

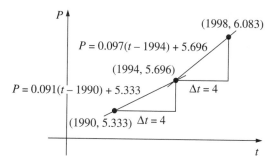

Figure 4.8 Using two steps to approximate 1998 world population

The process of estimating a new population from a known population can be represented with recursive equations. When we substitute $t = 1998$ into equation (9), we obtain the new approximate population value by adding the old population value (5.696) to the product of the slope of the new line (0.097) and the change in t (1998 − 1994). We can write this in equation form as

$$P_{new} = P_{old} + slope \cdot (t_{new} - t_{old}).$$

The slope is determined by evaluating the derivative $\frac{dP}{dt} = 0.017P$ at the previous value P_{old}. Using Δt for $t_{new} - t_{old}$, the change in t, we can write the system of recursive equations

$$t_{new} = t_{old} + \Delta t$$
$$P_{new} = P_{old} + \frac{dP}{dt}(P_{old}) \cdot \Delta t, \tag{10}$$

where $\frac{dP}{dt}(P_{old})$ is the value of the function $\frac{dP}{dt}$ evaluated at P_{old}.

The recursive equations in (10) allow us to determine a new ordered pair (t_{new}, P_{new}) based on knowledge of an old pair, (t_{old}, P_{old}), and this in turn allows us to make predictions about the population in future years. Whereas we used two steps in Figure 4.8 to go from a known value of P in 1990 to a prediction in 1998, additional steps with a smaller Δt can be taken using the recursive equations in (10). We expect that the smaller Δt is, the better the successive approximations will be, and the more confidence we can have in our predictions. In Table 4.9 are the calculations that use the tangent line at $(1990, 5.333)$ to approximate the population in 1992, and then use a new line at $t = 1992$ to approximate the 1994 population. This process is continued using a step size of $\Delta t = 2$. We expect that an approximation that results from iterating four times with $\Delta t = 2$ will generally be closer to the actual population value than a single estimate made with $\Delta t = 8$ or a two-step estimate made with $\Delta t = 4$. Table 4.9 shows an estimate of the 1998 population level at 6.096 billion.

t_{old}	P_{old}	$t_{new} = t_{old} + \Delta t$	$P_{new} = P_{old} + 0.017 \cdot P_{old} \cdot \Delta t$
1990	5.333	$1992 = 1990 + 2$	$5.514 = 5.333 + 0.017(5.333)(2)$
1992	5.514	$1994 = 1992 + 2$	$5.702 = 5.514 + 0.017(5.514)(2)$
1994	5.702	$1996 = 1994 + 2$	$5.896 = 5.702 + 0.017(5.702)(2)$
1996	5.896	$1998 = 1996 + 2$	$6.096 = 5.896 + 0.017(5.896)(2)$
1998	6.096		

Table 4.9 Population values generated with $\Delta t = 2$

Our estimate of the 1998 population can be further improved by decreasing the size of Δt to 1 year. With $\Delta t = 1$, we need eight iterations of the recursive equations in (10) to reach $t = 1998$. Table 4.10 shows that by using $\Delta t = 1$, our estimate of the population level in 1998 is 6.103 billion.

t_{old}	P_{old}	$t_{new} = t_{old} + \Delta t$	$P_{new} = P_{old} + 0.017 \cdot P_{old} \cdot \Delta t$
1990	5.333	$1991 = 1990 + 1$	$5.424 = 5.333 + 0.017(5.333)(1)$
1991	5.424	$1992 = 1991 + 1$	$5.516 = 5.424 + 0.017(5.424)(1)$
1992	5.516	$1993 = 1992 + 1$	$5.610 = 5.516 + 0.017(5.516)(1)$
1993	5.610	$1994 = 1993 + 1$	$5.705 = 5.610 + 0.017(5.610)(1)$
1994	5.705	$1995 = 1994 + 1$	$5.802 = 5.705 + 0.017(5.705)(1)$
1995	5.802	$1996 = 1995 + 1$	$5.901 = 5.802 + 0.017(5.802)(1)$
1996	5.901	$1997 = 1996 + 1$	$6.001 = 5.901 + 0.017(5.901)(1)$
1997	6.001	$1998 = 1997 + 1$	$6.103 = 6.001 + 0.017(6.001)(1)$
1998	6.103		

Table 4.10 Population values generated with $\Delta t = 1$

We have used a sequence of approximations to estimate the world population in 1998. In trying to improve our estimate, we used the recursive equations in (10) with step sizes of $\Delta t = 8$, $\Delta t = 4$, $\Delta t = 2$, and $\Delta t = 1$. In a situation in which we know the rate of change of a function, we can use this same method to generate any number of points that approximate the function. This strategy is called ***Euler's method*** ("Euler" is

Euler's method

pronounced "oiler"), named for Leonhard Euler (1707–1783), the most prolific writer of mathematics in the history of the subject.

As we decreased the step size Δt, our estimates for the world population in 1998 increased. How do we know that our estimates actually got closer to the actual value predicted by the differential equation $\frac{dP}{dt} = 0.017 \cdot P$ to which we applied Euler's method? We can verify that a function of the form $P(t) = Ae^{0.017t}$, where A is a constant, is a solution to this differential equation. The particular solution with the initial population (at time $t = 0$) of 5.333 billion is $P(t) = 5.333e^{0.017t}$, where t is in years since 1990 and P is in billions. According to this solution, the predicted population in 1998 (at time $t = 8$) is approximately 6.110 billion. Therefore, decreasing Δt in Euler's method improved our estimate for the world population in 1998.

Whenever we know a point of a function and an expression for the derivative of the function, we can use Euler's method to approximate function values. In general, if we know a starting point (x_0, y_0) and a formula for $\frac{dy}{dx}$, then we can choose a value for Δx and generate values of x and y using the recursive equations

$$x_i = x_{i-1} + \Delta x$$

$$y_i = y_{i-1} + \frac{dy}{dx}\bigg|_{(x_{i-1}, y_{i-1})} \cdot \Delta x. \tag{11}$$

The notation

$$\frac{dy}{dx}\bigg|_{(x_{i-1}, y_{i-1})}$$

indicates that the derivative $\frac{dy}{dx}$ is evaluated at the point (x_{i-1}, y_{i-1}). Each iteration of the equations in (11) adds Δx to the previous x value and adds Δy to the previous y value, where Δy is the product of Δx and the slope of a tangent line containing (x_{i-1}, y_{i-1}). This slope is equal to the derivative function evaluated at (x_{i-1}, y_{i-1}).

Since Euler's method can be used to approximate values of a function when we know an initial point and a rate of change, this method will be particularly useful when we are unable to determine the equation of the function from its rate of change. At this point in our study of calculus, constrained exponential growth is an example of a phenomenon for which we can describe the rate of change of the function, but for which we cannot find a closed form equation for the function. Given a constant of proportionality equal to 0.1 and a maximum population equal to 2000, we know that the rate of change of a population that experiences constrained growth is described by the differential equation

$$\frac{dP}{dt} = 0.1 \cdot P \cdot \left(1 - \frac{P}{2000}\right). \tag{12}$$

We do not know a function whose rate of change is described by the differential equation in (12), but Euler's method allows us to generate approximate values of such a function. Let the initial point be designated as (t_0, P_0). Then $t_1 = t_0 + \Delta t$, which simply means that we are estimating a point Δt units away from (t_0, P_0). The P value of this new point is given by

$$P_1 = P_0 + \frac{dP}{dt}(t_0, P_0)\Delta t.$$

In this case, the derivative is given in equation (12), so

$$P_1 = P_0 + 0.1 \cdot P_0 \cdot \left(1 - \frac{P_0}{2000}\right) \cdot \Delta t.$$

If we assume that $P_0 = 10$ and $t_0 = 0$, and we choose a step size $\Delta t = 1$, we find that $P_1 \approx 11.00$. We now use the point $(t_1, P_1) = (1, 11.00)$ to calculate the next point (t_2, P_2) using the equation

$$P_2 = P_1 + 0.1 \cdot P_1 \cdot \left(1 - \frac{P_1}{2000}\right) \cdot \Delta t,$$

which gives $P_2 = 12.09$ so that $(t_2, P_2) = (2, 12.09)$. We can generate as many points as we wish using the recursive equations

$$t_i = t_{i-1} + \Delta t$$
$$P_i = P_{i-1} + 0.1 \cdot P_{i-1} \cdot \left(1 - \frac{P_{i-1}}{2000}\right) \cdot \Delta t. \tag{13}$$

Figure 4.11 shows the first three points generated recursively using these equations.

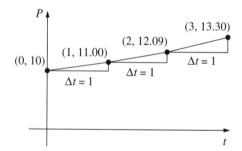

Figure 4.11 Three points generated recursively using equations in (13)

What is the meaning of the line segments that we follow from one point to the next in Figure 4.11? The first line segment has slope equal to the derivative of the particular solution at (0, 10), which is a point of the particular solution, so the first segment is tangent at (0, 10) to the curve we wish to estimate. The second line segment has slope equal to the derivative of the particular solution at (1, 11.00). This point is an approximation for a point on the curve, and it is not actually on the curve. Therefore, the second line segment is an approximation for a line segment tangent to the curve. The same is true for successive approximations after the first two. Each line segment, other than the segment containing the starting point, is an approximation for a line segment that is tangent to the curve. Since the slope of each segment is determined by the derivative of the particular solution we wish to approximate, each segment is tangent to a curve in the family of curves with the given derivative; however, the segments are only approximations for tangents to the particular curve with the given derivative and the given starting point.

Figure 4.12 shows the result of iterating the equations in (13) one hundred times with $\Delta t = 1$. Notice that the shape of the graph agrees with logistic curve A drawn in the slope field in Figure 4.6. The graph of P increases gradually at first, then increases rapidly, passes a point of inflection, increases less rapidly, and finally levels off when the P values approach 2000, the maximum sustainable population. Remember that the values of P that we generate using Euler's method are approximations. When we cannot find the particular solution of a differential equation with given initial conditions, Euler's method enables us to generate points that are close to those on the graph of the particular solution.

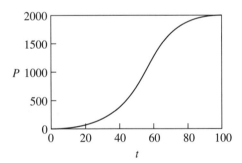

Figure 4.12 Logistic curve generated using equations in (13)

Lab 7 *Euler's Method*

In this lab, use the technology of your choice to execute Euler's method in the following investigations. Your choice of technology should allow input of the following:

– the derivative function $\frac{dP}{dt}$,

– the initial values for the variables t and P,

– the largest value of t for which the value of P is to be estimated, and

– the step size Δt.

The technology should graph ordered pairs (t_i , P_i) and should give the estimates for values of P.

1. The growth of a population that increases by 10% per year without constraint can be represented by the differential equation $\frac{dP}{dt} = 0.1 \cdot P$. Assume the initial population is 1000.

    a. Solve this differential equation analytically. Find the particular solution whose graph contains the initial point (0, 1000).

    b. According to the population function P you found in part a, how big is the population when $t = 100$ years?

    c. Use Euler's method and the given initial population with $\Delta t = 1$ year to estimate the population size when $t = 100$ years.

    d. Compare your population sizes in parts b and c. What is the absolute error in your Euler's method estimate? What is the relative error?

    e. How small must Δt be so that the relative error after 100 years is less than 10%?

2. The following table gives the U.S. population in millions according to census data collected every 10 years from 1790 through 1990.

    a. Assume the population in the table can be modeled using the derivative for unconstrained growth, $\frac{dP}{dt} = kP$. Use the data to estimate a value of k for the years 1790 through 1860. Write recursive equations for Euler's method, and use your equations to estimate the population in each of these years, assuming the initial population is 3.939 million and $\Delta t = 1$. For each year, examine the differences between the actual population and the Euler's method estimate. Find the value of k that minimizes the differences between the actual and the estimated population.

    b. Iterate the Euler's method equations with the final k value determined in part a to estimate the population for the years 1860–1990. Construct a table that compares the estimated population values with the actual population values for the years 1860–1990. What factors in the growth of the U.S. population might account for discrepancies between the predicted values and the actual values for these later years?

Year	U.S. Population (millions)	Year	U.S. Population (millions)
1790	3.939	1900	75.995
1800	5.308	1910	91.972
1810	7.240	1920	105.711
1820	9.638	1930	122.775
1830	12.866	1940	131.669
1840	17.069	1950	151.326
1850	23.192	1960	179.323
1860	31.443	1970	203.302
1870	39.818	1980	226.546
1880	50.156	1990	248.710
1890	62.948		

Reference

U.S. Bureau of the Census, *Statistical Abstract of the United States: 1992* (112th edition). Washington, DC, 1992.

3. a. Now assume the population data can be modeled using the derivative for constrained growth, $\frac{dP}{dt} = kP(1 - \frac{P}{M})$. Examine the data and estimate values for k and M for the years from 1790 through 1910. Write recursive equations for Euler's method, and use your equations to estimate the population in each of these years, assuming the initial population is 3.939 million and $\Delta t = 1$. Examine the differences between the actual population and the Euler's method estimate for each year. Find values for k and M that minimize the differences between the actual and the estimated populations.

   b. Iterate the Euler's method equations with the final k and M values determined in part a to estimate the population for the years 1910–1990. Construct a table that compares the estimated population values with the actual population values for the years 1910–1990. What factors in the growth of the U.S. population might account for discrepancies between the predicted values and the actual values for these later years?

Write a report summarizing your investigations.

Exercise Set 4.3

Use Euler's method and the technology of your choice to solve each problem.

1. For an object falling at a relatively slow speed over a short distance, the rate of change of the velocity v is described by the differential equation $\frac{dv}{dt} = g - kv$,

where g is the constant acceleration due to gravity and k is a constant of proportionality. This equation incorporates the assumption that the falling object is acted on by a force due to air resistance that is proportional to the velocity of the object. Assume that the initial velocity is 0 m/sec, $g = 9.8 \text{ m/sec}^2$, $k = 0.02$, and $\Delta t = 2$ sec. Graph an approximation for the velocity function over the first 50 seconds. What is the object's velocity at $t = 50$ seconds?

2. When a 12 volt battery is connected in a series circuit in which the inductance is 0.5 henry and the resistance is 10 ohms, the rate of change of the current in the circuit is described by $0.5 \cdot \frac{di}{dt} + 10i = 12$ where i is in amperes (amps) and t is in seconds. Assuming the initial current is 0 amps, approximate the current at each tenth of a second for the first two seconds. What happens if you approximate the current at each one hundredth of a second for the first two seconds? Which result is more accurate? Why?

3. When chemicals A and B combine in a first order reaction, the rate of change in the amount of the new compound C being formed is proportional to the remaining concentrations of each of the original two chemicals. Chemists measure quantities of substances in terms of the number of atoms or molecules. A mole is the amount of a substance that has the same number of elementary units (atoms or molecules) as there are in 12 grams of carbon-12, or 6.02×10^{23} units. Suppose it takes 1 mole of A and 2 moles of B to produce 1 mole of C and the rate constant is 0.06. If we start with 20 moles of chemical A and 50 moles of chemical B, then the rate of change of the amount of compound C is

$$\frac{dc}{dt} = k(20 - c)(50 - 2c)$$

where c is the amount of compound C. Use Euler's method to generate estimates for the moles of C produced over time. Obtain a graph of these values.

4. Consider the function that satisfies the differential equation $\frac{dy}{dx} = 0.2y$ and contains the point $(0, 1)$.

    a. Use Euler's method to approximate the value of y when $x = 20$.

    b. Find the particular solution of this differential equation analytically. What is the actual value of y when $x = 20$?

    c. Discuss the error in the estimation you made in part a. Was your estimate too large or too small? Explain why.

5. Consider the function that satisfies the differential equation $\frac{dy}{dx} = \frac{1}{2\sqrt{x}}$ and contains the point $(1, 2)$.

    a. Use Euler's method to approximate the value of y when $x = 10$.

b. Find the particular solution of this differential equation analytically. What is the actual value of y when $x = 10$?

c. Discuss the error in the estimation you made in part a. Was your estimate too large or too small? Explain why.

6. A certain function has $\frac{dy}{dx} = x - y$ as its derivative, and the function contains the ordered pair (0, 3). Use Euler's method with $\Delta x = 1$ to estimate the value of y when $x = 5$.

*The **immigration model** for population growth assumes that the population grows in proportion to the size of the current population and that the population also grows because of immigration. This model can be represented by the differential equation $\frac{dP}{dt} = kP + (immigration\ rate)$. If we assume that immigration occurs at a constant rate of m, the differential equation becomes $\frac{dP}{dt} = kP + m$.*

immigration model

7. Suppose two factors are known to contribute to the growth of a country's population. The population grows in proportion to the size of the current population with constant of proportionality equal to 0.03. The population also grows because of immigration that occurs at the rate of 100,000 people per year. The country's present population is 10 million.

a. Write a differential equation that describes the population growth in this country.

b. Use Euler's method to study the long-term population trend. Assuming that the growth rate and the immigration rate remain the same, find the number of years required for the population to reach 1 billion (1,000 million).

4.4 Falling Objects: In the Water and in the Air

Atomic Waste Disposal

For several years the United States disposed of radioactive waste by filling 55 gallon barrels with waste and dropping them into the ocean. The barrels were dropped in areas where the ocean floor is roughly 300 feet below the surface. Engineers were satisfied that the barrels would not deteriorate on the ocean floor, but they questioned whether the barrels would rupture from the impact of landing.

When engineers performed controlled tests, they found that any barrel that landed with a velocity greater than 40 feet per second was susceptible to rupture. Therefore, to determine whether the barrels of radioactive waste were susceptible to rupture when they landed on the ocean floor, the engineers needed to determine the velocity of the barrels after they had fallen to a depth of 300 feet. Since the engineers could not measure directly a barrel's velocity, they needed to find a way to predict this velocity based on other information.

How can we determine a barrel's velocity without being able to measure it? As in any situation involving the motion of an object, we know that the net force acting on an object is equal to the object's mass times its acceleration. This is Newton's second law of motion, and it is written as

$$F = ma.$$

In addition, we know that the acceleration of any moving object is equal to the rate of change of its velocity, which we write as

$$a = \frac{dv}{dt}.$$

Combining these two equations gives us the differential equation

$$\frac{dv}{dt} = \frac{F}{m} \tag{14}$$

To determine F, we need to identify the individual forces that act on a barrel as it falls. These forces are illustrated in Figure 4.13. The force that causes a barrel to descend is gravity. The magnitude of this force is the *weight* of the barrel. The weight W of a 55 gallon barrel filled with radioactive waste is estimated at 527 pounds. Two other forces act in opposition to gravity to retard the descent of a barrel. The first force is *buoyancy*, which is the upward force of the water acting on the barrel. The magnitude of the buoyancy force is equal to the weight of the water displaced by the barrel. The weight of 55 gallons of sea water is approximately 470 pounds. The second upward force is the *drag* force of the water, which is comparable to the force of air resistance that acts on objects moving through the air. The magnitude of the drag force depends upon the velocity of an object moving through the water so that the faster an object moves, the greater the drag. The engineers studying the barrels used towing experiments to determine that the drag force D is proportional to the velocity v of the barrel with proportionality constant 0.08, so that $D = 0.08v$. The orientation of the barrel as it falls through the water did not appear to affect significantly the relationship between velocity and drag.

Since buoyancy and drag act in opposition to gravity, the net effect F of the forces of weight W, buoyancy B, and drag D can be represented as follows:

$$F = W - B - D.$$

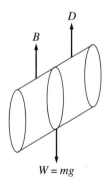

Figure 4.13 Forces on falling barrel

By substitution, we have

$$F = 527 - 470 - 0.08v,$$
$$= 57 - 0.08v.$$

Because we treat the force of gravity as a positive quantity and the forces of buoyancy and drag are subtracted, the positive direction of motion is towards the ocean floor.

Substituting the expression for F into differential equation (14) yields

$$\frac{dv}{dt} = \frac{1}{m}(57 - 0.08v).$$

We now need a numerical value for m, the mass of a barrel. By Newton's second law we know that $W = mg$, where g is the acceleration due to gravity (32.3 feet/sec^2). This implies that $m = \frac{W}{g} = \frac{527}{32.3}$.

Putting this all together gives the differential equation

$$\frac{dv}{dt} = (57 - 0.08v) \cdot \frac{32.3}{527}. \tag{15}$$

At this point in the course, we cannot solve differential equation (15) analytically. That is, we cannot determine the equation of a function v that satisfies this differential equation; however, we can use Euler's method to find approximate values for $v(t)$. The recursive equations

$$t_i = t_{i-1} + \Delta t,$$
$$v_i = v_{i-1} + \frac{dv}{dt}(v_{i-1}) \cdot \Delta t \tag{16}$$

can be used to generate ordered pairs (time, velocity).

The graph in Figure 4.14 shows ordered pairs (t, v) which were generated using the recursive equations in (16) with $t_0 = 0$, $v_0 = 0$, and $\Delta t = 0.1$. Notice that the velocity levels off over time. This occurs because the forces of buoyancy and drag work against the acceleration due to gravity. Although Figure 4.14 gives us important information about a barrel's motion, it does not tell us what we really want to know. We want to know velocity in terms of *distance* so that we can determine the value of v when d is 300.

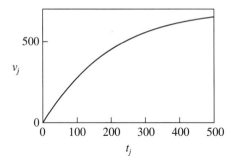

Figure 4.14 Velocity versus time for barrels of radioactive waste

Since velocity v is the derivative of distance d with respect to time, we know that $\frac{d}{dt}(d) = v$. This differential equation describes the rate of change of d and allows us to generate values of d with the recursive equation

$$d_i = d_{i-1} + \frac{dd}{dt}\bigg|_{(t_{i-1}, d_{i-1})} \cdot \Delta t,$$

or

$$d_i = d_{i-1} + v_{i-1} \cdot \Delta t. \tag{17}$$

The recursive equation in (17) uses the values generated previously for v to generate values for d. Note that we do not know an equation for velocity in terms of time, but we do know numerical values of velocity that have been generated previously using (16). Figure 4.15 shows ordered pairs (t, d) and allows us to see how the barrel's distance from the ocean surface changes over time.

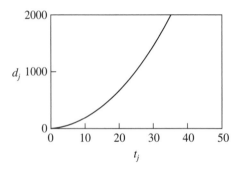

Figure 4.15 Distance of barrel from surface versus time

We have used recursive equations to generate values for t, d, and v. We have already graphed v versus t as well as d versus t and have observed how velocity and distance are related to time. We can also graph v versus d, and this graph will show how velocity is related to distance. Figure 4.16 shows ordered pairs (d, v) that have been generated using the recursive equations in (16) and (17) with $v_0 = 0$, $d_0 = 0$, and $\Delta t = 0.1$.

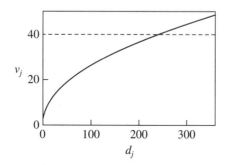

Figure 4.16 Velocity versus distance for a barrel

We still need to determine how fast a barrel is moving when it reaches a depth of 300 feet below the surface. The numerical values used to generate Figures 4.14, 4.15, and 4.16 indicate that after 132 iterations of the recursive equations, 13.2 seconds have elapsed, the barrel is about 295.74 feet below the surface, and the barrel's velocity is about 44.7 feet/sec. After 133 iterations, 13.3 seconds have elapsed, the barrel is about 300.2 feet below the surface, and the barrel's velocity is about 45.0 feet per second. These values tell us that the barrel reaches the ocean floor between 13.2 and 13.3 seconds after it is released, and when it hits the ocean floor its velocity is about 45 feet/sec.

Recall that when engineers performed controlled tests, they found that any barrel that landed with a velocity greater than 40 feet per second was susceptible to rupture. The fact that the barrel's velocity at impact with the ocean floor is greater than 40 feet per second means that the practice of disposing of radioactive waste in this way is not safe. Indeed the United States now forbids dumping radioactive waste into the ocean.

Reference

Braun, Martin. *Differential Equations and Their Applications*, 3rd Ed. Springer-Verlag New York, Inc.: New York, 1983.

Parachuting

When a parachutist falls freely through the air, there are several forces in action. Gravity is pulling downward, and air resistance is an upward force since it resists the parachutist's downward motion. We will be concerned only with the parachutist's vertical position, so any horizontal forces in action, such as wind or thermal currents, will be ignored. Suppose the parachutist falls from a height above ground h of 2000 meters. The parachutist's height decreases over time. This means that $\frac{dh}{dt}$, which is equal to the parachutist's velocity v, is negative. As the altitude continues to decrease, the parachutist falls faster so that the velocity becomes more negative. This means that v is decreasing over time, so $\frac{dv}{dt}$ is negative. Since $\frac{dv}{dt}$ is equal to the acceleration, we know that the acceleration is negative.

In our initial analysis of the parachutist's motion we will ignore air resistance. While this makes the mathematics easier to deal with, this assumption renders the parachute useless. In the absence of air resistance, the only force acting on the parachutist is gravity, which causes a negative acceleration equal to $-g$, where $g = 9.8 \text{ m}/\sec^2$. Since acceleration is simply the derivative of velocity, we can write

$$\frac{dv}{dt} = -g.$$

A function whose derivative with respect to t is equal to the constant $-g$ is the linear function

$$v(t) = -gt + c_1.$$

If v_0 symbolizes the initial velocity at $t = 0$, then $c_1 = v_0$, and this equation becomes

$$v(t) = -gt + v_0.$$

Since velocity is the derivative of height with respect to time, we can write

$$\frac{dh}{dt} = -gt + v_0.$$

A function whose derivative with respect to t is equal to the linear function $-gt + v_0$ is the quadratic function

$$h(t) = -\frac{g}{2}t^2 + v_0 t + c_2.$$

The initial height above ground at time $t = 0$ is 2000 meters, which implies that $c_2 = 2000$, and the height function becomes

$$h(t) = -\frac{g}{2}t^2 + v_0 t + 2000.$$

This final function allows us to determine the height above ground of the parachutist at any instant of time if we know the value of v_0. The function v allows us to determine the velocity of the parachutist at any instant of time if we know v_0. By

ignoring air resistance, we are able to model the motion of the parachutist with differential equations that we can solve analytically, and we can find closed form expressions for the position and velocity of the supplies over time. In the following lab, we
will include air resistance in our investigation of the motion of a parachutist in free
fall.

Lab 8 *Parachuting*

The simplifying assumption we made about air resistance in the previous section is
seriously flawed. In fact, a parachute is useful only because of the existence of air
resistance. The force that air resistance exerts on the parachutist opposes, or acts in
the opposite direction to, the force of gravity. The amount of force exerted by air
resistance depends on how fast the parachutist is falling. Researchers have estimated
that the force is proportional to the square of the parachutist's velocity.

This information leads us to the replace the differential equation $\frac{dv}{dt} = -g$ with

$$\frac{dv}{dt} = -g + kv^2.$$

(18)

The constant of proportionality k is given by

$$k = \frac{C\rho A}{2m},$$

where C is the coefficient of air resistance, ρ is the density of air, A is the surface area
of the parachutist, and m is the mass of the parachutist. Equation (18) implies that, at
any given time, $\frac{dv}{dt}$ will not be as negative as it would be in the absence of air resistance. Therefore the parachutist's velocity will not decrease (increase in the negative
direction) as rapidly as it would in the absence of air resistance.

Introducing air resistance into the model yields a differential equation that we are
not able to solve analytically. We can, however, use Euler's method to find numerical
solutions for the velocity and the height above ground as functions of time.

1. Write the recursive equation to generate values with Euler's method for the ve-
    locity of the free-falling parachutist. Use the following values of the constants:

    $g = 9.8$ m/sec^2 $C = 0.57$

    $\rho = 1.3$ kg/m^3 $A = 0.7$m^2

    $m = 75$ kg

2. Assume that the parachutist jumps from a plane so that the initial downward ve-
    locity is zero. Use the recursive equation developed in part 1 to generate values
    of v for every half second from 0 to 45 seconds. Graph the velocity versus time.
    Explain why the graph has the shape it does.

3. Write the recursive equation you would use to generate values for height with Euler's method.

4. Use the recursive equation developed in part 3 and the velocity values generated in part 2 to find the altitude of the parachutist every half second from 0 to 45 seconds. Assume the parachutist jumped from a plane so that the initial height above ground was 2000 meters. Graph the height versus time. Explain why the graph has the shape it does.

5. When the parachutist opens the parachute, the surface area A abruptly increases to 25 m^2. Assume that the chute is opened 30 seconds after the fall begins. Generate graphs of velocity and altitude until the parachutist reaches the ground. What happens immediately after the chute opens? What adjustments, if any, were necessary with Euler's method to achieve accurate results? How do the new velocity and height graphs compare to the graphs without opening the parachute?

6. How long can the parachutist wait before opening the parachute and still achieve a safe landing? What is the least height at which the parachutist can open the chute and still land safely? Support your answers.

Summarize your results in a report that includes graphs of velocity and height for the entire jump. What weaknesses are there in the method? What relevant questions have we not considered?

References
Kincanon, Eric. Skydiving as an aid to physics, *Physics Education* 25 (December 1990), 267–269.
Serway, R. A. *Physics for Scientists and Engineers.* Saunders College: London, 1990.
Symon, K. R. *Mechanics.* Addison-Wesley: London, 1971.

4.5 Coupled Differential Equations and an Improved Euler's Method

Harmonic Motion

Figure 4.17 shows a spring with an object attached to one end. When the object, with the spring attached, is moved to the right, the spring is stretched beyond its equilibrium length. The spring exerts a force on the object that pulls on the object toward the left, back toward the spring's equilibrium position. The magnitude of this force is related to the distance that the spring has been stretched so that the greater the distance is, the greater the force is. If we now release the object, it will move to the left past the equilibrium position, so that the spring is compressed. This in turn exerts a force that causes the object to move to the right, and the process of stretching and compressing is repeated. Assuming that the object slides without friction, how can we describe the motion of the object over time?

Actual data from springs reveal that the relationship between distance stretched and force is nearly linear for distances that are not too large. Let x represent the distance between the object and the position where the spring is at equilibrium. The force F exerted by the spring is proportional to x, that is,

$$F = -kx, \tag{19}$$

where k is a positive constant. The negative sign is necessary because the force is exerted in the direction opposite to the object's displacement from equilibrium. The relationship in (19) between F and x is known as **Hooke's law**.

Hooke's law

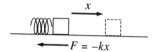

Figure 4.17 Spring system

From Newton's second law, we know that $F = ma$, which implies that

$$F = m \cdot \frac{d^2 x}{dt^2}, \tag{20}$$

since acceleration is the second derivative of position. Combining (19) and (20) gives us the differential equation

$$m \cdot \frac{d^2 x}{dt^2} = -kx,$$

or

$$\frac{d^2 x}{dt^2} = -\frac{k}{m} \cdot x. \tag{21}$$

Equation (21) is an example of a **second order differential equation**. It is called *second order* because it includes information about the second derivative of x with respect to t. Analytical techniques for solving most second order differential equations are beyond the scope of this text. However, knowing the acceleration allows us to use Euler's method to generate values of the velocity over time. These values can then be used to generate values of position over time.

second order differential equation

Since the velocity v of the object is the derivative of position with respect to time, we can write equation (21) as $\frac{dv}{dt} = -\frac{k}{m} x$. With Euler's method we can therefore generate v values using the recursive equation

$$v_i = v_{i-1} + \frac{dv}{dt}\bigg|_{(t_{i-1}, v_{i-1})} \cdot \Delta t,$$

or

$$v_i = v_{i-1} + \left(-\frac{k}{m} \cdot x_{i-1}\right) \cdot \Delta t. \tag{22}$$

Each v value generated by equation (22) allows us to generate an x value with the recursive equation

$$x_i = x_{i-1} + \frac{dx}{dt}\bigg|_{(t_{i-1}, x_{i-1})} \cdot \Delta t,$$

or

$$x_i = x_{i-1} + v_{i-1} \cdot \Delta t. \tag{23}$$

Using equations (22) and (23) to generate values of v and x presents an interesting problem. Given initial values v_0 and x_0, you can find v_1 with (22); however, you cannot find v_2 until you know x_1, so you must first use (23) to find x_1. Once you know v_1 and x_1, you can use (23) to find x_2, but then you must know v_2 in order to find x_3. In other words, you must generate values for x and v in pairs: x_1 and v_1, then x_2 and v_2, then x_3 and v_3.

Equations (22) and (23) are linked; that is, neither equation can be iterated without using values generated by the other. These equations were constructed from the two differential equations

$$\frac{dx}{dt} = v \text{ and } \frac{dv}{dt} = -\frac{k}{m} \cdot x. \tag{24}$$

These equations imply that the rate of change of x depends on v and the rate of change of v depends on x. For this reason the differential equations in (24) are called ***coupled differential equations***.

coupled differential equations

Figure 4.18 shows graphs of velocity versus time and of position versus time that were generated by using Euler's method and the recursive equations in (22) and (23). These graphs were generated using $k = 1$, $m = 1$, $x_0 = 10$, $v_0 = 0$, and $\Delta t = 0.1$. Note that the differential equations we have used do not take into account forces such as friction or air resistance that would retard the motion of the object. The motion shown in Figure 4.18 is an example of ***simple harmonic motion***.

simple harmonic motion

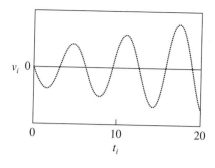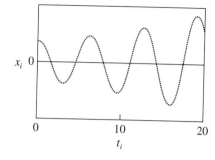

Figure 4.18 Velocity versus time and position versus time for object on a spring

Both the velocity and position functions in Figure 4.18 appear to be sinusoidal. Notice that differential equation (21) implies that the position function x has a second derivative which is proportional to itself, and this is a characteristic of the sine and cosine functions. For example, if $x = \cos t$, then $\frac{dx}{dt} = -\sin t$, and $\frac{d^2x}{dt^2} = -\cos t = -x$.

Quadratic Approximation and Euler's Method

One of the striking features about Figure 4.18 is that the amplitude of each graph appears to get larger as time passes. We know that this does not correspond to what actually happens when an object oscillates on a spring, so we suspect that this error is a consequence of the numerical method we used to solve the coupled differential equations. We know that some error is inherent in Euler's method. This is because Euler's method is based on a series of linear approximations, and each linear approximation is somewhat inaccurate (unless the function we are approximating is itself linear). The error in Euler's method is more pronounced the more a function curves away from a tangent line. The graphs of harmonic motion in Figure 4.18 have many turning points, and tangent line approximations deviate from the graphs significantly, even over small intervals near the turning points.

A tangent line is a good approximation for a function about a point because it has the same function value and the same slope as the function at that point. However, since a line has no concavity, a linear approximation will diverge from the function values as the distance from the point of tangency increases. If a function is concave up, a linear approximation will underestimate a function value. If a function is concave down, a linear approximation will overestimate a function value. These errors are illustrated in Figure 4.19.

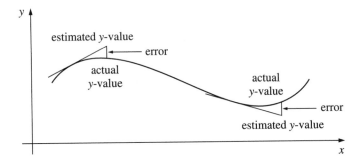

Figure 4.19 Errors in linear approximations

How can we improve Euler's method to make this problem less significant? To correct partially these errors, we can approximate a function with a simple curve that matches both the slope and the concavity of the function at a point. By using a quadratic approximation, we are able to bend the approximating curve to match the concavity of the function. A quadratic approximation, or "tangent parabola," must have the same function value, the same first derivative, and the same second derivative as the function at that point. Having the same first derivative as the function we are approximating guarantees that the parabola has the same slope at the point as the original function. Having the same second derivative as the function we are approximating guarantees that the parabola has the same concavity at the point as the original function.

Example 1

Write the equation of the quadratic approximation to $f(x) = \cos x$ about $x = 0$.

Solution   We are seeking an equation of the form $q(x) = ax^2 + bx + c$ that satisfies each of three conditions.

1. The value of $q(x)$ at $x = 0$ is the same as the value of $f(x)$ at $x = 0$, which is

$$f(0) = \cos(0) = 1.$$

2. The slope of the graph of $q(x)$ at $x = 0$ is the same as the slope of the graph of $f(x)$ at $x = 0$, which is

$$f'(0) = -\sin(0) = 0.$$

3. The concavity of the graph of $q(x)$ at $x = 0$ is the same as the concavity of the graph of $f(x)$ at $x = 0$, which is

$$f''(0) = -\cos(0) = -1.$$

The values of the constants a, b, and c in the equation for q must be chosen so that

$$q(0) = 1, \quad q'(0) = 0, \quad \text{and} \quad q''(0) = -1.$$

Values for a, b, and c are determined as shown below.

$q(x) = ax^2 + bx + c \quad \Rightarrow \quad q(0) = a \cdot 0^2 + b \cdot 0 + c = 1 \quad \Rightarrow \quad c = 1.$

$q'(x) = 2ax + b \quad \Rightarrow \quad q'(0) = 2a \cdot 0 + b = 0 \quad \Rightarrow \quad b = 0.$

$q''(x) = 2a \quad \Rightarrow \quad q''(0) = 2a = -1 \quad \Rightarrow \quad a = -\frac{1}{2}.$

These values yield the quadratic function $q(x) = -\frac{1}{2}x^2 + 1$. Graphs of the quadratic function q and the trigonometric function $f(x) = \cos x$ are shown on the same axes in Figure 4.20. The linear approximation to $f(x) = \cos x$ about $x = 0$, which has equation $y = 1$, is also shown on the graph. Notice that the parabola better approximates the cosine curve over a larger interval of x values than the line does. ∎

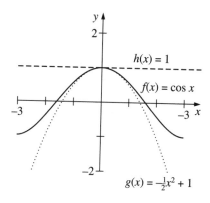

Figure 4.20 Quadratic and linear approximations to $f(x) = \cos x$ about $x = 0$

We can generalize the process used in the preceding example to write the equation of a quadratic approximation for any function. Suppose we want the quadratic function $q(x) = ax^2 + bx + c$ to approximate $f(x)$ about $x = 0$. We want the function value of q at $x = 0$ to be equal to the function value of f at $x = 0$, so $q(0) = f(0)$. Since $q(0) = c$, this implies

$$c = f(0).$$

We want the slope of q at $x = 0$ to be equal to the slope of f at $x = 0$, so $q'(0) = f'(0)$. Since $q'(x) = 2ax + b$, we have $q'(0) = b$. This implies

$$b = f'(0).$$

We want the concavity of q at $x = 0$ to be equal to the concavity of f at $x = 0$, so $q''(0) = f''(0)$. Since $q''(x) = 2a$, we have $q''(0) = 2a$. Thus, $2a = f''(0)$, or

$$a = \frac{f''(0)}{2}.$$

Therefore, the quadratic approximation for $f(x)$ about $x = 0$ is

$$q(x) = \frac{f''(0)}{2} x^2 + f'(0)x + f(0). \tag{25}$$

When we use equation (25) to write the equation of the quadratic approximation about $x = 0$, we usually know the equation of the function f we are trying to approximate. Thus, we would be able to find numerical values for the coefficients $\frac{1}{2} f''(0)$, $f'(0)$, and $f(0)$. For example, if $f(x) = \ln(x+1)$ then we know $f'(x) = \frac{1}{x+1}$ and $f''(x) = -\frac{1}{(x+1)^2}$. Thus $f(0) = 0$, $f'(0) = 1$ and $f''(0) = -1$. According to equation (25), the parabola that approximates this $f(x)$ about $x = 0$ is $q(x) = -\frac{1}{2} x^2 + x$.

Of course, we also want to be able to write a quadratic approximation to a function about points other than $x = 0$. Suppose we want q to estimate f at $x = a$. In this case, equation (25) must be modified in two ways. First, the graph of q must be shifted so that what originally occurred at $x = 0$ now occurs at $x = a$. This can be accomplished by replacing x with $x - a$ in the right side of equation (25). Second, the function values, first derivatives, and second derivatives must be determined at $x = a$, rather than at $x = 0$. This means that $f(a)$, $f'(a)$, and $f''(a)$ must be substituted for $f(0)$, $f'(0)$, and $f''(0)$, respectively. Therefore, equation (25) becomes

$$q(x) = \frac{f''(a)}{2}(x - a)^2 + f'(a)(x - a) + f(a), \tag{26}$$

which is the quadratic approximation for $f(x)$ about $x = a$.

Example 2

Write the equation of the quadratic approximation for $f(x) = \ln x$ about $x = 1$.

Solution   We can use equation (26) with $a = 1$, $f(x) = \ln x$, $f'(x) = \frac{1}{x}$, and $f''(x) = -\frac{1}{x^2}$. Evaluating $f(x) = \ln x$ at $x = 1$ gives $\ln 1 = 0$. Evaluating $f'(x) = \frac{1}{x}$ at $x = 1$ gives $\frac{1}{1} = 1$. Evaluating $f''(x) = -\frac{1}{x^2}$ at $x = 1$ gives $-\frac{1}{1^2} = -1$. Therefore, the approximating quadratic is $q(x) = -\frac{1}{2}(x - 1)^2 + (x - 1) + 0$. Figure 4.21 shows the graph of $f(x) = \ln x$ together with the graph of the quadratic approximation $q(x)$ about $x = 1$. ∎

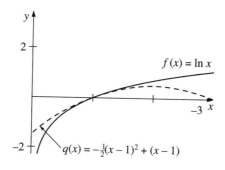

Figure 4.21 Graph of $f(x) = \ln x$ with quadratic approximation

We are now prepared to combine the power of quadratic approximations with the recursive power of Euler's method. Let us first review the connection between tangent lines and Euler's method. Figure 4.22 shows the point $(a, f(a))$ on the graph of the function $f(x)$ and the line tangent to $f(x)$ at $x = a$. The equation of the line tangent to this function through $(a, f(a))$ is

$$y = f'(a) \cdot (x - a) + f(a).$$

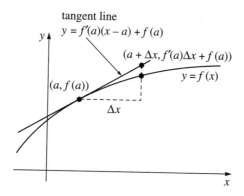

Figure 4.22 Linear approximation to $f(x)$ about $x = a$

This tangent line allows us to estimate the y-coordinate of a point near $(a, f(a))$. At a point with an x-coordinate of $a + \Delta x$, the actual y-coordinate $f(a + \Delta x)$ is estimated by y_{new}.

$$y_{new} = f'(a) \cdot (a + \Delta x - a) + f(a)$$
$$= f'(a) \cdot \Delta x + f(a) \qquad (27)$$

The expression in (27) corresponds to the iterative equation we use in Euler's method, $y_{new} = y_{old} + \frac{dy}{dx} \cdot \Delta x$, with $f(a) = y_{old}$ and $f'(a) = \frac{dy}{dx}$.

We can iterate the quadratic approximation to a function in a way that is comparable to the way we iterate linear approximations in Euler's method. Figure 4.23 shows the point $(a, f(a))$ on the graph of the function f, together with the quadratic approximation to $f(x)$ about $x = a$. The equation of the tangent parabola for this function at $(a, f(a))$ is

$$q(x) = \frac{f''(a)}{2}(x-a)^2 + f'(a)(x-a) + f(a). \tag{28}$$

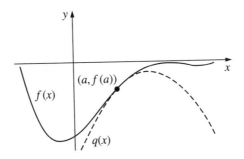

Figure 4.23 Quadratic approximation $q(x)$ to $f(x)$ about $x = a$

To estimate the y-coordinate of a point Δx units to the right of $(a, f(a))$, we can substitute $a + \Delta x$ for x in equation (28). The new y value on the tangent parabola is given by

$$y_{new} = \frac{f''(a)}{2}(a + \Delta x - a)^2 + f'(a) \cdot (a + \Delta x - a) + f(a),$$

which can be written as

$$y_{new} = f(a) + f'(a) \cdot \Delta x + \tfrac{1}{2} f''(a) \cdot \Delta x^2. \tag{29}$$

Equation (29) provides a means of approximating function values based on quadratic approximation rather than on linear approximation. If we want to generate a series of x and y values that approximate the behavior of a function, we can use equation (29) recursively, which gives

$$x_i = x_{i-1} + \Delta x$$
$$y_i = y_{i-1} + \frac{dy}{dx}\bigg|_{(x_{i-1}, y_{i-1})} \cdot \Delta x + \frac{1}{2} \cdot \frac{d^2y}{dx^2}\bigg|_{(x_{i-1}, y_{i-1})} \cdot \Delta x^2. \tag{30}$$

The technique of using the recursive equations in (30) to generate values for x and y is

quadratic Euler's method   called the **quadratic Euler's method**.

Example 3

Suppose we know that $\frac{dy}{dx} = 2y$ and that $(x_0, y_0) = (0,1)$. Use the linear Euler's method to approximate the value of y when $x = 5$, then use the quadratic Euler's method to approximate the value of y when $x = 5$. Determine the particular solution of $\frac{dy}{dx} = 2y$ that contains the point $(0, 1)$. Compare the y value approximations generated by each Euler's method with the actual value of y when $x = 5$.

Solution   To use the linear Euler's method to generate values of x and y we use the system of recursive equations

$$x_i = x_{i-1} + \Delta x$$
$$y_i = y_{i-1} + \frac{dy}{dx} \cdot \Delta x,$$

which becomes

$$x_i = x_{i-1} + \Delta x$$
$$y_i = y_{i-1} + 2y_{i-1} \cdot \Delta x \qquad (31)$$

by substitution for $\frac{dy}{dx}$. Figure 4.24 shows the result of iterating (31) fifty times using $\Delta x = 0.1$ with $(x_0, y_0) = (0,1)$. Our calculations find that $y \approx 9100$ when $x = 5$.

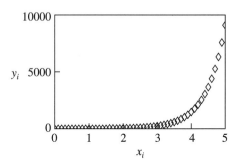

Figure 4.24 Linear Euler's method with $\frac{dy}{dx} = 2y$

To use the quadratic Euler's method to generate values of y, we use the recursive equation

$$y_i = y_{i-1} + \frac{dy}{dx} \cdot \Delta x + \frac{1}{2} \cdot \frac{d^2y}{dx^2} \cdot \Delta x^2.$$

The second derivative of y with respect to x is calculated as follows:

$$\frac{d^2y}{dx^2} = \frac{d}{dx}\left(\frac{dy}{dx}\right)$$

$$= \frac{d}{dx}(2y)$$

$$= 2 \cdot \frac{dy}{dx}$$

$$= 4y.$$

The recursive equations for the quadratic Euler's method are

$$x_i = x_{i-1} + \Delta x$$

$$y_i = y_{i-1} + 2y_{i-1} \cdot \Delta x + \frac{1}{2} \cdot 4y_{i-1} \cdot \Delta x^2. \tag{32}$$

Figure 4.25 shows the result of iterating (32) fifty times using $\Delta x = 0.1$ with $(x_0, y_0) = (0,1)$. Now our calculations find that $y \approx 20797$ when $x = 5$.

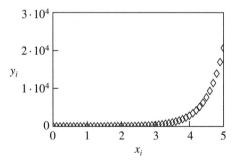

Figure 4.25 Quadratic Euler's method on $\frac{dy}{dx} = 2y$

The particular solution of $\frac{dy}{dx} = 2y$ that contains $(0, 1)$ is $y = e^{2x}$. This solution and the Euler's method approximations are shown in Figure 4.26. On the particular solution, $y \approx 22026$ when $x = 5$. The relative error in the linear Euler approximation is $(22026 - 9100)/22026 \approx 0.59$, or 59%. The relative error in the quadratic Euler approximation is $(22026 - 20797)/22026 \approx 0.06$, or 6%. Note that the two approximations are obtained with the same size Δx and the same number of iterations, yet the quadratic approximation is much more accurate, which is a consequence of taking both slope and concavity into account in the quadratic calculations. ∎

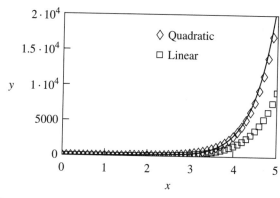

Figure 4.26 Linear and quadratic Euler's approximations
compared to a particular solution for $\frac{dy}{dx} = 2y$

Exercise Set 4.5

1. Write the equation of the quadratic approximation to each function about $x = 0$.

    a. $y = e^x$ b. $y = \sqrt{x+1}$

    c. $y = \dfrac{1}{x+1}$

2. Write the equation of the quadratic approximation to each function about the given point.

    a. $y = e^x$ about $(1, e)$ b. $y = \dfrac{1}{x}$ about $(-1, -1)$

3. When we studied the motion of a spring, we used the recursive equations $v_i = v_{i-1} + \frac{dv}{dt} \cdot \Delta t$ and $x_i = x_{i-1} + \frac{dx}{dt} \cdot \Delta t$, and we assumed that $\frac{dv}{dt} = -\frac{k}{m} \cdot x$. Modify these recursive equations as necessary to use the quadratic Euler's method to study the spring's motion. Use the modified recursive equations to graph velocity versus time and position versus time. Compare your graphs to those in Figure 4.18. Explain why using the quadratic Euler's method rather than the linear Euler's method causes your graphs to differ from Figure 4.18.

4. Suppose you know that a function goes through $(0, 5)$ and satisfies the differential equation $\frac{dy}{dx} = -0.1y$.

    a. Use ten iterations of the linear Euler's method to estimate the value of y when $x = 10$.

    b. Use ten iterations of the quadratic Euler's method to estimate the value of y when $x = 10$.

    c. Find the equation of the particular solution of $\frac{dy}{dx} = -0.1y$ that goes through $(0, 5)$. According to your equation, what is the value of y when $x = 10$?

    d. Compare values in parts a, b, and c, and explain your observations.

5. Find the equation of a cubic approximation to $f(x)$ about $x = a$ by extending the concept of a quadratic approximation. That is, find the values of the coefficients in $p(x) = q(x-a)^3 + r(x-a)^2 + s(x-a) + t$ so that $p(x)$ and $f(x)$ have the same values of the function, the first derivative, the second derivative, and the third derivative at $x = a$.

4.6   Parametric Equations

parametric equations

The position of an object in a coordinate plane can be described by specifying its x- and y-coordinates. If we have information about how the x- and y-coordinates are related to time, we can write equations such as $x = g(t)$ and $y = h(t)$ that describe these relationships. Equations such as these are called **parametric equations.**

After a space shuttle lifts off from Cape Canaveral, NASA employees describe the position of the shuttle by stating its vertical and horizontal distances from the launching pad. The vertical and horizontal components of the motion can be described separately. Using parametric equations like $x = g(t)$ and $y = h(t)$, it is convenient to determine the position of the shuttle at any instant in time. In contrast, if the position of the shuttle is described solely with the Cartesian form $y = f(x)$, then the altitude of the shuttle can be determined only if one knows the shuttle's horizontal distance from the launch pad.

parameter

The variable t is called the **parameter** of the parametric equations $x = g(t)$ and $y = h(t)$. Recall that in Chapter 3 we encountered problems in which the equation of a function to be maximized or minimized included parameters. For example, we found that the surface area S of a cylindrical can is given by

$$S(r) = \frac{2V}{r} + 2\pi r^2,$$

where r is the radius of the can and V is the volume. In this expression, r is a variable and V is a parameter. In the context of minimizing surface area for a given volume, the parameter V is treated as a constant. In contrast, in the context of parametric equations, a parameter is not treated as a constant, but rather as a variable upon which the values of x and y depend. However, the term *parameter* actually is not being used in two different ways, since the value of r that gives the minimum surface area depends on V.

Suppose a projectile is shot into the air and follows a path described by the parametric equations

$$x = 19.28t$$
$$y = -4.9t^2 + 22.98t.$$

In these equations t is measured in seconds and x and y are measured in meters. Graphs of x versus t and y versus t are shown in Figure 4.27. The position of the object at any time can be found by evaluating each of the parametric equations separately, as shown in Table 4.28.

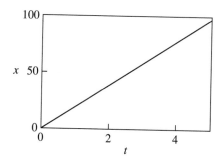
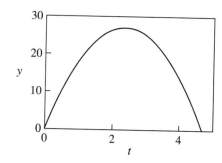

Figure 4.27 Graphs of $x = 19.28t$ and $y = -4.9t^2 + 22.98t$

t	x	y
0.00	0.00	0.00
0.50	9.64	10.27
1.00	19.28	18.08
1.50	28.92	23.45
2.00	38.56	26.36
2.50	48.20	26.83
3.00	57.84	24.84
3.50	67.48	20.41
4.00	77.12	13.52
4.50	86.76	4.19
5.00	96.40	-7.60

Table 4.28 Coordinates determined by $x = 19.28t$ and $y = -4.9t^2 + 22.98t$

The actual path of the projectile can be seen by graphing each y value on the vertical axis and the corresponding x value on the horizontal axis, as shown in Figure 4.29. Each ordered pair (x, y) on the graph corresponds to the x and y values for a particular t value. Figure 4.29 shows ordered pairs (x, y) for t values from 0 to 4.7.

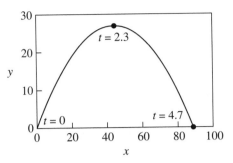

Figure 4.29 Path of projectile with equations $x = 19.28t$ and $y = -4.9t^2 + 22.98t$

We can write an equation of the form $y = f(x)$ for the graph in Figure 4.29 by solving $x = 19.28t$ for t, which gives

$$t = \frac{x}{19.28},$$

and then substituting for t in the equation $y = -4.9t^2 + 22.98t$.

$$y = -4.9\left(\frac{x}{19.28}\right)^2 + 22.98\left(\frac{x}{19.28}\right)$$

$$= -0.013x^2 + 1.19x. \tag{33}$$

This function describes the path of the projectile in Cartesian form, and has the same graph as that shown in Figure 4.29. The process of changing from parametric equations to an equation in Cartesian form, as we have just done, is called *eliminating the parameter.*

eliminating the parameter

Example 1

At what angle with the horizontal is the projectile traveling at the instant when it has been airborne for 1 second?

Solution   To answer this question we need to know the slope of the line tangent to the curve in Figure 4.29 at the point corresponding to $t = 1$. Then we can determine the angle θ that the tangent line makes with the horizontal at this point. The relationship between angle θ and the slope of the line is based on the right triangle definition of the tangent function. Referring to Figure 4.30, we can pick a point on the tangent line and drop a perpendicular to form the right triangle shown. If the vertical distance is Δy and the horizontal distance is Δx, we can use right triangle trigonometry to write

$$\tan\theta = \frac{\Delta y}{\Delta x}.$$

By the definition of slope, we can also write

$$\text{slope of tangent line} = \frac{\Delta y}{\Delta x}.$$

Thus we can conclude that the slope of the tangent line equals the tangent of the angle we wish to find.

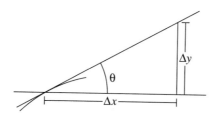

Figure 4.30 Relationship between angle and tangent slope

The slope of a line tangent to the graph in Figure 4.29 is given by the derivative of the function in (33), which is

$$y' = -0.026x + 1.19.$$

The x-coordinate of the projectile is given by the parametric equation $x = 19.28t$, which yields an x-coordinate of 19.28 when $t = 1$. Since $y' \approx 0.69$ when $x = 19.28$, we know that $\tan \theta \approx 0.69$, and therefore $\theta \approx 0.60$. ∎

We already know how to find and interpret the derivative of the Cartesian function for the path of the projectile. For instance, by setting the derivative equal to 0, we can find that the projectile reaches a maximum height when $x \approx 45.21$. We can also use the derivative of the Cartesian function to determine the angle at which the projectile is traveling, as in Example 1. In many cases, however, it will be difficult or impossible to eliminate the parameter from parametric equations to obtain a function in Cartesian form. In these situations, the parametric equations can be used to obtain the same information that the Cartesian form gives, as we will show in the rest of this section. First we will gain additional familiarity with parametric equations by studying several more examples.

Example 2

The parametric equations $x = t^3$ and $y = t^2$, with $t > 0$, are graphed in Figure 4.31. (Notice that the corresponding t values are not shown on this graph.) Write an equation for this graph in Cartesian form.

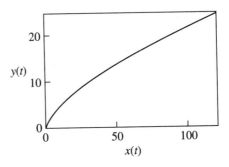

Figure 4.31 Graph of $x = t^3$ and $y = t^2$, with $t > 0$

Solution   Solving $x = t^3$ for t yields $t = x^{1/3}$. Substituting $t = x^{1/3}$ into $y = t^2$ yields $y = (x^{1/3})^2 = x^{2/3}$. The restriction that $t > 0$ implies that $x > 0$. Therefore, the Cartesian equation for these parametric equations is $y = x^{2/3}$, $x > 0$, which has the same graph as shown in Figure 4.31. ■

Example 3

Write parametric equations to represent the function $y = x^4$, for $x > 0$.

Solution   The simplest parametric form for this function is $x = t$, $y = t^4$, $t > 0$. The ordered pairs (x, y) generated by $x = t$, $y = t^4$, $t > 0$ are identical to the ordered pairs (x, y) generated by $y = x^4$, $x > 0$. There are many other parametric representations of this function, two examples of which are

$$x = \sqrt{t}, \; y = t^2, \; t > 0$$

and

$$x = t^2, \; y = t^8, \; t > 0.$$

Each of these parametric equations will produce a graph identical to the one shown in Figure 4.32. The only difference between these representations lies in the t values associated with each point in the xy-coordinate plane. If t ranges from 0 to 2, the graph of $x = t$, $y = t^4$ will have ordered pairs that extend from $(0, 0)$ to $(2, 16)$, the graph of $x = \sqrt{t}$, $y = t^2$ will extend from $(0, 0)$ to $(\sqrt{2}, 4)$, and the graph of $x = t^2$, $y = t^8$ will extend from $(0, 0)$ to $(4, 256)$. ■

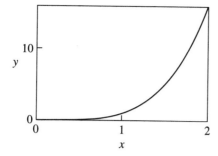

Figure 4.32 Graph of $x = t$, $y = t^4$, $t > 0$ or $y = x^4$, $x > 0$

Using parametric equations allows us to write formulas for many curves that are not the graphs of functions. The curve defined by

$$x = \cos t - 3 \sin t$$
$$y = 2 \cos t \cdot \sin t \tag{34}$$

is shown in Figure 4.33. This curve cannot be represented by a function of the form $y = f(x)$, yet we would like to be able to apply the tools of calculus to this curve. In particular, each point on the curve has a unique tangent line, and we want to determine the slope of the tangent line at various points on the curve. This brings us back to the question of how we can use the parametric equations in (34) to determine the slope of a tangent line at any point on the curve.

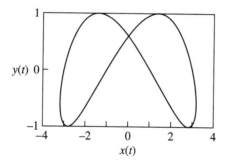

Figure 4.33 Curve defined by $x = \cos t - 3 \sin t$, $y = 2 \cos t \cdot 3 \sin t$

Suppose $x = g(t)$ and $y = h(t)$ are parametric equations defining a graph in the xy-coordinate plane. Because each t value corresponds to a unique ordered pair (x, y), there exists a relationship between x and y. Sometimes we can express y explicitly as a function of x. When this is difficult or impossible to do, we can assume that y is an implicitly defined function of x. Since y depends on x and x depends on t, the chain rule allows us to write

$$\frac{dy}{dt} = \frac{dy}{dx} \cdot \frac{dx}{dt},$$

which we can rewrite by solving for $\frac{dy}{dx}$:

$$\frac{dy}{dx} = \frac{dy/dt}{dx/dt}. \tag{35}$$

This implies that the rate of change of y with respect to x is equal to the rate of change of y with respect to t divided by the rate of change of x with respect to t. We will explore this concept in the context of the function $y = x^4$ that we investigated in Example 3.

Consider the parametric representation $x = t$ and $y = t^4$ for this function. Since $\frac{dx}{dt} = 1$ and $\frac{dy}{dt} = 4t^3$, according to equation (35) we have

$$\frac{dy}{dx} = \frac{4t^3}{1} = 4t^3.$$

Since $t = x$, $\frac{dy}{dx} = 4x^3$, which we recognize as the derivative with respect to x of $y = x^4$.

Now consider the parametric representation $x = \sqrt{t}$ and $y = t^2$ for the function $y = x^4$. Since $\frac{dx}{dt} = \frac{1}{2\sqrt{t}}$ and $\frac{dy}{dt} = 2t$, according to equation (35) we have

$$\frac{dy}{dx} = \frac{2t}{\frac{1}{2\sqrt{t}}} = 4t^{3/2}.$$

Since $t = x^2$, $\frac{dy}{dx} = 4(x^2)^{3/2} = 4x^3$, which we once more recognize as the derivative with respect to x of $y = x^4$.

Finally consider the parametric representation $x = t^2$ and $y = t^8$ for this function. Since $\frac{dx}{dt} = 2t$ and $\frac{dy}{dt} = 8t^7$, we have

$$\frac{dy}{dx} = \frac{8t^7}{2t} = 4t^6.$$

Since $t = \sqrt{x}$, we have $\frac{dy}{dx} = 4(\sqrt{x})^6 = 4x^3$, which we again recognize as the derivative with respect to x of $y = x^4$.

The relationship in equation (35) means that if x and y are defined parametrically by the functions $x = g(t)$ and $y = h(t)$ and we graph y versus x, then the slope of the line tangent to this graph at any point is equal to $\frac{dy/dt}{dx/dt}$ evaluated at that point.

Example 4

A projectile is following a path described by the parametric equations $x = 19.28t$ and $y = -4.9t^2 + 22.98t$. Without eliminating the parameter, find the angle that the path of the projectile makes with the horizontal at the instant when $t = 1$.

Solution   Differentiating x and y with respect to t gives $\frac{dx}{dt} = 19.28$ and $\frac{dy}{dt} = -9.8t + 22.98$. Applying equation (35), we have

$$\frac{dy}{dx} = \frac{-9.8t + 22.98}{19.28}.$$

This means that when $t = 1$, $\frac{dy}{dx} = \frac{-9.8(1) + 22.98}{19.28} \approx 0.68$. Since $\tan^{-1} 0.68 \approx 0.60$, the path of the projectile makes an angle of approximately 0.60 radians with the horizontal at this time. This result, obtained using only the parametric representation of the projectile's path, agrees to two decimal places with the result we obtained in Example 1 using the Cartesian function $y = -0.013x^2 + 1.19x$. ∎

Example 5

Suppose an object is traveling along the curve defined by the parametric equations $x = 3\cos t + 5$ and $y = 3\sin t + 9$. Determine the position of the object and describe the direction in which it is moving at the instant when $t = 2$.

Solution   We can find the position of the object by evaluating the parametric equations at $t = 2$, which gives

$$x = 3\cos 2 + 5 \approx 3.75$$

and

$$y = 3\sin 2 + 9 \approx 11.73.$$

At time $t = 2$, the object is located at the point with coordinates $x = 3.75$, $y = 11.73$.

Figure 4.34 shows a graph of ordered pairs (x, y) generated by $x = 3\cos t + 5$ and $y = 3\sin t + 9$. In addition, the location of the object at several t values is indicated. These values reveal that the object is moving counterclockwise around a circle with center (5, 9) and radius 3.

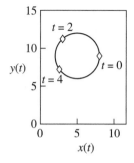

Figure 4.34 Graph of $x = 3\cos t + 5$ and $y = 3\sin t + 9$

We can describe the direction of the object's motion by finding the slope of the line tangent to the object's path. Since $\frac{dy}{dx} = \frac{dy/dt}{dx/dt}$, we have

$$\frac{dy}{dx} = \frac{3\cos t}{-3\sin t} = -\frac{\cos t}{\sin t}.$$

Evaluating $\frac{dy}{dx}$ at $t = 2$ gives

$$\frac{dy}{dx} = -\frac{\cos 2}{\sin 2} \approx 0.46.$$

At the instant when $t = 2$, the slope of the line tangent to the object's path is about 0.46. Since the object is moving in a counterclockwise direction, it is moving to the left and down. Notice that the slope of the tangent line by itself does not tell you whether the object is moving left or right along the curve. ∎

Exercise Set 4.6

1. Use the technology of your choice to graph each of the following.

   a. $x = 2\cos\frac{t}{3}$ and $y = 2\sin\frac{t}{3}$ for

       i. $0 \le t \le 2\pi$ ii. $0 \le t \le \pi$

       iii. $-\pi \le t \le \pi$ iv. $0 \le t \le 3\pi$

   b. $x = 6\cos 2t$ and $y = 6\sin 2t$ for

       i. $0 \le t \le \frac{\pi}{2}$ ii. $0 \le t \le \pi$

       iii. $0 \le t \le 2\pi$

2. Use the preceding exercise to predict what the graph of $x = r \cdot \cos kt$ and $y = r \cdot \sin kt$ looks like. How does the interval of t values affect the graph?

3. A projectile's path from its point of origin is described by the parametric equations $x = 59.09t$ and $y = -4.9t^2 + 10.42t$. Determine the angle that the projectile's path makes with the horizontal at $t = 1$, $t = 2$, and $t = 5$ seconds.

4. The path of an object thrown into the air with an initial velocity v_0 and at an angle of θ degrees with the horizontal can be described by the parametric equations $x = (v_0 \cdot \cos\theta)t$, $y = -\frac{g}{2}t^2 + (v_0 \cdot \sin\theta)t$, where g is the force of gravity (−9.8 m/sec^2). Write an expression for $\frac{dy}{dx}$ as a function of t in terms of v_0, θ, and g.

5. Use the parametric equations $x = t^2$ and $y = t^3$ to determine $\frac{dy}{dx}$ in terms of x. Compare your result to $\frac{dy}{dx}$ determined by differentiating $y = x^{3/2}$.

6. Refer to the curve defined by $x = \cos t - 3\sin t$, $y = 2\cos t \cdot \sin t$, shown here.

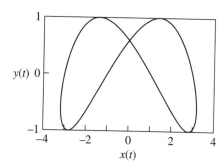

a. For what points on the curve is $\frac{dy}{dt} = 0$? What difficulties, if any, does this cause when writing the equation of the tangent line at these points?

b. For what points on the curve is $\frac{dx}{dt} = 0$? What difficulties, if any, does this cause when writing the equation of the tangent line at these points?

7. The ellipse $\frac{x^2}{16} + \frac{y^2}{9} = 1$ can be described parametrically as $x = 4\cos t$ and $y = 3\sin t$. Use parametric equations to determine the equations of the two lines tangent to the curve at the points with $x = 0.5$.

8. The ellipse $\frac{x^2}{16} + \frac{y^2}{9} = 1$ can also be described by piecing together the two functions $f(x) = \frac{3}{4}\sqrt{16 - x^2}$ and $g(x) = -\frac{3}{4}\sqrt{16 - x^2}$. Use these Cartesian equations to determine the equations of the two lines tangent to the curve at the points with $x = 0.5$.

9. Consider the parametric equations $x = \sin t + t$ and $y = t \cdot \cos t$.

a. Graph the curve defined by these equations for $0 \le t \le 10$.

b. Locate on the graph the points where it appears that $\frac{dy}{dt} = 0$. Locate the points where it appears that $\frac{dx}{dt} = 0$.

c. Use the parametric equations to determine the coordinates of the points where $\frac{dy}{dt} = 0$ and where $\frac{dx}{dt} = 0$.

d. Your answers to part c probably do not agree with your answers to part b. Sketch a graph to show more clearly what is happening close to the point $(\pi, -\pi)$.

4.7   Predator/Prey Models

For many years the Hudson Bay Company kept records of the number of pelts they traded, including pelts of the Canadian lynx and the snowshoe hare. Figure 4.35 shows a graph of the number of pelts of the lynx and the hare in two different regions from 1850 to 1900. If we assume that the number of pelts of a species varies directly with the population of that species, then the graph in Figure 4.35 illustrates how the populations of these two species changed over time. The lynx is a small cat whose primary food source is the snowshoe hare, and the cyclic behavior shown in the graph is typical for species that are in a predator/prey relationship. Notice that the numbers of both types of pelts oscillate with roughly the same period. In the next lab we will study the way predator/prey populations change, and we will try to account for this periodic behavior.

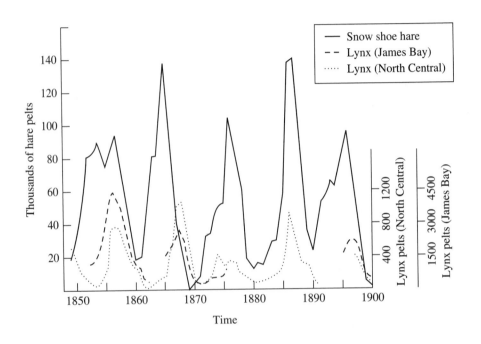

Figure 4.35 Lynx and hare pelts [Source: Martin Braun, ed., *Differential Equation Models*, Springer Verlag, New York, 1983]

The Italian mathematician Volterra was one of the first people to study predator/prey systems. One of Volterra's colleagues, the biologist D'Ancona, was investigating data on the catch of several species of fish in the Mediterranean during the years around World War I. In particular, D'Ancona noticed peculiar figures for the catch of selachians (sharks, skates, rays, *etc.*) as a percentage of the total catch of fish. The percentage of selachians increased dramatically during the years of the war, then returned to pre-war levels after the war. During the war, fishing was greatly restricted in the Mediterranean. Reduced fishing would increase the populations of all species of fish, so why did selachians thrive more than the other species? Selachians are predators, and they prey on other fish for their sustenance. Why would reduced levels of fishing benefit the predators more than the prey? When researching competing species, D'Ancona was perplexed by this question and turned to Volterra for assistance. When modeling populations of species which interact, such as sharks and fish, the rates of change of the species depend on the populations of both species. The differential equations Volterra formulated to assist D'Ancona's studies therefore must be treated as coupled differential equations.

There are many instances of the predator/prey relationship, and to model one, we must use coupled differential equations similar to Volterra's. For our discussion of predator/prey phenomena, we will use yet another example—rabbits and foxes.

What differential equation describes the rate at which the rabbit population grows? If the food supply was abundant and there were no foxes present, the rabbit population would grow at a rate proportional to their numbers, which results in unconstrained exponential growth. The differential equation describing this growth rate is

$$\frac{dr}{dt} = a \cdot r \tag{36}$$

where a is the intrinsic growth rate of the rabbits in the absence of foxes ($a > 0$) and r is the number of rabbits at time t. However, the rabbit population does not actually grow at this rate because there are predators present. How is the rate of change of the rabbit population influenced by the interaction of foxes and rabbits? Each contact between a rabbit and a fox results in an opportunity for a fox to kill a rabbit. The number of contacts per unit time between rabbits and foxes is proportional to the product of the sizes of the populations, $r \cdot f$ (where f is the number of foxes at time t). To understand why this is true, consider that if you double r, you will double the chance of contact between a rabbit and a fox. The same occurs if you double f. Furthermore, the number of rabbits killed by foxes is proportional to the number of contacts between rabbits and foxes. This means that the number of rabbit deaths caused by foxes is proportional to $r \cdot f$. Thus, equation (36) should be modified to

$$\frac{dr}{dt} = a \cdot r - b \cdot r \cdot f \tag{37}$$

where b is the rate at which foxes kill rabbits ($b > 0$). Equation (37) describes how the rate of change of the rabbit population depends on the current levels of the rabbit and fox populations.

What differential equation describes the rate at which the fox population grows? If the food supply of the foxes consisted solely of rabbits and there were no rabbits present, the fox population would die off at a rate proportional to their numbers (exponential decay). However, the fox population does not actually decline at this rate because there are rabbits present. The fox population will increase at a rate proportional to the number of contacts between foxes and rabbits. These two influences on the growth of the fox population are represented in the differential equation

$$\frac{df}{dt} = -c \cdot f + d \cdot r \cdot f, \tag{38}$$

where c is the intrinsic rate of decline of the fox population in the absence of rabbits ($c > 0$) and d measures the effect that the catching of rabbits by foxes has on increasing the fox population ($d > 0$).

The system of differential equations consisting of equations (37) and (38) is difficult to solve using analytic methods. With further modifications to the model (such as the ones we will investigate in Lab 9), the system may even be impossible to solve analytically. However, Euler's method can be used to generate values for the two populations over time. Since each differential equation refers to values for both populations, we must work with (37) and (38) as coupled differential equations.

Lab 9 *Predator/Prey Models*

In population models in which foxes (predators) feed on rabbits (prey), changes in the size of each population depend on the number of individuals in each species. The differential equation that describes the rate of change of the rabbit population is

$$\frac{dr}{dt} = a \cdot r - b \cdot r \cdot f,$$

where a and b are positive constants, r is the number of rabbits at time t, and f is the number of foxes at time t. The differential equation that describes the rate of change of the fox population is

$$\frac{df}{dt} = -c \cdot f + d \cdot r \cdot f,$$

where c and d are positive constants. We are not able to solve these equations analytically for r and f as functions of time, but we can determine values for both populations numerically using Euler's method.

1. Write the recursive equations that use the linear Euler's method to generate values for each population over time. Note that since the equations depend on each other, you must generate values of r and f simultaneously, as opposed to doing all the r values first and then all the f values.

2. Let the initial rabbit population be 200 and the initial fox population be 50. Define the four constants in the differential equations as $a = 0.05$, $b = 0.001$, $c = 0.03$, and $d = 0.0002$. Examine a 200 month cycle, letting Δt be 1 month. Investigate the behavior of the populations by making several graphs. First plot the number of rabbits versus time and the number of foxes versus time on the same axes. Then make a separate plot of the number of foxes versus the number of rabbits.

3. Use your graphs to investigate how the populations are related. What is the relationship between the peaks and valleys on the fox population graph and the

peaks and valleys on the rabbit population graph, and vice versa? What are the average population levels about which the rabbit and fox populations oscillate?

4. The graphs produced in part 2 may show the fox and rabbit populations experiencing larger swings from maximum size to minimum size. These increasing amplitudes are due to overshoot error in Euler's method near the turning points of the graphs. These errors can be minimized if we make Δt smaller, but this requires more iterations. An alternative to shrinking Δt is to use the quadratic Euler's method. Modify the equations you wrote in part 1 to implement the quadratic Euler's method. Generate new graphs of foxes versus time, rabbits versus time, and foxes versus rabbits. How do these graphs compare with the graphs in part 2?

5. Investigate the effects produced by changing the values of the parameters a, b, c, and d. Change the parameters one at a time, by small amounts. Note in particular how such changes influence the average population levels. Explain in terms of the predator/prey relationship why these changes in average values occur.

6. Values for r and f that cause both of the rates of change $\frac{df}{dt}$ and $\frac{dr}{dt}$ to be zero at the same time produce populations that do not oscillate, but rather remain constant over time. These values for r and f are called **stable points.** The ordered pair (0, 0) is one stable point; if $r = 0$ and $f = 0$, both $\frac{df}{dt}$ and $\frac{dr}{dt}$ will be equal to zero and both populations will remain at zero. Solve $\frac{df}{dt} = 0$ and $\frac{dr}{dt} = 0$ to find the coordinates of another stable point. How are the coordinates of this stable point related to the average population levels about which the rabbit and fox populations oscillate?   *stable points*

7. From the section on parametric equations, we know that $\frac{df}{dr} = \frac{df/dt}{dr/dt}$, which by substitution can be written as $\frac{df}{dr} = \frac{-cf+drf}{ar-brf}$. This derivative represents the rate of change of the number of foxes with respect to the number of rabbits. Use technology to draw a slope field for the differential equation $\frac{df}{dr} = \frac{-cf+drf}{ar-brf}$ with the values for the parameters a, b, c, and d given in part 2. Use the information the slope field gives you to describe the relationship between f and r.

8. This model has many different extensions that can be investigated with Euler's method. Some suggestions are given below.

   a. Introduce a change in the value of a parameter that occurs after a certain time. For example, suppose after 50 months a food shortage causes the rabbits' growth rate to change from 0.05 to 0.03. Investigate the effects of these changes if they are implemented at different times within the population cycles.

   b. Implement a logistic constraint on the rabbit population using an upper limit of 500. (See "Constrained Exponential Growth" in section 4.2.) Model the rabbit and fox populations using the linear Euler's method. (The quadratic method becomes too complicated.) What effect does the logistic constraint produce in the model? How does the introduction of a logistic constraint influence the average values about which the populations oscillate? Is the model more realistic? How does changing the upper limit of the rabbit population influence the long term behavior of the rabbit and fox populations?

c. Return to the original question investigated by Volterra: What effect does fishing have on the model of interaction between selachians and food fish? In terms of rabbits and foxes, we need to investigate the effect of hunting on the rabbit and fox populations. To account for hunting, introduce the term $-k \cdot r$ into the right side of the equation for $\frac{dr}{dt}$ and the term $-k \cdot f$ into the right side of the equation for $\frac{df}{dt}$, where k is a constant of proportionality that represents the intensity of hunting. We assume that hunters will kill rabbits and foxes at the same rate. This is parallel to the Volterra problem where fishing would affect selachians and food fish at the same rate. What are the effects of increasing the level of hunting on the predator and prey populations? What are the effects of reducing the level of hunting? How are the coordinates of the non-zero stable points changed by the cessation of hunting? Use your results to explain why the cessation of fishing in the Mediterranean during World War I led to an increase in the percentage of sharks caught. Note that prior to World War I the status quo was that fishing occurred; the change you are to explain involves going from fishing to absence of fishing.

4.8 Taylor Polynomials

If a function f is differentiable, then a tangent line can be used as an approximation for f close to the point of tangency. We know that the linear approximation for $f(x)$ about $x = a$ has the equation

$$y = f'(a) \cdot (x - a) + f(a).$$

Recall that the linear approximation and $f(x)$ have the same y value and the same value for the first derivative at $x = a$. In Section 4.5 we introduced a quadratic approximation, which is generally an improvement over a linear approximation because it takes into account the concavity of f. The quadratic approximation for $f(x)$ about $x = a$ has the equation

$$y = \frac{1}{2} f''(a) \cdot (x - a)^2 + f'(a) \cdot (x - a) + f(a).$$

Recall that the quadratic approximation has the same y value and same values for the first and second derivatives at $x = a$ as f does. Is a cubic approximation better than a quadratic approximation, and if so, is a fourth-degree polynomial even better? We will now investigate a method to find approximating polynomials of degree greater than 2.

Generalizing from a quadratic approximation, a cubic approximation for $f(x)$ about $x = a$ has the same y value and the same values for the first, second, and third derivatives at $x = a$ as f does. For example, $c(x) = \frac{1}{6}x^3 + \frac{1}{2}x^2 + x + 1$ is a cubic approximation to $f(x) = e^x$ about $x = 0$ with the same y value and values of the derivatives. This is true because

$$c(0) = 1 = f(0)$$
$$c'(0) = 1 = f'(0)$$
$$c''(0) = 1 = f''(0), \text{ and}$$
$$c'''(0) = 1 = f'''(0).$$

What fourth-degree polynomial will approximate $f(x) = e^x$ about $x = 0$? This polynomial is of the form $q(x) = a_1 x^4 + a_2 x^3 + a_3 x^2 + a_4 x + 1$, since $q(0)$ must equal $f(0)$, which is 1. We must find values for the coefficients a_1, a_2, a_3, and a_4 so that $q(x)$ and $f(x)$ have the same values for the first, second, third, and fourth derivatives at $x = 0$. We know that

$$q'(x) = 4a_1 x^3 + 3a_2 x^2 + 2a_3 x + a_4$$

and that $q'(0)$ must equal $f'(0)$. Since $f'(0) = e^0 = 1$, we have $a_4 = 1$. We know that

$$q''(x) = 4 \cdot 3a_1 x^2 + 3 \cdot 2a_2 x + 2a_3$$

and that $q''(0)$ must equal $f''(0)$. Since $f''(0) = e^0 = 1$, we have $2a_3 = 1$, so $a_3 = \frac{1}{2}$. We know that

$$q'''(x) = 4 \cdot 3 \cdot 2a_1 x + 3 \cdot 2a_2$$

and that $q'''(0)$ must equal $f'''(0)$. Since $f'''(0) = e^0 = 1$, we have $3 \cdot 2 \cdot a_2 = 1$, so $a_2 = \frac{1}{6}$. Finally, we know that

$$q^{(4)}(x) = 4 \cdot 3 \cdot 2 \cdot a_1$$

and that $q^{(4)}(0)$ must be equal to $f^{(4)}(0)$. Since $f^{(4)}(0) = e^0 = 1$ then $4 \cdot 3 \cdot 2 \cdot a_1 = 1$, so $a_1 = \frac{1}{24}$. Therefore, the equation of the fourth-degree polynomial that approximates $f(x) = e^x$ about $x = 0$ is

$$q(x) = \frac{1}{24}x^4 + \frac{1}{6}x^3 + \frac{1}{2}x^2 + x + 1.$$

Based on this pattern, we can predict the fifth-degree polynomial that approximates $f(x) = e^x$ about $x = 0$. From the pattern of coefficients we can infer that $\frac{1}{2 \cdot 3 \cdot 4 \cdot 5} = \frac{1}{120}$ is the coefficient of x^5, so

$$e^x \approx \frac{1}{120}x^5 + \frac{1}{24}x^4 + \frac{1}{6}x^3 + \frac{1}{2}x^2 + x + 1 \tag{39}$$

near $x = 0$. The polynomial on the right side of equation (39) is an example of a ***Taylor polynomial***. A Taylor polynomial approximates a function about $x = a$ by the method we have been using here: equating the y values and derivative values at $x = a$.

Taylor polynomial

Suppose we want to find the fifth-degree Taylor polynomial p for some arbitrary function g about $x = 0$. The coefficients in $p(x) = a_1x^5 + a_2x^4 + a_3x^3 + a_4x^2 + a_5x + a_6$ must be determined so that $p(0) = g(0)$, $p'(0) = g'(0)$, $p''(0) = g''(0)$, and so on. To find the appropriate values for the coefficients, we need to determine expressions for the derivatives of p, and then we need to evaluate these expressions at $x = 0$. The first through fifth derivatives of p are as follows.

$$p'(x) = 5a_1x^4 + 4a_2x^3 + 3a_3x^2 + 2a_4x + a_5$$
$$p''(x) = 5 \cdot 4a_1x^3 + 4 \cdot 3a_2x^2 + 3 \cdot 2a_3x + 2a_4$$
$$p'''(x) = 5 \cdot 4 \cdot 3a_1x^2 + 4 \cdot 3 \cdot 2a_2x + 3 \cdot 2a_3$$
$$p^{(4)}(x) = 5 \cdot 4 \cdot 3 \cdot 2a_1x + 4 \cdot 3 \cdot 2a_2$$
$$p^{(5)}(x) = 5 \cdot 4 \cdot 3 \cdot 2a_1$$

Evaluating each derivative at $x = 0$ yields the following values.

$$p'(0) = a_5$$
$$p''(0) = 2a_4$$
$$p'''(0) = 3 \cdot 2a_3$$
$$p^{(4)}(0) = 4 \cdot 3 \cdot 2a_2$$
$$p^{(5)}(0) = 5 \cdot 4 \cdot 3 \cdot 2a_1$$

factorial notation   **Factorial notation** provides a compact way to write many of the coefficients we encounter in Taylor polynomials. The symbol $n!$, read "n factorial," represents the product of the positive integers less than or equal to n, where n is a positive integer, so that
$$n! = n(n-1)(n-2)(n-3)\ldots(3)(2)(1).$$

Thus, $p^{(4)}(0) = 4 \cdot 3 \cdot 2 \cdot a_2$ can be written as $p^{(4)}(0) = 4!a_2$ and $p^{(5)}(0) = 5 \cdot 4 \cdot 3 \cdot 2 \cdot a_1$ can be written as $p^{(5)}(0) = 5!a_1$.

Using this information to rewrite the coefficients of the approximation for the function g, we have

$$p(0) = g(0) \quad\Rightarrow\quad a_6 = g(0);$$
$$p'(0) = g'(0) \quad\Rightarrow\quad a_5 = g'(0);$$
$$p''(0) = g''(0) \quad\Rightarrow\quad 2!a_4 = g''(0) \quad\text{or}\quad a_4 = \tfrac{1}{2!}g''(0);$$
$$p'''(0) = g'''(0) \quad\Rightarrow\quad 3!a_3 = g'''(0) \quad\text{or}\quad a_3 = \tfrac{1}{3!}g'''(0);$$
$$p^{(4)}(0) = g^{(4)}(0) \quad\Rightarrow\quad 4!a_2 = g^{(4)}(0) \quad\text{or}\quad a_2 = \tfrac{1}{4!}g^{(4)}(0);$$
$$p^{(5)}(0) = g^{(5)}(0) \quad\Rightarrow\quad 5!a_1 = g^{(5)}(0) \quad\text{or}\quad a_1 = \tfrac{1}{5!}g^{(5)}(0).$$

Thus, the fifth-degree Taylor polynomial for $g(x)$ about $x = 0$ is

$$p(x) = \frac{1}{5!} g^{(5)}(0)x^5 + \frac{1}{4!} g^{(4)}(0)x^4 + \frac{1}{3!} g'''(0)x^3 + \frac{1}{2!} g''(0)x^2 + g'(0)x + g(0). \quad (40)$$

The coefficient of x^i in the Taylor polynomial $p(x)$ in (40) is equal to $\frac{1}{i!} g^{(i)}(0)$. In general, the nth-degree Taylor polynomial approximation about $x = 0$ for the specific function $g(x) = e^x$ is

$$e^x \approx 1 + x + \frac{1}{2!} x^2 + \frac{1}{3!} x^3 + \frac{1}{4!} x^4 + \frac{1}{5!} x^5 + \frac{1}{6!} x^6 + \frac{1}{7!} x^7 + \cdots + \frac{1}{n!} x^n,$$

since all the derivatives of e^x equal e^x, which has a value of 1 at $x = 0$. As we have done here, it is customary to write the equations of Taylor polynomials with the powers of x in ascending order.

Example 1

Write the equation of the twelfth-degree Taylor polynomial for $\cos x$ about $x = 0$.

Solution   We need to find the derivatives of $f(x) = \cos x$ and evaluate these derivatives at $x = 0$.

$$
\begin{aligned}
f(x) &= \cos x &\Rightarrow\quad & f(0) = 1 \\
f'(x) &= -\sin x &\Rightarrow\quad & f'(0) = 0 \\
f''(x) &= -\cos x &\Rightarrow\quad & f''(0) = -1 \\
f'''(x) &= \sin x &\Rightarrow\quad & f'''(0) = 0 \\
f^{(4)}(x) &= \cos x &\Rightarrow\quad & f^{(4)}(0) = 1 \\
f^{(5)}(x) &= -\sin x &\Rightarrow\quad & f^{(5)}(0) = 0
\end{aligned}
$$

The values of $f^{(n)}(0)$ form the pattern of 1, 0, –1, 0, 1, 0, –1, 0 and so on. The first few terms of a Taylor polynomial for $f(x) = \cos x$ about $x = 0$ are

$$p(x) = 1 + 0x - \frac{1}{2!} x^2 + 0x^3 + \frac{1}{4!} x^4 + 0x^5 - \frac{1}{6!} x^6,$$

so the twelfth-degree Taylor polynomial for $f(x) = \cos x$ about $x = 0$ is

$$p(x) = 1 - \frac{1}{2!} x^2 + \frac{1}{4!} x^4 - \frac{1}{6!} x^6 + \frac{1}{8!} x^8 - \frac{1}{10!} x^{10} + \frac{1}{12!} x^{12}.$$

The values of x must be in radians for this Taylor polynomial to approximate $\cos x$. ∎

Notice that only terms with even powers appear in a Taylor polynomial for $\cos x$. Therefore, when graphed in the xy-plane, the Taylor polynomial for $\cos x$ is symmetric with respect to the y-axis, as is the graph of $y = \cos x$. Functions that are symmetric with respect to the y-axis are known as *even functions*, a term that is reflected in the nature of the exponents in the Taylor polynomials for these functions.

Based on the Taylor polynomials we have constructed so far, we can see that the general equation for the nth-degree Taylor polynomial that approximates $f(x)$ about $x = 0$ is

$$p(x) = f(0) + f'(0) \cdot x + \tfrac{1}{2!} f''(0) \cdot x^2 + \tfrac{1}{3!} f'''(0) \cdot x^3 + \cdots + \tfrac{1}{n!} f^{(n)}(0) \cdot x^n. \quad (41)$$

The function f must be differentiable n times at $x = 0$ to construct a Taylor polynomial of degree n about $x = 0$.

Of course, we may want to use a polynomial to approximate a function about a point other than $x = 0$. Suppose we want a polynomial p that approximates the function f about $x = a$. In this case, equation (41) must be modified in two ways. First, the graph of p must be shifted horizontally so that what originally occurred at $x = 0$ now occurs at $x = a$. This is accomplished by replacing x with $(x - a)$ in the right side of equation (41). Second, the function value and the derivative values must be evaluated at $x = a$ rather than at $x = 0$. This means that $f(a)$ must be substituted for $f(0)$, $f'(a)$ must be substituted for $f'(0)$, $f''(a)$ must be substituted for $f''(0)$, and so on. Therefore, equation (41) becomes

$$p(x) = f(a) + f'(a) \cdot (x - a) + \tfrac{1}{2!} f''(a) \cdot (x - a)^2 + $$
$$\tfrac{1}{3!} f'''(a) \cdot (x - a)^3 + \ldots + \tfrac{1}{n!} f^{(n)}(a) \cdot (x - a)^n. \quad (42)$$

Equation (42) is the nth-degree Taylor polynomial that approximates the function f about $x = a$. The function f must be differentiable n times at $x = a$ to construct this polynomial. Notice that the function p and the first n derivatives have the same values at $x = a$ as f and the derivatives of f.

Example 2

Write the expression for the fifth-degree Taylor polynomial for $\ln x$ about $x = 1$.

Solution   We can use equation (42) with $a = 1$ and $f(x) = \ln x$. To determine the coefficients of the fifth-degree Taylor polynomial we need to evaluate $f(x)$ and the first five derivatives of $f(x) = \ln x$ at $x = 1$.

$$
\begin{aligned}
f(x) &= \ln x && \Rightarrow & f(1) &= 0 \\
f'(x) &= x^{-1} && \Rightarrow & f'(1) &= 1 \\
f''(x) &= -x^{-2} && \Rightarrow & f''(1) &= -1 \\
f'''(x) &= 2x^{-3} && \Rightarrow & f'''(1) &= 2 \\
f^{(4)}(x) &= -6x^{-4} && \Rightarrow & f^{(4)}(1) &= -6 \\
f^{(5)}(x) &= 24x^{-5} && \Rightarrow & f^{(5)}(1) &= 24
\end{aligned}
$$

Using these numerical values in equation (42) yields

$$p(x) = (x-1) - \frac{1}{2}(x-1)^2 + \frac{1}{3}(x-1)^3 - \frac{1}{4}(x-1)^4 + \frac{1}{5}(x-1)^5$$

as the fifth-degree Taylor polynomial approximation for $f(x) = \ln x$ about $x = 1$. ∎

What is the advantage of using a fifth-degree polynomial over a second- or third-degree polynomial? We have seen that a quadratic approximation p to a function f about $x = 1$ carries information about the concavity of f at $x = a$ that is missing in the linear approximation. What effect does adding additional terms to the polynomial have on the accuracy of the approximation? The next two examples help to answer this question.

Example 3

Investigate the effect of increasing the number of terms in the Taylor polynomial for $f(x) = \sin x$ about $x = 0$.

Solution   We first need to write a Taylor polynomial. We can determine the coefficients by evaluating $f(0)$ and the derivatives of $f(x) = \sin x$ at $x = 0$.

$$\begin{aligned}
f(x) &= \sin x & \Rightarrow & & f(0) &= 0 \\
f'(x) &= \cos x & \Rightarrow & & f'(0) &= 1 \\
f''(x) &= -\sin x & \Rightarrow & & f''(0) &= 0 \\
f'''(x) &= -\cos x & \Rightarrow & & f'''(0) &= -1 \\
f^{(4)}(x) &= \sin x & \Rightarrow & & f^{(4)}(0) &= 0
\end{aligned}$$

The values of $f^{(n)}(0)$ form the pattern $0, 1, 0, -1, 0, 1, 0, -1, 0$, and so on. Using these numerical values in equation (41), we can write the eleventh-degree Taylor polynomial for $f(x) = \sin x$ about $x = 0$:

$$p(x) = x - \frac{1}{3!}\cdot x^3 + \frac{1}{5!}\cdot x^5 - \frac{1}{7!}\cdot x^7 + \frac{1}{9!}\cdot x^9 - \frac{1}{11!}x^{11}.$$

Graphs of $f(x) = \sin x$ with Taylor polynomials using terms through degree 3, 5, 7, 9, and 11 in $p(x)$ are shown in Figure 4.36. These graphs show that as the degree of the Taylor polynomial increases, there is a corresponding increase in the set of x values over which the polynomial is a good approximation for the function $f(x) = \sin x$ For example, the third-degree Taylor polynomial is accurate to within 10% of $f(x) = \sin x$ on the interval $[-1.6, 1.6]$, whereas the fifth-degree Taylor polynomial is accurate within 10% of $f(x) = \sin x$ on the interval $[-2.3, 2.3]$. ∎

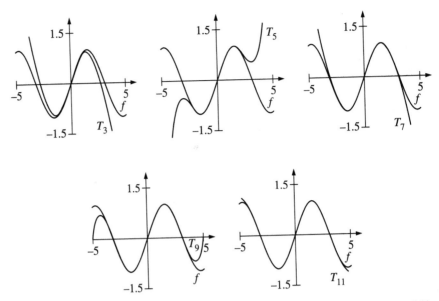

Figure 4.36 Graphs of $f(x) = \sin x$ and Taylor polynomials of degree 3, 5, 7, 9, and 11

Notice that only terms with odd powers appear in a Taylor polynomial for sin x. Therefore, when graphed in the xy-plane, the Taylor polynomial for sin x is symmetric with respect to the origin, as is the graph of $y = \sin x$. Functions that are symmetric with respect to the origin are known as *odd functions*, a term that is reflected in the nature of the exponents in the Taylor polynomials for these functions.

Example 4

Investigate the effect of increasing the number of terms in the Taylor polynomial for $f(x) = \frac{1}{x+1}$ about $x = 0$.

Solution   We first need to write a Taylor polynomial. We can determine the coefficients by evaluating $f(0)$ and the derivatives of $f(x) = \frac{1}{x+1}$ at $x = 0$.

$$f(x) = \frac{1}{x+1} \qquad \Rightarrow \qquad f(0) = 1$$

$$f'(x) = -\frac{1}{(x+1)^2} \qquad \Rightarrow \qquad f'(0) = -1$$

$$f''(x) = \frac{2}{(x+1)^3} \qquad \Rightarrow \qquad f''(0) = 2$$

$$f'''(x) = \frac{-6}{(x+1)^4} \qquad \Rightarrow \qquad f'''(0) = -6$$

$$f^{(4)}(x) = \frac{24}{(x+1)^5} \qquad \Rightarrow \qquad f^{(4)}(0) = 24$$

Using these numerical values in equation (41), we can write the fourth-degree Taylor polynomial for $f(x) = \frac{1}{x+1}$ about $x = 0$:

$$p(x) = 1 - x + x^2 - x^3 + x^4.$$

Figure 4.37 shows graphs of $f(x) = \frac{1}{x+1}$ and Taylor polynomials of degrees 1 through 4. These graphs show behavior different than the graphs in Figure 4.36. In Figure 4.37, when the degree of the Taylor polynomial is increased there is only a small increase in the set of x values over which the polynomial is a good approximation for the function $f(x) = \frac{1}{x+1}$. The interval over which the Taylor polynomial seems to be a good approximation for the function $f(x)$ appears to be limited. However, for x values close to -1, the accuracy improves substantially over this small interval. ∎

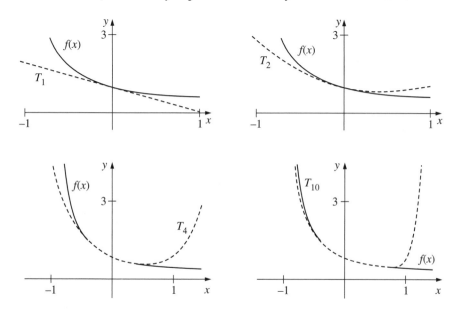

Figure 4.37 Graphs of $f(x) = \frac{1}{x+1}$ and Taylor polynomials of degree 1, 2, 4 and 10

In this chapter, we have introduced two methods for approximating a function, Euler's method and Taylor polynomials. Both approximate functions using our knowledge of the derivatives of the function; however, they accomplish the approximation in very different ways. If you know the derivative of a function, Euler's method allows you to use recursion to generate values that approximate the values of the function from a starting point to some other point. In contrast, a Taylor polynomial is not calculated recursively, but is determined by calculating the values of the function and its derivatives at one point.

Exercise Set 4.8

1. Write the first five non-zero terms in the Taylor polynomial about $x = 0$ for each function.

    a. $f(x) = \sqrt{x+1}$ b. $f(x) = \ln(x+1)$

2. a. Write the sixth-degree Taylor polynomial for the function $f(x) = \sqrt{x}$ about $x = 1$.

    b. Graph $f(x) = \sqrt{x}$ and its Taylor polynomials of degree 2, 4, and 6 on the same axes.

    c. Describe how increasing the degree of the polynomial affects the set of x values for which the polynomial does well approximating $f(x) = \sqrt{x}$.

3. a. Write the nth-degree Taylor polynomial for the function e^x about $x = 0$.

    b. Graph $f(x) = e^x$ and its Taylor polynomials of degree 1, 2, 3, and 4 on the same axes.

    c. Describe how increasing the degree of the polynomial affects the set of x values for which the polynomial is a good approximation for $f(x) = e^x$.

4. a. Write the Taylor polynomial of degree $2n$ for the function $f(x) = \cos x$ about $x = 0$.

    b. Determine the minimum degree Taylor polynomial needed to approximate the value of $\cos 1$ with relative error less than 1%.

    c. Determine the minimum degree Taylor polynomial needed to approximate the value of $\cos 3$ with relative error less than 1%.

5. Consider the function that satisfies the differential equation $f'(x) = \frac{1}{x+1}$ and contains the point $(0, 1)$. We can approximate the solution to this differential equation using either Taylor polynomials or Euler's method.

    a. Write the fifth-degree Taylor polynomial for f about $x = 0$.

    b. Use the Taylor polynomial you wrote in part a to estimate the value of $f(0.75)$.

    c. Use three iterations of the linear Euler's method to estimate the value $f(0.75)$.

    d. Verify that $f(x) = \ln(x+1)+1$ satisfies the differential equation. Use this function to find $f(0.75)$. Compare this value with the estimates you made in parts b and c.

    e. How could you improve the accuracy of your estimates using Taylor polynomials? using the linear Euler's method?

6. Consider the function that satisfies the differential equation $f'(x) = \ln x$ and contains the point $(1, 1)$. Use the linear Euler's method and a Taylor polynomial to approximate $f(1.75)$. Improve both methods so that the approximations agree to the second decimal place. What is your best estimate of $f(1.75)$?

END OF CHAPTER EXERCISES

Review and Extensions

1. For each of the following differential equations, find the particular solution that includes the given ordered pair.

    a. $\dfrac{dy}{dx} = 2 + \cos 3x;\ (0,1)$ b. $\dfrac{dP}{dt} = 0.085P;\ (0,10)$

2. Assume that the following differential equation models the rate of change in the fish population, f, of a new pond on a local farm:

    $$\frac{df}{dt} = 0.05f\left(1 - \frac{f}{800}\right).$$

    a. Suppose the farmer proposes to stock the pond initially with 100 fish. Use linear Euler's method to estimate the fish population after two years. Let $\Delta t = 0.5$ years for these calculations.

    b. Explain to the farmer what will happen to the number of fish over time. Be as specific and thorough as possible.

    c. Suppose the farmer initially stocks the pond with 1200 fish. Describe what will happen to the number of fish over time.

    d. Adjust the differential equation model to reflect the fact that each year 10% of the fish will be removed by fishing.

3. Suppose your friend missed class the day we discussed slope fields. Write a few sentences to describe the method we used to draw a slope field. Also explain what information a slope field provides.

4. Given that $\dfrac{dy}{dx} = 2x + y$ and that $y = 1$ when $x = 0$, use quadratic Euler's method with $\Delta x = 0.5$ to approximate two additional values of y.

5. Write the equation of the quadratic approximation to $y = x \ln x + x$ about the point (1, 1).

6. Consider the curve defined by the parametric equations $x = t - 3\sin t$ and $y = t \sin t - \cos t$.

    a. Find the slope of the tangent line at $t = 1$.

    b. Determine the coordinates of a point where the tangent line is vertical.

7. a. Find the third degree Taylor polynomial for $f(x) = e^{-x/2}$ about $x = 0$.

    b. For what values of $x > 0$ is the approximation from the Taylor polynomial within 0.1 of the actual value of the function?

8. a. Find the fourth degree Taylor polynomial for $f(x) = \sqrt{x}$ about $x = 9$.

    b. Use this polynomial to estimate $\sqrt{8}$ and determine the relative error associated with the estimate.

CHAPTER 5

Integrals and the Fundamental Theorem

5.1 The Fundamental Theorem of Calculus

In Lab 8, we used Euler's method to generate approximate values for the height above ground of a parachutist. The values for the altitude were plotted versus time to obtain a graph that approximated the height above ground of a parachutist during the time of the descent. The model we obtained in this manner is an example of the following general principle used throughout Chapter 4.

Given the derivative $\frac{dy}{dx}$ of a function with equation $y = f(x)$, we can generate approximate values for $f(x)$ by iterating the recursive system

$$x_i = x_{i-1} + \Delta x$$

$$y_i = y_{i-1} + \left.\frac{dy}{dx}\right|_{(x_{i-1},\,y_{i-1})} \cdot \Delta x. \tag{1}$$

Given a starting value y_0 equal to $f(x_0)$, equation (1) allows us to generate y_i values that are approximations for $f(x_i)$. In many situations, however, we do not need to know the intermediate y_i values, but instead we are primarily interested in the net change of $f(x)$ between an initial x value, x_0, and a final x value, x_n. In this section we will examine how Euler's method can be used to answer questions related to the net change of a function.

Suppose a projectile is fired straight up from the ground. We may or may not know a formula for the function H that describes exactly how the height of the projectile depends on time. For example, such a formula is particularly difficult to obtain if air resistance has an effect on the flight of the projectile. Assume, however, that we know a formula $H'(t)$ for the rate at which the height of the projectile is changing.

Euler's method allows us to generate a sequence of values h_n that approximate the height at time t_n by iterating the equation

$$h_n = h_{n-1} + H'(t_{n-1}) \cdot \Delta t. \tag{2}$$

The accompanying values of t_n are produced by incrementing time by Δt at each step, so that

$$t_n = t_{n-1} + \Delta t.$$

Using initial values $t_0 = 0$ and $h_0 = H(t_0)$, the first value h_1 in the sequence of approximate heights is produced by the first iteration of equation (2) with $n = 1$, which gives

$$h_1 = h_0 + H'(t_0) \cdot \Delta t. \tag{3}$$

We now use equation (2) with $n = 2$ to find h_2, which gives

$$h_2 = h_1 + H'(t_1) \cdot \Delta t. \tag{4}$$

Substituting the expression for h_1 from equation (3) into equation (4) yields

$$h_2 = h_0 + H'(t_0) \cdot \Delta t + H'(t_1) \cdot \Delta t. \tag{5}$$

Another iteration will help us see a pattern for h_n. The third value in the sequence of approximate heights is

$$h_3 = h_2 + H'(t_2) \cdot \Delta t,$$

which, by substitution from equation (5), is equivalent to

$$h_3 = h_0 + H'(t_0) \cdot \Delta t + H'(t_1) \cdot \Delta t + H'(t_2) \cdot \Delta t. \tag{6}$$

Observing a pattern in equations (3), (5), and (6), we can see that in general, after n iterations, an approximation h_n for the height of the projectile at time t_n is given by

$$h_n = h_0 + H'(t_0) \cdot \Delta t + H'(t_1) \cdot \Delta t + \cdots + H'(t_{n-1}) \cdot \Delta t. \tag{7}$$

summation notation

A sum like the one on the right side of equation (7) can be written in a compact form using ***summation notation***. For example, the sum

$$x_1 + x_2 + x_3 + x_4 + x_5$$

can be written as

$$\sum_{i=1}^{5} x_i.$$

The Greek letter Σ (*sigma*) symbolizes the *sum*. In the given example, the letter i is the *index of the summation*, and 1 and 5 are the lower and upper limits of summation.

This summation includes one term for each integer from 1 to 5, all of which are added together to give the value of the summation.

We can rewrite equation (7) using summation notation as:

$$h_n = h_0 + \sum_{i=0}^{n-1} H'(t_i) \cdot \Delta t. \tag{8}$$

Subtracting h_0 from both sides of equation (8) gives

$$h_n - h_0 = \sum_{i=0}^{n-1} H'(t_i) \cdot \Delta t.$$

The difference $h_n - h_0$ is an approximation for the net change in height of the projectile from time t_0 to time t_n. If we know a formula for $H(t)$, then the actual net change in height is $H(t_n) - H(t_0)$, which is approximated by $h_n - h_0$, so

$$H(t_n) - H(t_0) \approx \sum_{i=0}^{n-1} H'(t_i) \cdot \Delta t. \tag{9}$$

We have used Euler's method to show that the net change in height from t_0 to t_n is approximated by a sum of n terms, each of which is a product of the form $H'(t_i) \cdot \Delta t$, where i is an integer varying from 0 to $n - 1$. As illustrated in Figure 5.1, each $H'(t_i) \cdot \Delta t$ term is the product of a slope and a change in t, which gives a vertical change. Thus, each term represents an approximation for the change in H corresponding to a change in time from t_i to t_{i+1}. The sum of these changes gives us an approximation for the net change in H from $t = t_0$ to $t = t_n$. This representation is consistent with our use of Euler's method in Chapter 4.

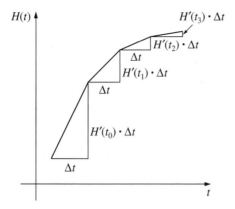

Figure 5.1 Net change in H approximated by Euler's method

The net change in a projectile's height from t_0 to t_n does not always equal the vertical distance it travels from t_0 to t_n. If the projectile rises and then falls, the total vertical distance traveled by the projectile is equal to the distance it rises plus the distance it falls, which is not equal to the net change in height. The net change in height is the difference between the final height and the initial height, which is always less than or equal to the total vertical distance traveled. For example, if you kick a ball on the ground so that it rises 10 ft. then falls back to the ground, the vertical distance traveled is 20 ft. but the net change in height is 0 ft.

From our previous work, we know that the Euler's method approximations for the height improve as Δt gets smaller. Note, however, that as Δt shrinks, more iterations are needed to approximate the height over the same interval of time. The sequence of h_i's does not stay the same as Δt shrinks. More values are generated in the sequence, and the values are closer together. As Δt goes to zero, there is improvement in our approximation for the net change in height given in equation (9). In fact, taking the limit as Δt goes to zero allows us to write "=" instead of "$\approx$" in equation (9). In doing so we are assuming that as Δt goes to 0, the sum gets closer to the actual value of $H(t_n) - H(t_0)$, so that

$$H(t_n) - H(t_0) = \lim_{\Delta t \to 0} \sum_{i=0}^{n-1} H'(t_i) \cdot \Delta t, \tag{10}$$

where $\Delta t = \frac{t_n - t_0}{n}$ and $t_i = t_{i-1} + \Delta t$. We have assumed that Δt is the constant $\frac{t_n - t_0}{n}$, which means that taking the limit as Δt goes to zero is equivalent to taking the limit as n, the number of iterations, goes to infinity. Therefore, the net change in height also can be written as

$$H(t_n) - H(t_0) = \lim_{n \to \infty} \sum_{i=0}^{n-1} H'(t_i) \cdot \Delta t, \tag{11}$$

where $t_i = t_{i-1} + \Delta t$.

Riemann sum

The sum on the right side of equation (10) is an example of a ***Riemann sum***, named for the 19th century mathematician Georg Friedrich Bernhard Riemann (1826–1866). A Riemann sum has the general form

$$\sum_{i=0}^{n-1} f(x_i) \cdot \Delta x,$$

where the terms that are added together in the sum are products of the form "function value" times "change in independent variable".

The right side of equation (10) is the limit of a Riemann sum. The special symbol used to represent this limit is

$$\int_a^b H'(t)dt, \tag{12}$$

where a is the initial time t_0 and b is the final time t_n. The $\int$ symbol is similar to an elongated S, which is consistent with the fact that we are dealing with a type of sum. The entire expression in (12) is called a **definite integral**. The function H' in the definite integral is called the **integrand**. The values a and b are called the **limits of integration**. Equating expression (12) with the right side of equation (10) gives the **definition of a definite integral**

definite integral
integrand
limits of integration
definition of a definite integral

$$\int_a^b H'(t)dt = \lim_{\Delta t \to 0} \sum_{i=0}^{n-1} H'(t_i) \cdot \Delta t, \qquad (13)$$

where $t_0 = a$, $t_n = b$, $\Delta t = \frac{t_n - t_0}{n}$, and $t_i = t_{i-1} + \Delta t$. Assuming Δt is constant, then the limit in (13) can be written with n approaching infinity as in equation (11). Also notice that on the right side of equation (13), the function H' is evaluated at the discrete valued, subscripted variable t_i, whereas on the left side of equation (13), the function H' is evaluated at the continuous variable t.

Since the right side of equation (13) is the net change in $H(t)$ from $t = a$ to $t = b$, which is $H(b) - H(a)$, the definite integral on the left side of equation (13) also equals the net change in $H(t)$, so that

$$\int_a^b H'(t)dt = H(b) - H(a). \qquad (14)$$

The relationship in (14) is the conclusion of the **Fundamental Theorem of Calculus**, which is stated as follows:

Fundamental Theorem of Calculus

If $H(x)$ is continuous and differentiable at all x values from a to b, inclusive, then

$$\int_a^b H'(t)dt = H(b) - H(a).$$

This result is *fundamental* because it links derivatives and integrals, the two central concepts of calculus.

We have interpreted the definite integral $\int_a^b H'(t)dt$ in the Fundamental Theorem as the net change in the function H. A second interpretation of the Fundamental Theorem is linked to the graph of the function H'. In the Riemann sum on the right side of equation (13), we are adding products of $H'(t_i)$ and Δt. If $H'(t)$ has positive values for $a \le t \le b$, then these products represent the areas of rectangles whose heights are $H'(t_i)$ and whose widths are Δt, as shown in Figure 5.2. The sum of the areas of these rectangles approximates the total area between the graph of $H'(t)$ and the t-axis from $t = a$ to $t = b$. As Δt approaches 0, the sum of the rectangular areas gets closer and closer to the actual area between the curve and the t-axis. The area between a graph that has positive values and the horizontal axis is also called the *area under the graph* or the *area under the curve*. Thus, Figure 5.2 shows that $\int_a^b H'(t)dt$ represents the area under the graph of $H'(t)$ from $t = a$ to $t = b$.

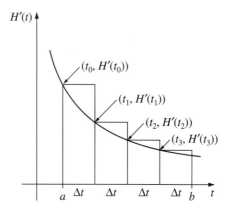

Figure 5.2 Approximating area under $H'(t)$ using rectangles

We now see that the definite integral $\int_a^b H'(t)dt$ can represent either the net change in the function H from $t = a$ to $t = b$ or, if $H'(t)$ is positive over this interval, the area under the graph of $H'(t)$ from $t = a$ to $t = b$.

Example

Evaluate $\int_2^5 3x^2 dx$. Interpret your answer in two different ways.

Solution   The Fundamental Theorem tells us that in general

$$\int_a^b f'(x)dx = f(b) - f(a).$$

The function f evaluated on the right side of this equation must have a derivative equal to the integrand f' on the left side. We can evaluate $\int_2^5 3x^2 dx$ by computing $f(5) - f(2)$, given that f is a function for which $f'(x) = 3x^2$. Clearly, *one* choice for f is $f(x) = x^3$; therefore, according to the Fundamental Theorem, the value of the integral is given by following the sequence of calculations:

$$\int_2^5 3x^2 dx = f(5) - f(2)$$
$$= 5^3 - 2^3$$
$$= 125 - 8$$
$$= 117.$$

The result 117 can be interpreted as the net change in $f(x) = x^3$ from $x = 2$ to $x = 5$ and as the area under the curve defined by $f'(x) = 3x^2$ from $x = 2$ to $x = 5$. ∎

In the example, we chose x^3 as the function whose derivative is equal to $3x^2$, yet any function of the form $g(x) = x^3 + C$ has a derivative equal to $3x^2$. The constant C results in a vertical shift of the graph of x^3. Such a shift does not affect the net change in $g(x)$ from $x = 2$ to $x = 5$, and therefore does not affect the value of the integral. This is because the integral equals the difference between $g(5)$ and $g(2)$, and this difference does not change when both $g(5)$ and $g(2)$ are shifted vertically by the same amount. Using $g(x) = x^3 + C$ in the Fundamental Theorem, the definite integral can be evaluated as follows:

$$\int_2^5 3x^2 dx = g(5) - g(2)$$
$$= (5^3 + C) - (2^3 + C)$$
$$= 125 + C - 8 - C$$
$$= 117.$$

Comparing this result with the result from the example, we see that the constant C does not affect the value of the definite integral. In fact, this is true for all definite integrals, not just $\int_2^5 3x^2 dx$. Thus, we can write

$$\int_2^5 3x^2 dx = x^3 \Big|_2^5,$$

which equals $5^3 - 2^3$, and we can evaluate this integral as we did in the previous example. The notation $x^3 \big|_2^5$ indicates that x^3 is evaluated at 5 and at 2 and the results are subtracted. This notation is a helpful shorthand for the evaluation of definite integrals.

Exercise Set 5.1

1. Evaluate each definite integral.

   a. $\displaystyle\int_0^{\pi/2} \cos x \, dx$

   b. $\displaystyle\int_1^3 x^2 dx$

   c. $\displaystyle\int_5^9 (x^2 + 6x) \, dx$

   d. $\displaystyle\int_{-1}^4 e^{x+2} dx$

   e. $\displaystyle\int_1^e \frac{3}{x} dx$

   f. $\displaystyle\int_0^4 \frac{1}{x+1} dx$

   g. $\displaystyle\int_0^1 \sqrt{x+4} \, dx$

   h. $\displaystyle\int_{-4}^4 |x| \, dx$

2. Write two interpretations for the value of the definite integral in part a above.

3. Use the Fundamental Theorem to show that $\displaystyle\int_a^b f(x) dx = -\int_b^a f(x) dx$.

5.2 Area Between Curves

We can use the Fundamental Theorem to determine the value of the definite integral $\int_0^3 x^2 dx$. To do so, we need to evaluate $f(3) - f(0)$, where f is a function whose derivative is $f'(x) = x^2$. One choice for $f(x)$ is $\frac{1}{3}x^3$, which allows us to write

$$\int_0^3 x^2 dx = \frac{1}{3}x^3\Big|_0^3$$
$$= \frac{1}{3}\cdot 3^3 - \frac{1}{3}\cdot 0^3$$
$$= 9.$$

The value 9 represents the area under the graph of $y = x^2$ between $x = 0$ and $x = 3$, as shown in Figure 5.3.

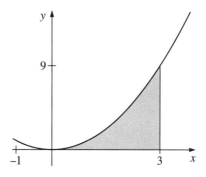

Figure 5.3 Area under the graph of $y = x^2$ between $x = 0$ and $x = 3$

Consider the integrals $\int_0^3 (2x - 4)dx$ and $\int_0^{2\pi} \cos x \, dx$. These integrals can be evaluated as follows:

$$\int_0^3 (2x - 4)\, dx = (x^2 - 4x)\Big|_0^3 = 9 - 12 = -3.$$
$$\int_0^{2\pi} \cos x \, dx = \sin x\Big|_0^{2\pi} = \sin(2\pi) - \sin(0) = 0.$$

Neither of these integrals represents an area, because we know that area must be a non-negative number. The integral $\int_0^3 x^2 dx$, the first one we evaluated, *does* represent an area because the integrand x^2 has non-negative values between the limits of integration $x = 0$ and $x = 3$. In contrast, the integrands $2x - 4$ and $\cos x$ have positive and negative values between the limits of integration. In general, $\int_a^b f(x)dx$ represents the area between the graph of $f(x)$, the x-axis, $x = a$ and $x = b$ provided $f(x) \geq 0$ for all x values between $x = a$ and $x = b$.

How does one find areas of regions that cannot be described in this way? In this first example, we will look at the area between two curves.

Example 1

Find the area of the region shown in Figure 5.4, which is bounded by the graphs of the functions $f(x) = 4 - x^2$ and $g(x) = -3x$ between their points of intersection.

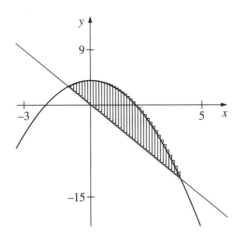

Figure 5.4 Area bounded by $f(x) = 4 - x^2$ and $g(x) = -3x$

Solution   The first step in finding the area is to determine the x-coordinates of the points where the graphs intersect. Setting $4 - x^2$ equal to $-3x$ and solving for x, we find that the points of intersection occur where $x = -1$ and $x = 4$. One way to approximate the area is to subdivide the region into rectangles of width Δx and heights calculated from the functions shown in Figure 5.4. The height of each rectangle can be obtained by subtracting the y-coordinate of the bottom of the rectangle from the y-coordinate of the top of the rectangle; therefore, the height of an arbitrary rectangle is given by $4 - x^2 - (-3x)$. The area of an arbitrary rectangle is $(4 - x^2 + 3x) \cdot \Delta x$, where Δx is the width of the rectangle. If we make all the rectangles the same width, and if n is the number of rectangles, then the width of each rectangle is $\frac{4-(-1)}{n}$, so that $\Delta x = \frac{5}{n}$. The sum of the areas of these rectangles is equal to the Riemann sum

$$\sum_{i=0}^{n-1} (4 - x_i^2 + 3x_i)\Delta x,$$

where $\Delta x = \frac{5}{n}$ and $x_0 = -1$.

The sum of the areas of the rectangles approaches the actual area between the graphs as n approaches infinity or, equivalently, as Δx approaches 0. The actual area is the limiting value of the Riemann sum, which is written as

$$\lim_{n \to \infty} \sum_{i=0}^{n-1} (4 - x_i^2 + 3x_i)\Delta x.$$

By definition, this limit is equal to the definite integral

$$\int_{-1}^{4} (4 - x^2 + 3x)dx,$$

and is evaluated using the Fundamental Theorem as follows:

$$\int_{-1}^{4} (4 - x^2 + 3x)dx = \left(4x - \frac{1}{3}x^3 + \frac{3}{2}x^2\right)\Big|_{-1}^{4} = \left(16 - \frac{64}{3} + 24\right) - \left(-4 - \frac{-1}{3} + \frac{3}{2}\right) = \frac{125}{6}.$$

The area of the region between the two graphs is $\frac{125}{6}$ square units. ■

In the next example, both functions have only positive values on the interval. Unlike Example 1, however, the first function is less than the second for part of the interval and greater than the second for the rest of the interval.

Example 2

Find the area bounded by $y = \frac{1}{x}$ and $y = x$ for $\frac{1}{4} \le x \le 4$, the shaded region shown in Figure 5.5.

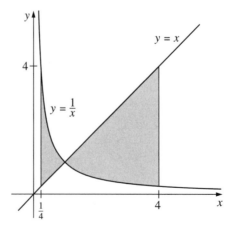

Figure 5.5 Area for Example 2

Solution   The boundaries of the shaded region are the graphs of $y = \frac{1}{x}$, $y = x$, and the vertical lines $x = \frac{1}{4}$ and $x = 4$. The graphs of $y = \frac{1}{x}$ and $y = x$ intersect at the point where $x = 1$. Between $x = \frac{1}{4}$ and $x = 1$, the graph of $y = \frac{1}{x}$ is above the graph of $y = x$, so when we subdivide the region between the graphs into narrow rectangles, the height of each rectangle is given by the difference $\frac{1}{x} - x$. This means that the area of the shaded region to the left of $x = 1$ is given by the value of the integral $\int_{1/4}^{1} (\frac{1}{x} - x)\,dx$. Between $x = 1$ and $x = 4$, the graph of $y = x$ is above the graph of $y = \frac{1}{x}$, so when we subdivide the region between the graphs into narrow rectangles, the height of each rectangle is given by the difference $x - \frac{1}{x}$. This means that the area of the shaded region to the right of $x = 1$ is given by the value of the integral $\int_{1}^{4} (x - \frac{1}{x})\,dx$. The values of these integrals can be determined as follows:

$$\int_{1/4}^{1} \left(\frac{1}{x} - x\right)dx = \left(\ln x - \frac{1}{2}x^2\right)\Big|_{1/4}^{1} = \left(\ln 1 - \frac{1}{2}\right) - \left(\ln \frac{1}{4} - \frac{1}{2} \cdot \frac{1}{16}\right) \approx 0.92,$$

$$\int_{1}^{4} \left(x - \frac{1}{x}\right)dx = \left(\frac{1}{2}x^2 - \ln x\right)\Big|_{1}^{4} = \left(\frac{1}{2} \cdot 16 - \ln 4\right) - \left(\frac{1}{2} - \ln 1\right) \approx 6.11.$$

The total area of the shaded region in Figure 5.5 is approximately 7.03 square units. ∎

Example 3

Find the area of the region bounded by the graph of $f(x) = \cos x$ and the x-axis between $x = 0$ and $x = 2\pi$.

Solution   The region between the curve and the x-axis, together with approximating rectangles, is shown in Figure 5.6.

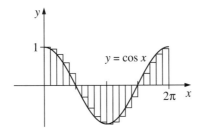

Figure 5.6   Area for Example 3

From $x = 0$ to $x = \frac{\pi}{2}$, the curve is above the x-axis, and the heights of the rectangles are given by $\cos x$. Thus, the area under the curve between $x = 0$ and $x = \frac{\pi}{2}$ is equal to the integral

$$\int_{0}^{\pi/2} \cos x \, dx. \tag{15}$$

From $x = \frac{\pi}{2}$ to $x = \frac{3\pi}{2}$, the curve is below the x-axis and the heights of the rectangles are obtained by subtracting the y-coordinate of the bottom of the rectangle from the y-coordinate of the top of the rectangle, which is $0 - \cos x$. Therefore, the area bounded by the curve and the x-axis between $x = \frac{\pi}{2}$ and $x = \frac{3\pi}{2}$ is equal to the integral

$$\int_{\pi/2}^{3\pi/2} -\cos x \, dx. \tag{16}$$

From $x = \frac{3\pi}{2}$ to $x = 2\pi$, the curve is above the x-axis, the heights of the rectangles are given by $\cos x$, and the area under the curve is equal to the integral

$$\int_{3\pi/2}^{2\pi} \cos x \, dx. \tag{17}$$

The total area between the graph of $y = \cos x$ and the x-axis from $x = 0$ to $x = 2\pi$ is the sum of the integrals in (15), (16), and (17),

$$\int_{0}^{\pi/2} \cos x \, dx + \int_{\pi/2}^{3\pi/2} -\cos x \, dx + \int_{3\pi/2}^{2\pi} \cos x \, dx,$$

which has the value

$$\sin x \Big|_{0}^{\pi/2} - \sin x \Big|_{\pi/2}^{3\pi/2} + \sin x \Big|_{3\pi/2}^{2\pi}.$$

This expression can be written as

$$\left(\sin \frac{\pi}{2} - \sin 0\right) - \left(\sin \frac{3\pi}{2} - \sin \frac{\pi}{2}\right) + \left(\sin 2\pi - \sin \frac{3\pi}{2}\right),$$

which equals

$$(1 - 0) - (-1 - 1) + [0 - (-1)],$$

or 4. The total area bounded by $y = \cos x$ and the x-axis between 0 and 2π is 4 square units. Because of the symmetry of the graph of $y = \cos x$, the area in question could also have been found by multiplying $\int_{0}^{\pi/2} \cos x \, dx$ by four. ∎

Exercise Set 5.2

1. Show that the area under one arch of the curve $y = \sin x$ is 2.

2. Find the area between the curve $y = x^2 - 4$ and the x-axis from $x = -2$ to $x = 2$.

3. Find the area of the region bounded by the graphs of $f(x) = 2 - x^2$ and $g(x) = x$.

4. Find the area of the region bounded by the graphs of $y = x^2 + 2$, $y = -x$, $x = 0$ and $x = 1$.

5. Find the area of the region bounded by the graphs of $y = \sin x$ and $y = \cos x$ between $x = 0$ and $x = \frac{\pi}{4}$.

6. Calculate the area bounded by $y = e^{2x}$ and the x-axis between $x = 0$ and $x = 2$.

7. Find the total area bounded by $y = x^3 - 4x$ and the x-axis.

8. a. Find the area under the curve $y = \frac{1}{x}$ from $x = 1$ to $x = 3$.

   b. Find the area under the curve $y = \frac{1}{x}$ from $x = 1$ to $x = b$ where $b > 1$.

   c. Use your results from parts a and b to help you express $\int_1^b \frac{1}{x} dx$ in terms of b, assuming $b > 1$.

9. Evaluate $\int_{-1}^3 f(x)dx$ if

$$f(x) = \begin{cases} 3 - x & \text{if } x \leq 2, \\ \frac{x}{2} & \text{if } x > 2. \end{cases}$$

5.3 Displacement and Distance Traveled

Our work in Section 5.2 has shown that the value of the definite integral $\int_a^b H'(t)dt$ is equal to the area under the curve $y = H'(t)$ from $t = a$ to $t = b$ provided $H'(t) > 0$. Our work with Euler's method and the Fundamental Theorem has shown that this definite integral is also equal to the net change in $H(t)$ from $t = a$ to $t = b$. We now apply these interpretations to determine how far an object travels during a certain interval of time.

Suppose a projectile is shot vertically into the air and its height $h(t)$ in meters at time t in seconds is given by $h(t) = -5t^2 + 1000t$. The velocity for this object is therefore $v(t) = -10t + 1000$. Graphs of $h(t)$ and $v(t)$ are shown in Figure 5.7.

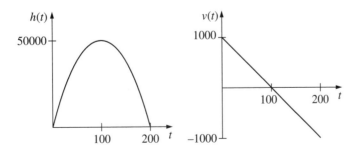

Figure 5.7 Graphs of height and velocity of projectile

Although the projectile is propelled straight up and falls straight down, the graph of height versus time shown in Figure 5.7 has the shape of a parabola. The vertex of the parabola, which corresponds to the highest point that the projectile reaches, has coordinates $(100, 50,000)$, which means that the maximum height of 50,000 m is reached after 100 seconds. At the instant when $t = 100$, the velocity is zero.

Since $h(200) = 0$, the projectile returns to the ground level after 200 seconds. What distance does the projectile travel during the 200 seconds it is in the air? The highest point the projectile reaches is 50,000 m, so the distance traveled by the projectile as it goes straight up and then falls straight down is 100,000 m. Note that we have solved this problem without calculus; however, not all distance problems can be solved so easily. We would like to see how calculus can be used in the solution so that we can then solve more complicated problems.

displacement

Because $v(t) = h'(t)$, the definite integral of the velocity function from $t = 0$ to $t = 200$ will give the projectile's net change in height, which is known as ***displacement***. We expect the displacement from $t = 0$ to $t = 200$ to be zero, since the projectile leaves the ground at time $t = 0$ and returns to the ground at time $t = 200$. Using a definite integral, we find the displacement is

$$\int_0^{200} (-10t + 1000)\,dt\,,$$

which is equal to

$$-5t^2 + 1000t \Big|_0^{200} = -5(200)^2 + 1000(200).$$

The value of this expression is 0, as we expected. The projectile's displacement from $t = 0$ to $t = 200$ is zero. This does not mean that the projectile is stationary during this time interval, but merely that its net change in height is zero.

In general, the integral of the velocity of an object from $t = a$ to $t = b$ gives the displacement of the object from $t = a$ to $t = b$. To determine the actual distance traveled rather than the displacement, we cannot merely evaluate a definite integral of velocity over the entire interval. We must determine distance traveled when the velocity is positive, and then separately determine the distance traveled when the velocity is negative.

What does this mean for the projectile whose height at time t is given by $h(t) = -5t^2 + 1000t$? Between $t = 0$ to $t = 100$ the projectile has positive velocity and is moving upward. The displacement from $t = 0$ to $t = 100$ is

$$\int_0^{100} (-10t + 1000)\,dt = \left(-5t^2 + 1000t\right)\Big|_0^{100} = 50,000.$$

Between $t = 100$ to $t = 200$ the projectile has negative velocity and is moving downward. The displacement from $t = 100$ to $t = 200$ is

$$\int_{100}^{200}(-10t+1000)dt = \left(-5t^2+1000t\right)\Big|_{100}^{200} = -50,000.$$

The negative displacement from $t = 100$ to $t = 200$ simply means that the height of the projectile at $t = 200$ is less than the height of the projectile at $t = 100$. The distance traveled over this interval of time is $|-50,000| = 50,000$. The total distance traveled between $t = 0$ to $t = 200$ is $50,000 + |-50,000| = 100,000$.

Note that the problem of finding the total distance traveled between $t = 0$ to $t = 200$ involves evaluating

$$\int_{0}^{100}(-10t+1000)dt + \left|\int_{100}^{200}(-10t+1000)dt\right|.$$

The same calculations would enable us to find the total area bounded by the graph of $v(t) = -10t + 1000$ and the t-axis between $t = 0$ to $t = 200$ shown in Figure 5.8.

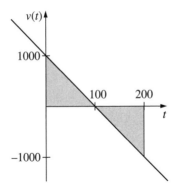

Figure 5.8 Area between $v(t) = -10t + 1000$ and the t-axis

We can now make the following observations about displacement and distance traveled:

–   Displacement is the difference between final position and initial position; we can find the displacement from $t = a$ to $t = b$ by evaluating $\int_{a}^{b}v(t)dt$.

–   When velocity is positive over an interval of time, the distance traveled and the displacement over the interval are the same.

–   Finding total distance traveled is equivalent to finding the total area bounded by the velocity curve and the horizontal axis over the interval.

Example

Suppose we begin recording the velocity of a hummingbird as it flies along the top of a fence. We model its velocity using the equation $v(t) = t^2 - 1$, where t represents time in seconds since we began watching the bird and velocity is measured in m/sec. Find the total distance the hummingbird travels from time $t = 0$ to $t = 4$.

Solution   A graph of the velocity function is shown in Figure 5.9. From the graph we see that the velocity is negative from $t = 0$ to $t = 1$ and positive from $t = 1$ to $t = 4$. To find the total distance the hummingbird travels, we need to find the displacements separately over each of these two intervals, then add the absolute values of the two displacements.

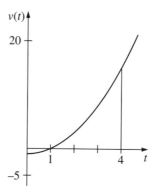

Figure 5.9 Graph of $v(t) = t^2 - 1$

The displacement when the velocity is negative is given by the definite integral

$$\int_0^1 (t^2 - 1)dt,$$

which has a value calculated as

$$\int_0^1 (t^2 - 1)dt = \left(\frac{1}{3}t^3 - t\right)\Big|_0^1 = \left(\frac{1}{3} - 1\right) - (0 - 0) = -\frac{2}{3}.$$

The displacement when the velocity is positive is given by the definite integral

$$\int_1^4 (t^2 - 1)dt,$$

which has a value calculated as

$$\int_1^4 (t^2 - 1)dt = \left(\frac{1}{3}t^3 - t\right)\Big|_1^4 = \left(\frac{64}{3} - 4\right) - \left(\frac{1}{3} - 1\right) = 18.$$

The distance traveled from $t = 0$ to $t = 1$ (when the velocity is negative) is $\left|-\frac{2}{3}\right| = \frac{2}{3}$ m. The distance traveled from $t = 1$ to $t = 4$ (when the velocity is positive) is 18 m. The total distance traveled is $\frac{56}{3}$ m. The value calculated for total distance traveled also represents the area bounded by the curve $v(t) = t^2 - 1$ and the horizontal axis between $t = 0$ to $t = 4$. Note that the definite integral $\int_0^4 (t^2 - 1)dt$ yields a positive number, $\frac{52}{3}$, but this number represents the displacement from $t = 0$ to $t = 4$, not the distance traveled. ∎

Exercise Set 5.3

1. In parts a–c, $v(t)$ represents the velocity (in meters/second) of a moving body as a function of the time t (in seconds). Sketch the graph of $v(t)$ versus t to find when the velocity is positive and when it is negative. Then find the displacement and the total distance traveled between the given values of t.

    a. $v(t) = t^2 - t - 2, \quad 0 \le t \le 3$

    b. $v(t) = t - \frac{8}{t}, \qquad 1 \le t \le 3$

    c. $v(t) = 2\cos 2t, \qquad 0 \le t \le \pi$

2. The function $a(t) = \sqrt{4t + 1}$ represents the acceleration (in m/s^2) of a moving body and $v_0 = -\frac{13}{3}$ is its velocity at $t = 0$. Find the distance traveled by the body between time $t = 0$ and $t = 2$.

3. Even though the graphs of $f(x) = x^4 - 2x^2 + 1$ and $g(x) = 1 - x^2$ intersect in three points, the area between the curves can be represented by a single definite integral. Explain why this is so and write an integral for this area.

4. The graphs of $f(x) = x^3$ and $g(x) = x$ intersect at three points, and the area between the curves *cannot* be represented by the single definite integral

    $$\int_{-1}^1 (x^3 - x)dx.$$

    Explain why this is so. Use symmetry to write a single integral that *does* represent the area.

5. An animal population is increasing at a rate of $r(t) = 200 + 50t$ animals per year (where t is measured in years). By how much does the animal population increase between the fourth and tenth years?

5.4   The Definite Integral Applied to Income Distribution

The distribution of income in our society is a concept of ongoing interest to economists, politicians, public policy analysts, and other concerned individuals. The data that economists use to quantify distribution of income is often presented in a form like Table 5.10, which in this case is constructed using data from 1992 (U.S. Bureau of the Census. Current population reports, series p60–p184. *Money income of households, families, and persons in the United States: 1992*, U.S. Government Printing Office, Washington, D.C., 1993).

Fifth of population	Percent of income
Lowest fifth	3.8
Second fifth	9.4
Third fifth	15.8
Fourth fifth	24.2
Highest fifth	46.9

Table 5.10 Percent distribution by household of aggregate income for 1992

Table 5.10 gives the percent of the total income of the United States earned by each fifth of the population, ordered by income. The procedure for determining the numbers in Table 5.10 can be thought as follows. Each family and each unattached individual counts as one household with one income level. Suppose all families and unattached individuals are lined up according to their earnings for the year, from least income to greatest income. Starting with those who earn the least, we count off one-fifth of the total number of households. All of the households in this group fall into the first (lowest) fifth in income level. We continue counting households in order of increasing income and divide the number of households into the second, third, fourth, and last (highest) fifth. The aggregate (total) income of each fifth, expressed as a percentage of the aggregate income of all households, gives the percentage distribution of aggregate income for each fifth.

The data in Table 5.10 mean that in 1992 the lowest fifth of all households earned 3.8% of all income for that year. The second fifth earned 9.4% of all income, whereas the highest fifth of all households earned 46.9% of all income.

The data in Table 5.10 can also be expressed cumulatively. To do so, we calculate the percent of income earned by all households in the lowest fifth, the lowest two-fifths, the lowest three-fifths, and the lowest four-fifths. The cumulative data can be obtained by adding successive entries in the right-hand column of Table 5.10, yielding the data shown in Table 5.11. Of course the lowest five-fifths of the population, which is all households, earns 100% of the income, although the numbers in Table 5.10 do not add to exactly 100% because of rounding.

Fifths of population	Percent of income
Lowest one-fifth	3.8
Lowest two-fifths	13.2
Lowest three-fifths	29.0
Lowest four-fifths	53.2
All fifths	100.0

Table 5.11 Cumulative percent distribution of aggregate income for 1992

A picture of the data in Table 5.11 is obtained by plotting the cumulative proportional distribution of aggregate income versus the proportion of the population, as shown in the graph given in Figure 5.12. The percentages have been converted to decimal numbers, that is, the lowest two-fifths earning 13.2% of the aggregate income is represented by the point (0.4, 0.132). The points (0, 0) and (1, 1) have been included because 0% of the households earn 0% of the income and 100% of the households earn 100% of the income.

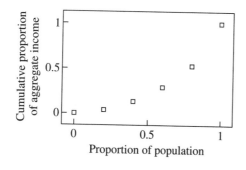

Figure 5.12 Graph of cumulative proportional distribution data

A curve that models cumulative data in both the dependent and independent variables as in Figure 5.12 is called a **Lorenz curve**. What would the Lorenz curve look like if income were distributed equally over the entire population? This means that each household earned the same amount, so each fifth of the population would have one-fifth of the income. The lowest fifth of the population has one-fifth of the income, the lowest two-fifths has two-fifths of the income, and so on. The corresponding points on a graph of cumulative proportional distribution lie along the line $y = x$, and

Lorenz curve

the Lorenz curve is this line. What would the Lorenz curve look like for perfect inequity of income distribution? This corresponds to a single household earning all the income in the society. A graph of the cumulative proportional distribution of aggregate income versus the proportion of the population has all the points on the horizontal axis from 0 to 1, except for the point corresponding to this single household, which is at (1, 1). Therefore, the Lorenz curve that models perfect inequity has a vertical coordinate of zero over the interval [0, 1], then makes a jump to the point (1, 1). These two situations are pictured in Figure 5.13.

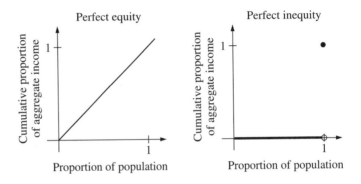

Figure 5.13 Lorenz curves for perfect equity and perfect inequity

Lorenz curves always lie somewhere between the two curves shown in Figure 5.13, since income distribution must fall somewhere between perfect equity and perfect inequity. Based on a graph of the Lorenz curve, we can make qualitative comparisons between one year and another regarding income distribution in a society. The closer a particular Lorenz curve is to the diagonal, the closer a society is to perfect equity of income distribution. The more bend there is in a Lorenz curve, the greater the inequity of income distribution. Although these qualitative comparisons are quite useful, we would like to develop a way to quantify the degree of inequity of income distribution in a society. For instance, given data about income distribution in two different years, we would like a numerical way to compare the extent to which one income distribution is more inequitable than another.

We start by finding the equation for the Lorenz curve based on the data from Table 5.11. A reasonable model for this data is a power function of the form $y = x^n$. The graph of this function contains the points (0, 0) and (1, 1) and, assuming $n > 1$, lies between the two Lorenz curves shown in Figure 5.13. We choose not to use a power least squares procedure to fit a power function to the data because a Lorenz curve *must* contain the point (1, 1). A power least squares curve has the form $y = ax^n$, and so it does not necessarily contain (1, 1). Instead, we will use the fact that a log-log re-expression linearizes data that is modeled by a power function. Since $y = x^n$, we take the logarithm of both sides of the equation to obtain

$$\ln y = \ln x^n,$$

which simplifies to

$$\ln y = n \ln x. \tag{18}$$

Equation (18) implies that the ordered pairs $(\ln x, \ln y)$ can be modeled by a line with slope n and intercept 0. We can solve equation (18) for n, which yields

$$n = \frac{\ln y}{\ln x}$$

We can obtain a reasonable estimate for n by forming the ratio $\frac{\ln y_i}{\ln x_i}$ for each point (x_i, y_i) in the data set and then averaging these ratios.

For the data given in Table 5.11, the average value of the ratios $\frac{\ln y_i}{\ln x_i}$ is approximately 2.38. We therefore choose the Lorenz curve with equation

$$y = x^{2.38}$$

as our model for the income distribution of Table 5.11.

What does a Lorenz curve with equation $y = x^{2.38}$ tell us about income distribution? How does it compare to other Lorenz curves such as $y = x^{2.8}$ or $y = x^{2.1}$? In general, the curve with equation $y = x^n$ is closer to the diagonal the closer n is to 1. This implies that x^n represents a more equitable income distribution the closer n is to 1. Conversely, as n increases, the distribution of income becomes more inequitable, approaching the case of perfect inequity displayed in Figure 5.13. So, we can compare one Lorenz curve to another by comparing the values of the exponents. We know that $y = x^{2.8}$ represents an income distribution that is less equitable than $y = x^{2.1}$, but how much less equitable is it? To answer this question, we wish to quantify the degree of inequity for a particular Lorenz curve.

Economists have derived a standard index of inequity of income distribution, called the ***Gini index***. The Gini index is based upon the area of the region bounded by the Lorenz curve and the line $y = x$, the line that corresponds to a perfectly equitable income distribution. In the case of perfect inequity of income distribution, the region bounded by the Lorenz curve and the line $y = x$ is a triangle with vertices $(0, 0)$, $(1, 0)$, and $(1, 1)$. This triangle has area 0.5. In general, the Gini index is the ratio of the area of the region bounded by the Lorenz curve and the $y = x$ line to the area of the corresponding triangle for perfect inequity of income distribution, that is,

Gini index

$$\text{Gini index} = \frac{\text{area bounded by Lorenz curve and } y = x}{\text{area of triangle with vertices } (0,0), \ (1,0), \text{ and } (1,1)}.$$

The numerator of the Gini index is always less than the denominator, so the Gini index always has a value between 0 and 1. Since the area of the triangle with vertices (0, 0), (1, 0) and (1, 1) is one-half, the Gini index is twice the area bounded by the Lorenz curve and $y = x$. A graphical representation of the Gini index is shown in Figure 5.14.

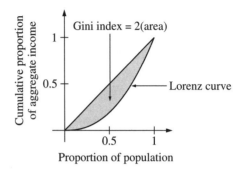

Figure 5.14 Graphical representation of Gini index

We can find the Gini index by using a definite integral to calculate the area of the region bounded by $y = x$ and $y = x^n$ (the Lorenz curve), and then multiplying the area by 2. For the year 1992, for which the Lorenz curve has power $n = 2.38$, the Gini index is approximately 0.41, which is calculated as follows:

$$\text{Gini index} = 2(\text{area between } y = x \text{ and } y = x^{2.38})$$

$$= 2\int_0^1 (x - x^{2.38})\,dx$$

$$= 2\left(\frac{1}{2}x^2 - \frac{1}{3.38}x^{3.38}\right)\Big|_0^1$$

$$= 2\left(\frac{1}{2} - \frac{1}{3.38}\right)$$

$$\approx 0.41.$$

Exercise Set 5.4

1. Determine the percent distribution of aggregate income by fifths (as in Table 5.10) for the following set of 25 incomes:

6,400	7,100	7,700	8,200	8,800
9,000	14,000	19,000	24,200	29,000
29,300	30,900	31,700	32,500	33,800
39,800	44,300	48,100	52,300	55,700
83,600	88,800	93,300	98,900	103,600

Your results should be similar to the data in Table 5.10.

2. Can a Lorenz curve applied to income distribution have an exponent less than 1? Why or why not?

3. As the exponent in the equation of the Lorenz curve approaches 1, what happens to the Gini index? What happens to the Gini index as the exponent approaches infinity?

4. Can different distributions of income have the same Gini index? Explain.

5. Determine a general expression for the Gini index in terms of n, the power in the equation for the Lorenz curve. Your expression should be based upon a definite integral used to represent area.

6. Most people would agree that perfect or close to perfect inequity of income distribution is not desirable in a just society. Depending on your political philosophy, however, perfect equity of income distribution may or may not be a desirable outcome in a society. Some would say that a society should allow economic inequity, perhaps as an incentive to work hard or to take financial risks. Consider, for example, the following income distribution: few people in society are poor, many people are in the middle class, and a few people are rich. This situation can be modeled with the triangular graph shown in Figure 5.15.

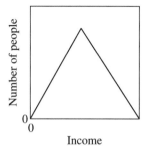

Figure 5.15 Distribution of income

    a.   Create a set of 25 incomes modeled by the graph shown in Figure 5.15.

    b.   Determine an equation for the Lorenz curve for your data.

    c.   Determine the Gini index for your data.

7.   The concept of a Gini index can be applied to situations other than income distribution. For example, one important aspect in determining the quality of life in a particular location is the population density. The population density, often measured in units such as people per square mile or 1000's of people per acre, is the average number of people that live on a given area of land. If the population density is 105 people per square mile, that does not mean that every square mile has 105 people living on it. A more informative measure might account for the distribution of the people on the land. A county with one large urban center surrounded by farmland may have the same population density as a county that is entirely a suburban area, but the population distributions will be very different.

    Consider a county divided into six townships with areas and populations given in Table 5.16. How has the population density of the county changed from 1980 to 1990? The total population has increased from 32,510 to 36,210, while the land area remains the same at 342.2 square miles. We can modify the concepts we used in the Gini index for the distribution of income to determine a Gini index for the distribution of people in the county. If the population were evenly distributed across the county, then 13% of the population would live on 13% of the land, 42% of the population on 42% of the land, and so on. If the population lives primarily in towns, then this distribution would be very uneven.

Township	1980 Pop. (1000s)	1990 Pop. (1000s)	Area (sq. miles)
River Walk	2.45	4.41	19.7
Rolesville	6.83	8.54	35.2
Grant	5.03	5.92	54.0
Hyatt	3.98	3.54	58.5
Burgess	6.78	7.10	64.4
West Hyatt	7.44	6.70	110.4
Total	32.51	36.21	342.2

Table 5.16 Township areas and populations for 1980 and 1990

Begin with the townships with the smallest areas and compute the cumulative proportional distributions of area and population for 1980 and 1990. Fit a Lorenz curve to the two cumulative data sets. For each year, determine the Gini index by calculating twice the area between the line $y = x$ and the Lorenz curve. Is the population becoming more unevenly distributed in this county? Explain how you know.

8. The locality of Yorkshire in England has three main townships: East Riding, which comprises 19% of the total land area of Yorkshire; North Riding, which comprises 35% of the land area; and West Riding, which comprises 46% of the total land area of Yorkshire. In 1801, the three townships had 13%, 18% and 69%, respectively, of the population. In 1961, they had 11%, 12%, and 77%, respectively, of the population. In 1961, was the population of Yorkshire more evenly distributed than in 1801? How do you know?

9. Research the population history of your state or region. Describe how the Gini index for population distribution for your state has changed since 1900.

Lab 10 *Gini Index*

The data in the following table is the income distribution by fifths in the United States for various years since 1935.

Percent Distribution by Household of Aggregate Income in the U.S.

Year	Lowest Fifth	Second Fifth	Third Fifth	Fourth Fifth	Highest Fifth
1935-36	4.1	9.2	14.1	20.9	51.7
1941	4.1	9.5	15.3	22.3	48.8
1946	5.0	11.1	16.0	21.8	46.1
1950	4.8	10.9	16.1	22.1	46.1
1955	4.8	11.3	16.4	22.3	45.2
1960	4.6	10.9	16.4	22.7	45.4
1964	4.2	10.6	16.4	23.2	45.5
1967	4.0	10.8	17.3	24.2	43.8
1970	4.1	10.8	17.4	24.5	43.3
1975	4.3	10.4	17.0	24.7	43.6
1979	4.1	10.2	16.8	24.7	44.2
1982	4.0	10.0	16.5	24.5	45.0
1985	3.9	9.8	16.2	24.4	45.6
1987	3.8	9.6	16.1	24.3	46.2
1989	3.8	9.5	15.8	24.0	46.8
1992	3.8	9.4	15.8	24.2	46.9

Reference

U.S. Bureau of the Census. Current population reports, series p60–p184. *Money income of households, families, and persons in the United States: 1992*, U.S. Government Printing Office, Washington, D.C., 1993.

1. Use the data in the table to find the cumulative percent distribution of aggregate income for each year.

2. For each year in the table, find the Lorenz curve $y = x^n$ that fits the data for proportion of population versus cumulative proportional distribution of aggregate income and calculate the Gini index.

3. Plot the Gini index against the year. What trends do you notice? To what historical events do these trends correspond? How reasonable is the Gini index as an indicator of the effects of public policy decisions? To what extent can Gini indices from different years be compared?

4. The data shown in the following table gives the distribution of household income for Colombia (1988), Poland (1987), and the United Kingdom (1979). Determine the Gini indexes for each country. How do these indexes compare to the Gini indexes in the appropriate year for the U.S.?

Fifth	Colombia	Poland	United Kingdom
Lowest fifth	4.0%	9.7%	5.8%
Second fifth	8.7	14.2	11.5
Third fifth	13.5	18.0	18.2
Fourth fifth	20.8	22.9	25.0
Highest fifth	53.0	35.2	39.5

Reference
World Development Indicators: 1992, Data on Diskette, World Bank, Washington, D.C., 1992.

Write a report summarizing your results, observations, and comments.

5.5   Antiderivatives and Indefinite Integrals

To use the Fundamental Theorem to evaluate $\int_a^b H'(x)dx$, we must know a function H that has the derivative H', so that

$$\int_a^b H'(x)dx = H(b) - H(a).$$

In general, therefore, to apply the Fundamental Theorem to an integral of the form $\int_a^b f(x)dx$, we must be able to find a function F whose derivative is f. The function F *antiderivative* is called an ***antiderivative*** of f. If we can find such a function F, then

$$\int_a^b f(x)dx = F(b) - F(a),$$

where $F'(x) = f(x)$.

We have been using the symbols $\frac{df}{dx}$ and $f'(x)$ to represent the derivative of f with respect to x. For cases where we want to find an antiderivative but do not want to evaluate it over an interval, it will be convenient to have a symbol for antiderivatives as well. We denote antiderivatives with the symbol

$$\int f(x)dx.$$

The symbol $\int f(x)dx$ is called the **indefinite integral** of f with respect to x. This indefinite integral of f represents a family of functions with each member having f as its derivative.   *indefinite integral*

An antiderivative of a function, if it exists, is not unique. Recall that the graph of a function can be shifted vertically and the derivative of the shifted function is the same as the derivative of the original function. In other words, adding a constant to a function does not change the derivative of that function. Therefore, knowing the derivative of a function enables us to determine the function only to within a constant. Thus, the notation $\int f(x)dx$ represents not a single function, but a *family* of functions in which the members differ by a constant. Each function in this family is an antiderivative of f, which means that each function has a derivative equal to f.

Why is the symbol for the definite integral $\int_a^b f(x)dx$ similar to the symbol for the indefinite integral $\int f(x)dx$? The former symbol represents a number that is the limit of a Riemann sum, while the latter symbol represents a family of functions. Even though there are fundamental differences between the definite and indefinite integral, they are closely related. The number represented by $\int_a^b f(x)dx$ can be found by evaluating an antiderivative of $f(x)$ at $x = a$ and at $x = b$ and subtracting. If we use an indefinite integral to represent the antiderivatives of the function f, then we can write

$$\int_a^b f(x)dx = \int f(x)dx\Big|_a^b.$$

In general, if $F(x)$ is an antiderivative of $f(x)$, then $F(x) + C$ represents a family of functions, each member of which is an antiderivative of $f(x)$. Using integral notation, we write

$$\int f(x)dx = F(x) + C,$$

where $F'(x) = f(x)$. The constant C is called the **constant of integration** for this antiderivative. In our previous work with area between curves, we found that introducing a constant of integration is not necessary when evaluating a definite integral.   *constant of integration*

In Chapter 1, we began this course by examining questions about change. Specifically, if we know how a phenomenon changes, how can we describe the phenomenon that has the observed rate of change? In Chapter 2, we then proceeded to investigate the reverse question; that is, given a function, how does it change? This led to expressions for the derivatives of functions. We now are discussing the problems we originally

integration

set out to solve; that is, we are finding functions from their derivatives, or in other words, finding antiderivatives. The process of finding antiderivatives is also called *integration*.

The operation of differentiation is symbolized by $\frac{d}{dx}[\ ]$, which tells us to differentiate with respect to x whatever expression is in the brackets. The operation of integration is symbolized by $\int [\ ]dx$, which tells us to integrate with respect to x whatever expression is in the brackets. With derivatives, the dx is necessary to indicate the variable with respect to which we are differentiating. Similarly, the dx is necessary in an indefinite integral to indicate the variable with respect to which we are integrating.

Example 1

Evaluate the indefinite integral $\int \frac{1}{\sqrt{x-1}}\,dx$.

Solution   We need to find a function such that its derivative with respect to x is equal to $\frac{1}{\sqrt{x-1}}$. The derivative of $\sqrt{x}$ is $\frac{1}{2\sqrt{x}}$, so the derivative of $2\sqrt{x}$ is $2\left(\frac{1}{2\sqrt{x}}\right)$, or $\frac{1}{\sqrt{x}}$. Shifting the function by one unit in the x direction also shifts the derivative. The derivative of $2\sqrt{x-1}$ is $\frac{1}{\sqrt{x-1}}$. The indefinite integral we are asked to evaluate represents the family of functions such that each member has $\frac{1}{\sqrt{x-1}}$ as its derivative, so

$$\int \frac{1}{\sqrt{x-1}}\,dx = 2\sqrt{x-1} + C.$$ ■

Example 2

Show that $\int \frac{1}{x}\,dx = \ln|x| + C$.

Solution   This equation is true if $\ln|x|$ is an antiderivative of $\frac{1}{x}$, so we must show that

$$\frac{d}{dx}\left[\ln|x|\right] = \frac{1}{x}.$$

If $x > 0$, then $\ln|x| = \ln x$ and $\frac{d}{dx}\left[\ln|x|\right] = \frac{d}{dx}\left[\ln x\right]$. We know that $\frac{d}{dx}\left[\ln x\right] = \frac{1}{x}$, so

$$\text{if } x > 0, \text{ then } \frac{d}{dx}\left[\ln|x|\right] = \frac{1}{x}.$$

If $x < 0$, then $\ln|x| = \ln(-x)$ and $\frac{d}{dx}\left[\ln|x|\right] = \frac{d}{dx}\left[\ln(-x)\right]$. By the chain rule, we know that $\frac{d}{dx}\left[\ln(-x)\right] = \frac{1}{-x}\cdot(-1)$, so $\frac{d}{dx}\left[\ln(-x)\right] = \frac{1}{x}$ for $x < 0$; thus,

$$\text{if } x < 0, \text{ then } \frac{d}{dx}\left[\ln|x|\right] = \frac{1}{x}.$$

We have shown that whether $x < 0$ or $x > 0$, the derivative with respect to x of $\ln|x|$ is equal to $\frac{1}{x}$; therefore, we have shown that

$$\int \frac{1}{x}\,dx = \ln|x| + C.$$

This result enables us to evaluate integrals that result in a natural logarithm without restricting the domain of the integrand to positive values. Note, however, that zero must still be considered and excluded from the domain. ∎

Example 3

Find the area under the graph of $y = \dfrac{2x+4}{x^2+4x+6}$ between $x = 1$ and $x = 7$.

Solution   The area we seek is shown in the graph in Figure 5.17. Since the y values are positive over the interval from $x = 1$ to $x = 7$, the area under the curve is given by the definite integral

$$\int_1^7 \frac{2x+4}{x^2+4x+6}\,dx.$$

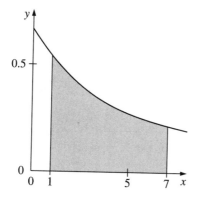

Figure 5.17 Area under the curve with equation $y = \dfrac{2x+4}{x^2+4x+6}$

To evaluate this integral using the Fundamental Theorem, we need to find an antiderivative of $\dfrac{2x+4}{x^2+4x+6}$. To find an antiderivative, we begin with the chain rule for derivatives, which states that

$$\frac{d}{dx}(f(g(x))) = f'(g(x)) \cdot g'(x). \tag{19}$$

If two quantities are equal, then their integrals are equal, so integrating both sides of equation (19) yields

$$\int \frac{d}{dx}(f(g(x)))dx = \int f'(g(x)) \cdot g'(x)dx. \tag{20}$$

The left side of equation (20) is an antiderivative of a derivative, which is just the original function plus a constant (to denote a family of functions). Substituting $f(g(x))$ for $\int \frac{d}{dx}(f(g(x)))dx$ in equation (20) and switching the left and right sides of the equation gives

$$\int f'(g(x)) \cdot g'(x)\, dx = f(g(x)) + C. \tag{21}$$

According to equation (21), the indefinite integral of a function of the form $f'(g(x)) \cdot g'(x)$ is simply $f(g(x)) + C$.

Does our function $\frac{2x+4}{x^2+4x+6}$ have the form $f'(g(x)) \cdot g'(x)$? Although the correspondence may not be obvious, the answer is yes. First of all, we need to think of $\frac{2x+4}{x^2+4x+6}$ as

$$\frac{1}{x^2+4x+6} \cdot (2x+4).$$

The fraction $\frac{1}{x^2+4x+6}$ can be considered $f'(x) = \frac{1}{x}$ evaluated at $g(x) = x^2 + 4x + 6$. Also, $2x+4$ can be considered $g'(x)$. Thus,

$$\frac{2x+4}{x^2+4x+6} = f'(g(x)) \cdot g'(x),$$

where $f'(x) = \frac{1}{x}$, $g(x) = x^2 + 4x + 6$, and $g'(x) = 2x+4$. Since $f'(x) = \frac{1}{x}$, we know from Example 2 that an antiderivative of f' is $f(x) = \ln|x|$. According to equation (21), an antiderivative of $(2x+4) \cdot \frac{1}{x^2+4x+6}$ is thus $f(g(x))$, which is $\ln|x^2 + 4x + 6|$. Since $x^2 + 4x + 6$ is always positive, we can drop the absolute value symbols, so that

$$\int \frac{2x+4}{x^2+4x+6}\, dx = \ln(x^2 + 4x + 6) + C.$$

We can check this result by taking the derivative of $\ln(x^2 + 4x + 6) + C$, which by the chain rule is $\frac{1}{x^2+4x+6} \cdot (2x+4)$. Notice how lucky we are that $2x+4$ is in the numerator of the integrand; otherwise, we would have had a difficult, if not impossible, time determining an antiderivative of this function.

The area shown in Figure 5.17 is equal to the value of the definite integral that is computed as follows:

$$\int_1^7 \frac{2x+4}{x^2+4x+6}\, dx = \ln(x^2 + 4x + 6) + C \Big|_1^7$$

$$= \ln(49 + 28 + 6) + C - \big(\ln(1 + 4 + 6) + C\big)$$

$$= \ln(49 + 28 + 6) - \ln(1 + 4 + 6)$$

$$= 2.02. \qquad \blacksquare$$

An alternate notation can be used to organize the process of evaluating

$$\int \frac{2x+4}{x^2+4x+6} dx.$$

If we define a new variable u that is equal to x^2+4x+6, then $\frac{du}{dx} = 2x+4$, and

$$\int \frac{2x+4}{x^2+4x+6} dx = \int \frac{1}{u} \cdot \frac{du}{dx} dx.$$

According to the chain rule, an antiderivative of $\frac{1}{u} \cdot \frac{du}{dx}$ with respect to x is $\ln|u|$. Since $u = x^2 + 4x + 6$, the final result can be expressed in terms of x as

$$\int \frac{2x+4}{x^2+4x+6} dx = \ln(x^2+4x+6) + C.$$

The technique of replacing an expression involving x with a new variable such as u in an integral is called **substitution of variables**.

substitution of variables

Example 4

Use substitution of variables to evaluate

$$\int \frac{x \ln(x^2+1)}{x^2+1} dx.$$

Solution   This integral is so complicated that we hope that a substitution of variables will simplify it. Let us try the substitution $u = \ln(x^2+1)$. The derivative of u with respect to x is given by

$$\frac{du}{dx} = \frac{1}{x^2+1} \cdot 2x.$$

Notice that for this choice of u, the original integrand is almost equal to $u \cdot \frac{du}{dx}$. Multiplying $\frac{du}{dx}$ by $\frac{1}{2}$ gives

$$\frac{1}{2} \cdot \frac{du}{dx} = \frac{1}{x^2+1} \cdot x,$$

which shows that the original integrand can be written as $\frac{1}{2} u \cdot \frac{du}{dx}$ if $u = \ln(x^2+1)$. Thus, the original integral is equivalent to

$$\int u \cdot \frac{1}{2} \cdot \frac{du}{dx} dx.$$

A property of integrals is that the integral of a constant times a function equals the constant times the integral of the function, which is written in symbols as

$$\int kf(x)dx = k \int f(x)dx.$$

(You will verify this in Exercise Set 5.5.) This means that we can write

$$\int u \cdot \frac{1}{2} \cdot \frac{du}{dx}\, dx = \frac{1}{2} \int u \cdot \frac{du}{dx}\, dx.$$

An antiderivative with respect to x of $u \cdot \frac{du}{dx}$ is $\frac{1}{2}u^2$, so

$$\frac{1}{2}\int u \cdot \frac{du}{dx}\, dx = \frac{1}{2}\left(\frac{1}{2}u^2 + C\right),$$

where C is a constant. The right side of this equation is equal to

$$\frac{1}{4}u^2 + K,$$

where the new constant K is equal to $\frac{1}{2}C$. Substitution of $u = \ln(x^2 + 1)$ gives the answer

$$\int \frac{x\ln(x^2 + 1)}{x^2 + 1}\, dx = \frac{\left[\ln(x^2 + 1)\right]^2}{4} + K. \qquad \blacksquare$$

Exercise Set 5.5

1.   Evaluate $\int (3x + 2)e^{3x^2 + 4x}\, dx$ by making the substitution $u = 3x^2 + 4x$.

2.   Evaluate $\int \frac{x+1}{x^2 + 2x - 3}\, dx$ by making the substitution $u = x^2 + 2x - 3$.

3.   Evaluate $\int \sqrt{1 + f(x)}\, f'(x)\, dx$ by making the substitution $u = 1 + f(x)$.

Evaluate each integral in problems 4–20.

4. $\displaystyle\int_0^{3\pi/10} \cos(-5x)\, dx$ 5. $\displaystyle\int (x + 1)e^{x^2 + 2x}\, dx$

6. $\displaystyle\int t(1 - t^2)\, dt$ 7. $\displaystyle\int xe^{x^2}\, dx$

8. $\displaystyle\int \cot x\, dx$ 9. $\displaystyle\int_0^5 \frac{dx}{\sqrt{1 + x}}$

10. $\displaystyle\int \sec^2 w\, dw$ 11. $\displaystyle\int x\sqrt{4 - x^2}\, dx$

12. $\displaystyle\int_{-\pi/4}^{\pi/4} \tan t \sec^2 t\, dt$ 13. $\displaystyle\int \frac{x^2}{e^{x^3}}\, dx$

14. $\displaystyle\int x^2 \cos(x^3)\, dx$ 15. $\displaystyle\int \tan^2 t\, dt$

16. $\displaystyle\int_{-5}^5 \frac{x}{x^2 + 1}\, dx$ 17. $\displaystyle\int \ln(\sin x)\frac{1}{\sin x}\cos x\, dx$

18. $\int \cos(2t)\sin(2t)dt$

19. $\int \cos t \sin(2t)dt$

20. $\int_{0.1}^{1} \frac{e^{\sqrt{x}}}{\sqrt{x}} dx$

21. $\int x^n dx$

22. Find the area under the arch of the graph of $y = x\sin(x^2)$ with left endpoint at the origin (that is, $0 \le x \le \sqrt{\pi}$).

23. The function $v = \frac{t}{t^2+10}$ represents the velocity (in m/sec) of a moving object. Find the total distance that the object travels between $t = 0$ and $t = 3$.

24. Use derivatives of both sides of the equation to show that

$$\int [f(x)+g(x)]dx = \int f(x)dx + \int g(x)dx,$$

which states that the integral of a sum is equal to the sum of the integrals.

25. Use derivatives of both sides of the equation to show that

$$\int kf(x)dx = k\int f(x)dx,$$

which states that the integral of a constant times a function is equal to the constant times the integral of the function.

5.6   Solving Differential Equations

Separation of Variables

In Chapter 1 and again in Chapter 4 we saw that a population experiences unconstrained exponential growth when the rate of change of the population is proportional to the existing population. This observation is represented by the differential equation

$$\frac{dP}{dt} = kP, \tag{22}$$

where k is a constant of proportionality. Based on our knowledge of exponential functions, we can guess that a function that satisfies differential equation (22) is $P = e^{kt}$. This guess is easy to check: if $P = e^{kt}$, then $\frac{dP}{dt} = ke^{kt}$, which is equivalent to $\frac{dP}{dt} = kP$. This confirms that a solution to the differential equation $\frac{dP}{dt} = kP$ is $P = e^{kt}$.

Recall from Chapter 4 that the general solution of a differential equation is a family of functions, each member of which satisfies the differential equation. For example, the general solution of $\frac{dy}{dx} = \cos x$ is $y = \sin x + C$ because any function with a derivative equal to $\cos x$ is a solution. In other words, any antiderivative of cos x is a solution. As we have seen in this chapter, the indefinite integral $\int \cos x \, dx$ represents the family of functions whose members are the antiderivatives of cos x and that can be written as sin $x + C$. The family of functions $y = \sin x + C$ therefore represents the general solution to the differential equation $\frac{dy}{dx} = \cos x$.

Can we write the general solution to differential equation (22) by adding a constant to a particular solution? If so, then the function $Q = e^{kt} + C$ would satisfy the differential equation $\frac{dQ}{dt} = kQ$. We can verify that $Q = e^{kt} + C$ is *not* a solution of the differential equation as follows. First find the derivative of Q with respect to t, which is $\frac{dQ}{dt} = ke^{kt}$. Second, find an expression for kQ, which is $k(e^{kt} + C)$. When we compare $\frac{dQ}{dt} = ke^{kt}$ with $kQ = ke^{kt} + kC$, we see that $\frac{dQ}{dt}$ does not equal kQ.

In Section 4.2, when we looked at the slope field for a differential equation of the form given in (22), we saw that tangent segments were parallel to other tangent segments with the same y-coordinates. This gave us solution curves that were horizontal shifts of each other. Changing the value of C in the function $Q = e^{kt} + C$ gives curves that are vertical shifts of each other; however, the slope field shows that the solution curves of (22) are not vertical shifts of each other. Therefore, we have shown both analytically and graphically that adding an arbitrary constant to a particular solution of (22) does not produce the general solution of (22).

To understand what happens to the constant of integration in the general solution of $\frac{dP}{dt} = kP$, we need to study an analytical technique for solving differential equations called ***separation of variables***. In this method, we try to separate the variables in a differential equation to opposite sides of the equal sign in the hope that the expressions on each side will be antiderivatives that we recognize.

separation of variables

We can rewrite the equation $\frac{dP}{dt} = kP$ in the form

$$\frac{1}{P} \cdot \frac{dP}{dt} = k,$$

so that all occurrences of P and its derivative are on the same side of the equal sign. Because the expressions on opposite sides of the equal sign are equal, their antiderivatives with respect to t differ at most by a constant and therefore the indefinite integrals are equal. Thus, we can write

$$\int \frac{1}{P} \cdot \frac{dP}{dt} \, dt = \int k \, dt. \tag{23}$$

The left side of equation (23) is equal to $\ln|P| + C_1$, since the derivative with respect to t of $\ln|P|$ is $\frac{1}{P} \cdot \frac{dP}{dt}$. The right side of equation (23) is equal to $kt + C_2$. Therefore, (23) can be written as

$$\ln|P| + C_1 = kt + C_2. \tag{24}$$

Since a constant minus a constant is a constant, the two constants of integration in equation (24) can be combined into a single constant, which gives us

$$\ln|P| = kt + C.$$

We can remove the logarithm on the left side by using the expression as an exponent on the base e, which is a process called **exponentiating**. Since $\ln|P| = kt + C$, we can write $e^{\ln|P|} = e^{kt+C}$. Thus,

exponentiating

$$|P| = e^{kt+C}$$
$$= e^C \cdot e^{kt},$$

or,

$$P = \pm e^C \cdot e^{kt}.$$

The coefficient $\pm e^C$ is a constant which we can relabel a, so that the solution is

$$P = a \cdot e^{kt}.$$

Thus, any function of the form $P = a \cdot e^{kt}$ satisfies the differential equation $\frac{dP}{dt} = kP$.

Example 1

In Section 4.2, we used slope fields to visualize solutions of differential equations. One of the solutions we investigated in Exercise Set 4.2 is the learning curve, which satisfies the differential equation

$$\frac{dy}{dt} = k(M - y),$$

where y is the amount learned at time t, k is a constant of proportionality, and M is the mastery level, or the maximum level of learning that can occur. Notice that when y is small, the growth rate is close to kM; however, as y increases toward M, the growth rate declines to 0. According to this analysis, the curve grows quickly for small y, but levels off as y approaches M. Solve the differential equation for this phenomenon to find an expression for y as a function of t.

Solution   First, separate the variables so the differential equation becomes

$$\frac{1}{M - y} \cdot \frac{dy}{dt} = k.$$

We then integrate both sides of the differential equation with respect to t, which gives

$$\int \frac{1}{M - y} \cdot \frac{dy}{dt}\, dt = \int k\, dt. \tag{25}$$

An antiderivative of $\frac{1}{M-y} \cdot \frac{dy}{dt}$ with respect to t is $-\ln|M - y|$, and an antiderivative of k with respect to t is kt. Thus, equation (25) becomes

$$-\ln|M - y| = kt + C,$$

so that

$$\ln|M - y| = -kt + C.$$

Notice that since the product of a constant and -1 is a constant, we simply redefined C to include the -1 coefficient. We next exponentiate both sides so that

$$|M - y| = e^{-kt+C},$$

or

$$M - y = a \cdot e^{-kt},$$

where a is a constant that replaces $\pm e^{C}$. We can now solve for y, which yields

$$y = M - a \cdot e^{-kt}.$$

The initial value of y, which occurs when $t = 0$, is $y_0 = M - a$. Since $a = M - y_0$, a represents the difference between the maximum amount learned and the initial amount learned. A graph of the solution function with $k = 0.2$, $M = 100$, and $a = 90$ (so that $y_0 = 10$) is shown in Figure 5.18. ■

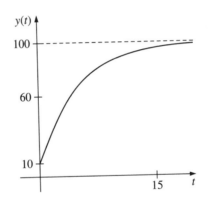

Figure 5.18 Graph of $y = 100 - 90e^{-0.2t}$

Exercise Set 5.6.A

1. Find the general solution of each differential equation.

    a. $\dfrac{dy}{dx} = x \cdot y^2$

    b. $\dfrac{dy}{dx} = \dfrac{x}{y}$

    c. $\dfrac{dy}{dx} = y \cdot \sin x$

    d. $\dfrac{dy}{dx} = \dfrac{x^2}{y^3}$

    e. $\dfrac{dy}{dx} = \dfrac{xy}{\ln y}$

    f. $\dfrac{dy}{dx} = x(1 + y)$

    g. $\dfrac{dy}{dx} = xy$

2. In Exercise Set 4.3, we introduced the differential equation for unconstrained population growth with immigration, which is

$$\frac{dP}{dt} = kP + m$$

where k is the natural growth rate of the population and m is the immigration rate.

a. Solve this differential equation for P as a function of t. (Hint: divide both sides of the equation by $kP + m$ to separate variables.)

b. How is the undetermined constant (not k or m) in your solution from part a related to the initial value of the population P_0?

c. Plot a graph of the solution with $k = 0.02$, $m = 25$, and $P_0 = 400$.

3. In Section 4.2, we introduced a differential equation for cooling

$$\frac{dT}{dt} = k(T - A),$$

where T is the temperature of the cooling substance at time t hours after cooling begins, k is a constant of proportionality, and A is the temperature of the surrounding environment (ambient temperature).

a. Solve this differential equation for T as a function of t.

b. How is the undetermined constant (not k or A) in your solution from part a related to the initial temperature T_0?

c. A cup of coffee is heated to $140°$ F and allowed to cool in a room in which the temperature is $70°$ F. After 30 minutes the temperature of the coffee is $100°$ F. Find an equation for the temperature of the coffee in terms of time. Plot a graph of temperature versus time.

4. For a student studying a particular subject, the number of facts F that have been memorized at time t can be modeled by the differential equation

$$\frac{dF}{dt} = a(M - F) - bF,$$

where M is the total number of facts to be memorized and a and b are positive constants. The first term in this differential equation has the same form as the differential equation for the learning curve. The second term is a forgetting term, and is based upon the assumption that facts are forgotten at a rate proportional to the number of facts that have been memorized.

a. Solve this differential equation for F in terms of t by separating variables.

b. Let $M = 100$, $F_0 = 10$, and $a = 0.2$. If $b = 0$, then the differential equation is identical to the learning curve differential equation in Example 1 with $k = 0.2$. Compare the learning curve graph with the graph from the forgetting model using the following values for b: 0.1, 0.2, and 0.3. How does increasing b affect the graph of the forgetting model? Why does this make sense? What happens to the graph for values of b that are much larger than a, such as 4 or 5 times a?

c. How is the undetermined constant (not M, a, or b) in your solution from part a related to the initial number of facts F_0?

d. In terms of the parameters M, a, and b, what level does the number of memorized facts approach in the long run?

5. During the pay-in phase of an annuity, you deposit the same amount of money each month at a fixed interest rate. The interest on the balance is compounded each month, just prior to the time at which the next deposit is made. Suppose you deposit $100 each month in an account that is paying 1% per month, compounded monthly, and that you do not withdraw the interest from the account. You continue with this plan for 20 years (240 months).

a. Write a recursive equation that will generate the amount in the account each month.

b. Write a differential equation that approximates the rate at which the account balance changes.

c. Solve the differential equation.

d. Compare the balance after 20 years which you would predict using your solution in part c with the balance found using the recursive equation in part a. Which is more realistic? Why would we ever use the solution of the differential equation rather than the value generated by the recursive equation?

Partial Fractions

In Chapter 4 we studied constrained exponential growth, also called logistic growth, which is described by the differential equation

$$\frac{dP}{dt} = k \cdot P \cdot \left(1 - \frac{P}{M}\right).$$ (26)

This equation describes a situation in which the rate of change of P with respect to t is proportional to both P and $\left(1 - \frac{P}{M}\right)$. This differential equation can be used to model the spread of a rumor. In this scenario, P represents the number of people who have heard the rumor at time t, and M is the total population available to hear the rumor. At first, as P increases, the rate $\frac{dP}{dt}$ also increases, which means that as more people hear the rumor it spreads at a faster rate. As P increases and approaches M, the rate $\frac{dP}{dt}$ decreases and approaches zero. This corresponds to the fact that if almost everyone has heard the rumor, then it spreads at a slower rate because there are few people left to hear it.

If we let $M = 1$ in equation (26), then P can be interpreted as the fraction of the total population that has heard the rumor at time t. If we separate the variables in equation (26) with $M = 1$, we get

$$\frac{1}{P(1-P)} \cdot \frac{dP}{dt} = k. \tag{27}$$

Unfortunately, the left side of equation (27) is not recognizable as an antiderivative of any function we have studied in this text. If we could express the fraction $\frac{1}{P(1-P)}$ as the sum of two fractions with denominators P and $(1 - P)$, then it may be possible to find an antiderivative of each of these fractions in terms of natural logarithms. The next example illustrates how this can be accomplished.

Example 2

Evaluate the indefinite integral $\int \frac{dx}{x^2 - 1}$.

Solution   We do not know a function whose derivative with respect to x is $\frac{1}{x^2 - 1}$. Although $f(x) = \ln(x^2 - 1)$ is a reasonable guess, $f'(x) = 2x \cdot \frac{1}{x^2 - 1}$, so f is not the function we are looking for. Notice that the denominator of the integrand, $x^2 - 1$, can be factored as

$$x^2 - 1 = (x + 1)(x - 1).$$

We will use this fact to help us write the fraction $\frac{1}{(x+1)(x-1)}$ as the sum of two other fractions $\frac{A}{x+1}$ and $\frac{B}{x-1}$, each of which has a natural logarithm function as an antiderivative. Doing so requires that we find constant values A and B so that $\frac{A}{x+1} + \frac{B}{x-1}$ is equal to $\frac{1}{x^2 - 1}$ as expressed in the equation

$$\frac{A}{x+1} + \frac{B}{x-1} = \frac{1}{x^2 - 1}. \tag{28}$$

Rewriting the left side of equation (28) with the common denominator $(x + 1)(x - 1)$ transforms the equation into

$$\frac{A(x-1) + B(x+1)}{x^2 - 1} = \frac{1}{x^2 - 1}. \tag{29}$$

For equation (29) to be true, the numerators must be equal for all values of x other than 1 and -1, so that

$$A(x - 1) + B(x + 1) = 1 \tag{30}$$

must be true for all x such that $x \neq 1$ and $x \neq -1$. If we take the limit of each side of equation (30) as x approaches 1, then

$$\lim_{x \to 1}[A(x - 1) + B(x + 1)] = \lim_{x \to 1} 1,$$

which is equivalent to $2B = 1$, so that $B = \frac{1}{2}$. Likewise, if we take the limit of each side of equation (30) as x approaches -1, then

$$\lim_{x \to -1}[A(x-1) + B(x+1)] = \lim_{x \to -1} 1,$$

which is equivalent to $-2A = 1$, so that $A = -\frac{1}{2}$. Substituting these values for A and B into equation (28) yields

$$\frac{-\frac{1}{2}}{x+1} + \frac{\frac{1}{2}}{x-1} = \frac{1}{x^2 - 1}.$$

Now we can rewrite the indefinite integral in a form we can evaluate, which is

$$\int \frac{1}{x^2 - 1}\, dx = \int \left[-\frac{1}{2}\left(\frac{1}{x+1}\right) + \frac{1}{2}\left(\frac{1}{x-1}\right) \right] dx. \tag{31}$$

Each term in the integrand on the right side of equation (31) is the derivative of a natural logarithm function, so that the integral equals

$$-\frac{1}{2}\ln|x+1| + \frac{1}{2}\ln|x-1| + C.$$

By using the division rule for logarithms, this answer also can be written as

$$\frac{1}{2}\ln\left|\frac{x-1}{x+1}\right| + C.$$

Therefore, we know that

$$\int \frac{1}{x^2 - 1}\, dx = \frac{1}{2}\ln\left|\frac{x-1}{x+1}\right| + C. \qquad\blacksquare$$

partial fractions

The technique used in Example 2 to write $\frac{1}{x^2 - 1}$ as $-\frac{1}{2}\left(\frac{1}{x+1}\right) + \frac{1}{2}\left(\frac{1}{x-1}\right)$ is called the method of ***partial fractions***. Example 2 prepares us to return to the differential equation for logistic growth, which we rewrote in the form

$$\frac{1}{P(1-P)} \cdot \frac{dP}{dt} = k. \tag{32}$$

(Recall that k is a constant, t is the independent variable, and P is the dependent variable.) To use the method of partial fractions, we must express $\frac{1}{P(1-P)}$ as a sum $\frac{A}{P} + \frac{B}{1-P}$, so we must find values for the constants A and B such that

$$\frac{A}{P} + \frac{B}{1-P} = \frac{1}{P(1-P)}. \tag{33}$$

As in Example 2, we find a common denominator for the left side of equation (33), which yields

$$\frac{A(1-P) + BP}{P(1-P)} = \frac{1}{P(1-P)},$$

which implies that

$$A(1-P)+BP=1. \tag{34}$$

Although P cannot equal 0 or 1 in equation (33), and thus in equation (34), we can equate the limits of both sides of equation (34) as P approaches 0 or 1. Taking the limit of each side of equation (34) as P approaches 0 yields $A=1$, and taking the limit of each side of equation (34) as P approaches 1 yields $B=1$.

Therefore, we can rewrite equation (33) as

$$\left(\frac{1}{P}+\frac{1}{1-P}\right)\cdot\frac{dP}{dt}=k. \tag{35}$$

Note that equation (32) involves a function of P for which we do not know an antiderivative, but equation (35) involves a sum of two functions of P, each of which has a natural logarithm function for an antiderivative. We write

$$\int\left(\frac{1}{P}+\frac{1}{1-P}\right)\cdot\frac{dP}{dt}\,dt=\int k\,dt$$

to symbolize integrating each side with respect to t. Evaluating these integrals gives us an equation with no derivatives, which is

$$\ln|P|-\ln|1-P|=kt+C. \tag{36}$$

Our task now is to solve equation (36) for P in terms of t. First we would like to remove the expressions involving P from the natural logarithms. To do so, we use the division rule for logarithms to write the equivalent equation

$$\ln\left|\frac{P}{1-P}\right|=kt+C.$$

Now we can exponentiate both sides of the equation so that

$$\left|\frac{P}{1-P}\right|=e^{kt+C},$$

or,

$$\frac{P}{1-P}=ae^{kt},$$

where $a=\pm e^C$. We next multiply both sides of the previous equation by $1-P$ to get

$$P=(1-P)\cdot ae^{kt}$$
$$=ae^{kt}-Pae^{kt}.$$

Collecting the terms containing P on one side of the equation gives us

$$P+Pae^{kt}=ae^{kt},$$

which can be factored as

$$P(1+ae^{kt})=ae^{kt}.$$

Solving for P results in the equation

$$P = \frac{ae^{kt}}{1 + ae^{kt}}. \tag{37}$$

Thus the function P with equation (37) is the general solution to differential equation (32). Multiplying both numerator and denominator of (37) by e^{-kt} allows us to write

$$P = \frac{a}{e^{-kt} + a} \tag{38}$$

as an equivalent form of the solution to (32) which is easier to use in some algebraic manipulations.

According to (38), the initial value of P (which occurs when $t = 0$) is

$$P_0 = \frac{a}{1 + a}. \tag{39}$$

We can solve equation (39) for a in terms of P_0. Multiplying both sides of (39) by $(1 + a)$ gives

$$(1 + a)P_0 = a,$$

or

$$P_0 + aP_0 = a.$$

Isolating the a terms on one side of the equation and factoring yields

$$P_0 = a - aP_0 = a(1 - P_0).$$

Thus we find that a is given by the equation

$$a = \frac{P_0}{1 - P_0}.$$

We can interpret a as the ratio of the proportion of the maximum population that you have initially to the proportion of the maximum population initially left to grow.

A graph of P versus t for equation (38) with $k = 0.05$ and $P_0 = 0.1$ (so that $a = \frac{0.1}{0.9} = \frac{1}{9}$) is shown in Figure 5.19. This curve has the S shape that is characteristic of logistic growth. In Lab 11 we will compare the analytic solution given in (38) with the numerical solution we will obtain by solving differential equation (32) using Euler's method.

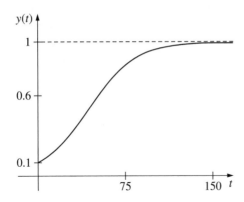

Figure 5.19 Graph of analytic solution of logistic differential equation

Exercise Set 5.6.B

1. Evaluate the following integrals using the method of partial fractions.

   a. $\int \dfrac{x+7}{x^2-x-6}\,dx$

   b. $\int \dfrac{2}{x^2-1}\,dx$

   c. $\int \dfrac{10x+3}{x^2+x}\,dx$

   d. $\int \dfrac{1}{2x^2+x}\,dx$

2. Solve the following differential equations using the method of partial fractions.

   a. $\dfrac{dP}{dt} = 0.1P\left(1-\dfrac{P}{100}\right)$

   b. $\dfrac{dP}{dt} = kP\left(1-\dfrac{P}{M}\right)$

3. The spread of a flu virus through a dormitory can be modeled by a logistic growth curve. Suppose a dorm has a total population of 500, and after a vacation 5 people return with the flu. Write a differential equation that describes the rate at which the flu spreads, assuming no one is immune. Solve the differential equation to find a function that gives s, the number of students who have had the flu, in terms of t, the number of days elapsed since the vacation. What additional information would you need to find the unknown constant in your function?

4. In a chemical reaction, substance A is transformed into substance B. The change in the concentration of B, $\dfrac{dx}{dt}$, is proportional to the concentration of substance A at that instant (concentration is measured in moles per liter). Suppose it takes 1 mole of A to form 1 mole of B. If the original concentration of substance A is α, then the concentration of A after x moles per liter of A have reacted is $\alpha - x$. Write a differential equation for the rate of change of the concentration of B with respect to time and find the general solution for this differential equation.

5. In a chemical reaction, two substances A and B combine to form substance C. The change in the concentration of C, $\frac{dx}{dt}$, is proportional to the product of the remaining concentrations of the two substances A and B (concentration is measured in moles per liter). Suppose it takes 1 mole of A and 2 moles of B to produce 1 mole of C. If the original concentration of substance A is a and the original concentration of substance B is b, write a differential equation for the rate of change of the concentration of C with respect to time and find the general solution for this differential equation.

6. The increase in the number of schools in which students use computers occurs at a rate that is proportional both to the number of schools currently using computers and to the number of schools that do not yet have computers. Accordingly, the percentage of schools with computers in the classroom has grown steadily over the years, but obviously this percentage can never exceed 100%. Write a differential equation for this phenomenon and then solve it. What information would you need to find the constant of proportionality?

7. Verify that the particular solution of the differential equation $\frac{dP}{dt} = kP\left(1 - \frac{P}{M}\right)$ that contains the point $(0, P_0)$ is

$$P = \frac{MP_0}{P_0 + (M - P_0)e^{-kt}}.$$

Lab 11 *The Discrete Logistic Model*

The purposes of this lab are to investigate the use of Euler's method to solve numerically the differential equation for logistic growth and to compare our results with the analytic solution to this differential equation.

The differential equation for logistic population growth is

$$\frac{dP}{dt} = kP\left(1 - \frac{P}{M}\right), \tag{40}$$

where P is the population at time t, k is a growth constant, and M is the limiting value of the population. We simplify equation (40) by letting $M = 1$. This changes the interpretation of P from the actual population to the *proportion* of the limiting value M of the population, and equation (40) becomes

$$\frac{dP}{dt} = kP(1 - P). \tag{41}$$

In the previous section and in problem 7 of Exercise Set 5.6.B, we found that solving (41) analytically using partial fractions yields

$$P(t) = \frac{P_0}{P_0 + (1 - P_0)e^{-kt}}. \tag{42}$$

Euler's method can be used to generate a numerical solution for (41) by iterating the equation

$$P_n = P_{n-1} + kP_{n-1}(1 - P_{n-1})\Delta t.$$

Defining $\Delta t = 1$ produces

$$P_n = P_{n-1} + kP_{n-1}(1 - P_{n-1}), \tag{43}$$

which can generate a sequence of P_n values that approximate the value of the population over time.

1. Generate a sequence of values for the discrete logistic model given by equation (43). Generate a sequence of values of time such that $t_n = t_{n-1} + 1$, since we are setting $\Delta t = 1$ in Euler's method. Assign the values $t_0 = 0$, $P_0 = 0.1$, and $k = 0.1$, and let n be the integer values from 1 to 50.

2. Graph the values of the continuous model $P(t)$ versus t using the same k and P_0 as in part 1. Graph the values of the discrete model P_n versus t_n on the same axes. Compare the graphs of the two models.

3. Increase the value of k by tenths. What is happening to the graphs? Increase k until reaching a value that causes the discrete model to oscillate between two different values. Such behavior is known as a **2-cycle**. For approximately what   *2-cycle* value of k does the discrete model first break into a 2-cycle? Above and below what value of the population does the cycle oscillate?

In part 3, we observed that the discrete model diverges from the analytic model as k increases, eventually breaking into a 2-cycle. The values of the discrete model initially increase toward 1, but for sufficiently large k, the values grow so quickly that they surpass 1. Once a population value is greater than 1, which is the maximum sustainable population, then population declines to a value below 1. Both of these behaviors, increasing population below 1 and decreasing population above 1, are caused by the product term $kP_{i-1}(1 - P_{i-1})$ in equation (43). For some values of k, the size of the oscillations about 1 decrease, and the values of the iterations approach 1. For other values of k, the oscillations do not vanish, but remain and settle into a 2-cycle. What is happening when the population is around 1 that causes a 2-cycle to occur for some values of k?

The values of the discrete model are obtained by iterating equation (43), which is equivalent to iterating the function

$$f(x) = x + kx(1 - x) \tag{44}$$

by letting $P_n = f(P_{n-1})$. If $P_n = 1$, then successive values of the iterations of f do not change because $f(1) = 1$. In other words, if the population is ever exactly 1, then it *fixed point*

will stay at exactly 1. The point with $x = 1$ is called a *fixed point* of the function f since the values of the iterations remain fixed at 1.

How is f changing at the fixed point $x = 1$? The derivative of f with respect to x is given by the equation

$$f'(x) = 1 + k - 2kx.$$

The value of the derivative at the fixed point $x = 1$ is $f'(1) = 1 + k - 2k(1) = 1 - k$.

4. What is the value of $f'(1)$ at the k value where you estimated in part 3 that the discrete model first yields iterations in the pattern of a 2-cycle? Why do you think this value of $f'(1)$ would cause the shift to a 2-cycle?

5. Find a value of k that yields a 4-cycle. Also try to find 8- and 16-cycles. You will need to increase k by increments smaller than one-tenth, otherwise you may pass over these cycles. You may also need to increase the number of iterations n so that you have enough iterations for the values of the population in the discrete model to settle into a pattern. In the continuum of k values, the places where the long-term behavior of the discrete model changes from a single limiting value to a 2-cycle, or from a 2-cycle to a 4-cycle, and so on, are called *bifurcation points*. The change of cycles from 1 to 2 to 4 to 8 and so on is called *period doubling*.

bifurcation points
period doubling

A bifurcation from a 2-cycle to a 4-cycle occurs for reasons similar to the reasons for the initial appearance of a 2-cycle. The points on a 2-cycle of the function f defined in equation (44) are actually fixed points of a function g defined as two iterations of f, so that $g(x) = f(f(x))$. The bifurcation of f from a 2-cycle to a 4-cycle is related to the derivative of g at its fixed points in the same way that the appearance of a 2-cycle of f is related to the derivative of f at its fixed point. We will not pursue this phenomenon further due to the difficulty of finding an analytic expression for the derivative of g at its fixed points.

Equation (42) gives an equation for a continuous function that comes from an analytic technique for finding an antiderivative. We can approximate values of this continuous function with a discrete set of values generated by equation (43), which comes from Euler's method. In reality, many populations do not change continuously, but change by discrete amounts at discrete times. A continuous model often gives reasonably accurate results; however, many situations dictate that we use a discrete model. The growth rates given by the k values used in this lab may seem unrealistically large in relation to human populations; however, these growth rates are not uncommon with other populations such as insects. In addition, the periodic and chaotic growth observed in the discrete model also occur in nature.

6. What characteristics of population growth exhibited in the discrete model are not present in the continuous model? What characteristic of Euler's method causes these differences between the discrete and continuous models? Discuss the strengths and the weaknesses of the discrete and continuous models.

Extensions

7. Not all k values lead to periodic behavior in the discrete model. Regions with no discernible pattern are classified as regions of **chaos**. One k value that leads to chaos is $k = 2.7$. Find several other k values that appear to produce chaotic iterations. What happened to the oscillations that are periodic with a period equal to 2^n, where n is a positive integer? We can gain insight into the answer to this question by observing what is happening to the spacing of the k values at which the bifurcations found in part 5 occur. How quickly with respect to k is the spacing decreasing? How could this account for the observation that no cycles with period 2^n occur for k values above a certain number?

chaos

8. Not all periodic behavior in the discrete model occurs with period 2^n. For example, period three behavior occurs for some values of k between 2.82 and 2.86. Find such a k value. Also find a k value in this interval that produces a 6-cycle. How does this relate to the period doubling observed in part 5?

9. In some situations, the limiting factor in logistic growth has a delayed effect, such as with environmental pollution; the impact of increased pollution usually is noticeable not immediately but after several years. Delay equations are difficult, if not impossible, to analyze using analytic techniques for continuous functions. On the other hand, delay equations can be investigated as discrete models. Suppose the limiting factor has a delay of 10 time units. This can be modeled by changing equation (43) to the piecewise-defined equation

$$P_n = \begin{cases} P_{n-1} + kP_{n-1} & \text{if } 1 \le n \le 10, \\ P_{n-1} + kP_{n-1}(1 - P_{n-11}) & \text{if } n \ge 11. \end{cases}$$

Starting with small values of k and gradually increasing k, compare the sequence of values produced by the delay equation for each k with the sequence of values produced by equation (43).

10. The various types of behavior observed in the discrete model for logistic growth occur with many other functions as well. Investigate iterations of the recursive equation $x_n = k \sin(x_{n-1})$ for k values between 1 and 3. Find k values that produce periodic behavior and chaotic behavior.

Write a report summarizing your results and explaining your method, observations, and conclusions.

5.7 Differentials and Tables of Integrals

In our study of calculus so far, we have learned two primary ways to solve differential equations.

– If we know $\frac{dy}{dx}$ and at least one point (x_0, y_0), we can build a sequence of approximate y values using Euler's method. This technique gives us information about the relationship between x and y, but it does not produce a closed form equation for y in terms of x.

– For certain types of differential equations, we can find an antiderivative analytically. In these cases, knowledge of $\frac{dy}{dx}$ allows us to write a closed form equation for y in terms of x.

The following examples illustrate three analytic techniques that we have developed in this chapter:

$$\frac{dy}{dx} = 2xe^{x^2} \qquad \Rightarrow \qquad y = e^{x^2} + C \qquad \text{(substitution of variables)}$$

$$\frac{dy}{dx} = xy \qquad \Rightarrow \qquad y = ae^{\frac{x^2}{2}} \qquad \text{(separation of variables)}$$

$$\frac{dy}{dx} = \frac{1}{x^2 - 1} \qquad \Rightarrow \qquad y = \frac{1}{2}\ln\left|\frac{x-1}{x+1}\right| + C \qquad \text{(partial fractions)}$$

One of the tools often used to find an antiderivative is an integral table, which contains formulas for indefinite integrals. One can spend much time finding integrals using various techniques. Tables and technology allow us to concentrate on the problems and the interpretation of their solutions rather than manipulating symbols. Of course, each of the integrals in a table were derived by mathematicians at some point, and the techniques they used are still useful.

A typical integral table is included in the appendix. This table includes many formulas that we already know, such as

$$\int e^u \, du = e^u + C.$$

Use of an integral table, however, often requires some manipulation of the integrand to make it match one of the formulas in the table. For example, earlier in this chapter we evaluated the integral

$$\int \frac{2x + 4}{x^2 + 4x + 6} \, dx$$

by letting $u = x^2 + 4x + 6$. Since $\frac{du}{dx} = 2x + 4$, we can construct a new integral in terms of u, which is

$$\int \frac{1}{u} \frac{du}{dx} \, dx. \tag{45}$$

The integral in (45) is often written simply as

$$\int \frac{1}{u}\,du\,.\tag{46}$$

The notation in (46) implies that we should integrate $\frac{1}{u}$ with respect to u, while (45) implies that we should integrate $\frac{1}{u}\frac{du}{dx}$ with respect to x. In both cases, the result of integration is equal to $\ln|u| + C$, which in terms of x is equal to $\ln\left|x^2 + 4x + 6\right| + C$.

Comparing (45) and (46) suggests that du was substituted for $\frac{du}{dx}\,dx$, as if dx represented a number that could be "canceled" by multiplication. Up until now, we have assumed that the notation $\frac{du}{dx}$ means the derivative of u with respect to x and that it does not represent a ratio of two numbers. Likewise, when dx is written next to an integral sign, we have interpreted it as an indicator of "with respect to x". Although we have not done so up to now, we also can think of du and dx as numbers, and this way of thinking can enhance our understanding of calculus.

A quantity such as du or dx is called a **differential.** To define du and dx as *differential* separate quantities where u is a function of x, we think of dx as a positive number that is smaller than any other positive number. If u is a function of x, $u(x)$, we define the differential of u with the equation

$$du = u'(x)dx.\tag{47}$$

Dividing both sides of the equation by the differential of x yields an equation with the ratio of two differentials

$$\frac{du}{dx} = u'(x).$$

Since by definition $u'(x) = \lim\limits_{\Delta x \to 0} \frac{\Delta u}{\Delta x}$, this ratio equals

$$\frac{du}{dx} = \lim\limits_{\Delta x \to 0} \frac{\Delta u}{\Delta x},$$

which is consistent with the notation we have used previously.

What does differential notation tell us? Assume $\frac{du}{dx}$ is the slope of a tangent line at a point with x-coordinate x_1, as indicated in Figure 5.20. Let Δx represent a small change in x from x_1 to x_2. Notice that near the point of tangency, if we let $dx = \Delta x$, then, since $du = u'(x)dx$, du is the vertical change in the tangent line over this interval and du approximates the actual change Δu in the function value over this same interval. The smaller the interval, the better the approximation, so we say

$$\frac{du}{dx} = \lim\limits_{\Delta x \to 0} \frac{\Delta u}{\Delta x}.$$

We therefore can think of the symbol $\frac{du}{dx}$ as the ratio of two differentials, which is

$$\frac{\text{small change in } u}{\text{small change in } x}.$$

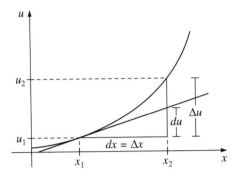

Figure 5.20 Differentials and deltas

Equation (47) will be useful to us in our work and can be applied to any function of a single variable. For example, if $v = f(t)$, then the differential of v is defined by $dv = f'(t)dt$, or $dv = \lim\limits_{\Delta t \to 0} \frac{\Delta v}{\Delta t}\, dt$. If this function has the equation $f(t) = t^2 - 7$, then $dv = 2t\, dt$.

We can now use the concept of differentials to evaluate the integral $\int \frac{2x+4}{x^2+4x+6}\, dx$ in terms of the variable u. As we look for an antiderivative, we substitute u for $x^2 + 4x + 6$. In differential notation, we write $du = (2x+4)dx$. The integral can be rewritten by substituting u for $x^2 + 4x + 6$ and du for $(2x+4)dx$, so that

$$\int \frac{2x+4}{x^2+4x+6}\, dx = \int \frac{1}{u}\, du,$$

which equals

$$\ln|u| + C.$$

Since we desire a final answer in terms of the original variable x, we can substitute $x^2 + 4x + 6$ for u to get

$$\int \frac{2x+4}{x^2+4x+6}\, dx = \ln\left|x^2 + 4x + 6\right| + C.$$

Example 1

Evaluate $\int (x^3 + 6x)^5 (6x^2 + 12)dx$.

Solution   The integrand has the structure (expression)^5 times something resembling the derivative of the expression in the parentheses. Let $u = x^3 + 6x$, so $\frac{du}{dx} = 3x^2 + 6$ and $du = (3x^2 + 6)dx$. The integrand contains u^5; note that $(6x^2 + 12)$ in the integral can be written as $2(3x^2 + 6)$. Rewriting the integral by substituting u for $x^3 + 6x$ and substituting $2\, du$ for $(6x^2 + 12)\, dx$ gives us

$$\int u^5 \cdot 2du.$$

An antiderivative with respect to u is $2 \cdot \frac{1}{6} u^6 + C$. Returning to our original variable x we have

$$\int (x^3 + 6x)^5 (6x^2 + 12)dx = \frac{1}{3}(x^3 + 6x)^6 + C.$$ ∎

Example 2

Evaluate the definite integral $\int_0^{\pi/2} \cos(3x)dx$.

Solution   We first find an antiderivative of the integrand. You may be able to do this without writing down any substitution; nevertheless, we will work through the substitution process to illustrate its effectiveness. We begin with $u = 3x$, which implies that $du = 3dx$. The original integral contains $\cos 3x$, for which we substitute $\cos u$. Since $du = 3dx$ we can substitute $\frac{1}{3} du$ for dx, so that

$$\int \cos(3x)dx = \int \cos u \cdot \frac{1}{3} du.$$

We know that $\int \frac{1}{3} \cos u \, du = \frac{1}{3} \sin u + C$, so that by substitution,

$$\int \cos(3x)dx = \frac{1}{3} \sin(3x) + C.$$

We can now evaluate the definite integral as follows:

$$\int_0^{\pi/2} \cos(3x)dx = \frac{1}{3} \sin(3x) \Big|_0^{\pi/2}$$

$$= \frac{1}{3} \left[\sin\left(\frac{3\pi}{2}\right) - \sin(0) \right]$$

$$= \frac{1}{3}.$$ ∎

The experience with differential notation provided by the preceding examples has prepared us to try a more difficult integral.

Example 3

Evaluate the definite integral $\int \frac{3}{\sqrt{5 - 9x^2}} dx$.

Solution   Suppose we let $u = 5 - 9x^2$. Then $\frac{du}{dx} = -18x$ and $du = -18x \, dx$. The equivalent of du, or $-18x \, dx$, does not appear in the integral, so this choice of substitution is not helpful. We might think of substituting $\frac{du}{-6x}$ for $3dx$ in the integrand, but then we would have an integral with a mix of u and x, and we would be worse off than when we started.

At this stage, we will consult a table of integrals. Notice that the integral looks similar to the general form $\int \frac{du}{\sqrt{a^2-u^2}}$, which appears as formula 16 in the table of integrals in the appendix. Our task is to make appropriate substitutions so that $\int \frac{3}{\sqrt{5-9x^2}}\,dx$ exactly matches $\int \frac{du}{\sqrt{a^2-u^2}}$

If we let $u = 3x$, then $\frac{du}{dx} = 3$ and $du = 3dx$. Substituting u for $3x$ and du for $3dx$ gives

$$\int \frac{3}{\sqrt{5-9x^2}}\,dx = \int \frac{1}{\sqrt{5-u^2}}\,du. \qquad (48)$$

This is $\int \frac{du}{\sqrt{a^2-u^2}}$ with $a = \sqrt{5}$. The table of integrals tells us that

$$\int \frac{1}{\sqrt{a^2-u^2}}\,du = \sin^{-1}\left(\frac{u}{a}\right) + C,$$

so we know that

$$\int \frac{1}{\sqrt{5-u^2}}\,du = \sin^{-1}\left(\frac{u}{\sqrt{5}}\right) + C. \qquad (49)$$

Combining equations (48) and (49) and substituting $3x$ for u yields

$$\int \frac{3}{\sqrt{5-9x^2}}\,dx = \sin^{-1}\left(\frac{3x}{\sqrt{5}}\right) + C. \qquad (50)$$

This result can be verified by differentiating the right side of (50) to check that it has a derivative with respect to x equal to the integrand of the left side of (50). ∎

Example 4

Evaluate $\int \sin^3 x\,dx$.

Solution   Suppose we let $u = \sin x$; then $\frac{du}{dx} = \cos(x)$ and $du = \cos x\,dx$. We can substitute u for $\sin x$, but the integral does not contain the $\cos x\,dx$ term that we need so that we can substitute du into the integral. Instead, notice that this integral has the form $\int \sin^n x\,dx$, where $n = 3$. Formula 23 in the integral table tells us that

$$\int \sin^n u\,du = -\frac{1}{n}\sin^{n-1} u \cos u + \frac{n-1}{n}\int \sin^{n-2} u\,du. \qquad (51)$$

Notice that the right side of equation (51) contains an indefinite integral, so this formula does not complete the job of evaluating the integral on the left side. The integral on the right side is simpler, however, than the one on the left side in terms of the power of the sine function. In general, equation (51) might be used several times to evaluate an integral of the form $\int \sin^n u\,du$, with each step yielding an integral with a lower power of n than in the previous step. An integral formula that yields a final answer by iterating an equation like (51) is called a **reduction formula**.

With $u = x$ and $du = dx$, and with $n = 3$, equation (51) becomes

$$\int \sin^3 x \, dx = -\frac{1}{3} \sin^{3-1} x \cos x + \frac{3-1}{3} \int \sin^{3-2} x \, dx,$$

or,

$$\int \sin^3 x \, dx = -\frac{1}{3} \sin^2 x \cos x + \frac{2}{3} \int \sin x \, dx. \tag{52}$$

Since $\int \sin x \, dx = -\cos x$, equation (52) is equivalent to

$$\int \sin^3 x \, dx = -\frac{1}{3} \sin^2 x \cos x - \frac{2}{3} \cos x + C,$$

and our work with this integral is complete. ∎

Exercise Set 5.7

Evaluate each integral. Use a table of integrals as needed.

1. $\displaystyle\int x^2 \sec^2(x^3) dx$

2. $\displaystyle\int \frac{4e^{1/x}}{x^2} dx$

3. $\displaystyle\int_{-3}^{5} \frac{e^x}{9 + 4e^{2x}} dx$

4. $\displaystyle\int \frac{3e^x}{\sqrt{1 - e^{2x}}} dx$

5. $\displaystyle\int \frac{1}{t\sqrt{4t^2 - 9}} dt$

6. $\displaystyle\int_0^{1.1} \frac{y}{\sqrt{16 - 9y^4}} dy$

7. $\displaystyle\int \frac{1}{9 - 16x^2} dx$

8. $\displaystyle\int \frac{1}{\sqrt{(x+1)^2 - 4}} dx$

9. $\displaystyle\int x^2 \sqrt{16x^2 + 25} \, dx$

10. $\displaystyle\int_0^{\pi/2} \sin(3x)\cos(5x) dx$

11. $\displaystyle\int \sin^2 x \, dx$

12. $\displaystyle\int \cos^3(2x) dx$

13. $\displaystyle\int \sin^4 x \, dx$

14. $\displaystyle\int \frac{4dx}{4x\sqrt{16x^2 - 4}}$

15. $\displaystyle\int \sin(4x)\sin(3x) dx$

16. $\displaystyle\int \frac{3\sqrt{9x^2 - 4}}{3x} dx$

17. $\displaystyle\int \frac{dx}{3x\sqrt{9x^4 - 256}}$

18. $\displaystyle\int_1^2 \frac{2\sqrt{9x^4 - 7}}{9x} dx$

19. $\displaystyle\int \frac{dx}{64x - x^3}$

20. $\displaystyle\int \frac{\cos(2x)}{[3 - 0.25\sin^2(2x)]^{3/2}} dx$

21. $\displaystyle\int \cos(3x^2)\sin(4x^2)2x\,dx$ 22. $\displaystyle\int \sin^3(49x)dx$

23. $\displaystyle\int \tan(3x)dx$ 24. $\displaystyle\int \frac{2x}{x^4\sqrt{36-x^4}}\,dx$

25. $\displaystyle\int_{4.5}^{8} \frac{x^2}{\sqrt{4x^2-64}}\,dx$ 26. $\displaystyle\int \frac{e^x}{16+2e^{2x}}\,dx$

27. Formula 14 in the table of integrals is

$$\int \sec u\,du = \ln|\sec u + \tan u| + C.$$

Multiply the integrand by 1 in the form of $\frac{\sec u+\tan u}{\sec u+\tan u}$, and then integrate to show that this formula is correct. A list of trigonometric identities is given in the appendix.

5.8 Numerical Methods of Integration

Even with the most clever substitutions, it may not be possible to evaluate an indefinite integral. For many functions, such as $f(x) = \sin(x^2)$ and $f(x) = e^{-x^2}$, we cannot find an antiderivative by analytical means. If we wish to evaluate a definite integral for such a function, we cannot use the Fundamental Theorem of Calculus to find the value of the integral. We can, however, use a numerical method to approximate the value of a definite integral without finding an antiderivative of the integrand. The methods introduced in this section are based upon interpreting a definite integral as the area bounded by the horizontal axis and the curve defined by the integrand.

Approximations with Rectangles

Recall that at the beginning of this chapter we defined a definite integral in terms of a Riemann sum, such as

$$\int_a^b h(t)dt = \lim_{\Delta t \to 0} \sum_{i=0}^{n-1} h(t_i) \cdot \Delta t,$$

where $t_0 = a$, $t_n = b$, and $\Delta t = \frac{b-a}{n}$. This definition says that the definite integral is equal to the limiting value of the sum as Δt approaches zero. As a consequence, if Δt is close to zero, the value of the sum will be close to the value of the definite integral, so we can write

$$\int_a^b h(t)dt \approx \sum_{i=0}^{n-1} h(t_i) \cdot \Delta t. \tag{53}$$

This means that if we cannot determine an antiderivative for $h(t)$, then we can use the sum to approximate the value of the definite integral. The approximation is closer to the exact value of the integral the closer Δt is to zero.

Figure 5.21 gives a geometric interpretation of the approximation in (53). Provided the integrand $h(t)$ has positive values between a and b, $\int_a^b h(t)dt$ represents the area under the graph of $h(t)$ between $t = a$ and $t = b$. Similarly, $\sum_{i=0}^{n-1} h(t_i) \cdot \Delta t$ represents the sum of the areas of n rectangles whose heights are $h(t_i)$ and whose widths are Δt. The sum of the areas of these rectangles approximates the actual area under the curve for small Δt.

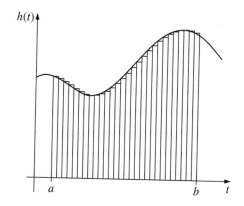

Figure 5.21 $\int_a^b h(t)dt$ approximated by $\sum_{i=0}^{n-1} h(t_i) \cdot \Delta t$

Example

Use a Riemann sum to approximate the value of $\int_1^2 3x^2 dx$.

Solution   We can find the exact value of this integral using the Fundamental Theorem and then compare this value with the Riemann sum approximations. The exact value is found by the calculations

$$\int_1^2 3x^2 dx = x^3 \Big|_1^2 = 2^3 - 1^3 = 7.$$

Using a Riemann sum, we know that

$$\int_1^2 f(x)dx \approx \sum_{i=0}^{n-1} f(x_i) \cdot \Delta x,$$

where $x_0 = 1$, $x_n = 2$, and $\Delta x = \frac{2-1}{n} = \frac{1}{n}$. If we use $n = 2$, then $\Delta x = 0.5$, and the values of the subscripted variables are $x_0 = 1$, $x_1 = 1.5$, and $x_2 = 2$. We have

$$
\begin{aligned}
\int_1^2 f(x)dx &\approx \sum_{i=0}^{1} f(x_i) \cdot \Delta x \\
&= f(x_0) \cdot 0.5 + f(x_1) \cdot 0.5 \\
&= f(1) \cdot 0.5 + f(1.5) \cdot 0.5 \\
&= 3 \cdot 0.5 + 6.75 \cdot 0.5 \\
&= 4.875.
\end{aligned}
$$

This means that $\int_1^2 3x^2 dx \approx 4.875$ using 2 rectangles, and this estimate is less than the actual value of 7. Figure 5.22 shows graphically why this approximation underestimates the area under the graph of $f(x)$.

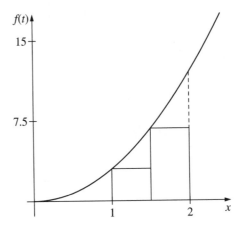

Figure 5.22 The value of $\int_1^2 3x^2 dx$ approximated by two rectangles

If we use $n = 4$, then $\Delta x = 0.25$, and the values of the subscripted variables are $x_0 = 1$, $x_1 = 1.25$, $x_2 = 1.5$, $x_3 = 1.75$, and $x_4 = 2$. We have

$$
\begin{aligned}
\int_1^2 f(x)dx &\approx \sum_{i=0}^{3} f(x_i) \cdot \Delta x \\
&= f(x_0) \cdot 0.25 + f(x_1) \cdot 0.25 + f(x_2) \cdot 0.25 + f(x_3) \cdot 0.25 \\
&\approx 3 \cdot 0.25 + 4.688 \cdot 0.25 + 6.75 \cdot 0.25 + 9.188 \cdot 0.25 \\
&\approx 5.907.
\end{aligned}
$$

This means that $\int_1^2 3x^2 dx \approx 5.907$ using 4 rectangles, as shown in Figure 5.23.

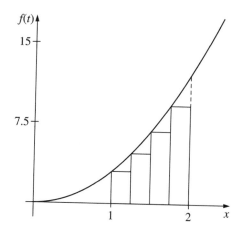

Figure 5.23 The value of $\int_1^2 3x^2 dx$ approximated by four rectangles

We can continue to increase n, which results in a smaller value for Δx. As Δx gets closer to zero, our approximation of the definite integral gets closer to the actual value of the integral. Using $n = 100$, we find that $\int_1^2 f(x)dx \approx \sum_{i=0}^{99} f(x_i) \cdot 0.01$, which is approximately 6.955. ∎

The integral we approximated with rectangles in the previous example has an integrand that is positive over the interval of integration. The summation formula for rectangles also approximates integrals in which the integrand has negative values over the interval of integration. In the exercise set at the end of this section, you will be asked to explain why you think this is true and to use technology to approximate values of definite integrals by using Riemann sums.

Approximations with Trapezoids

In the preceding example we approximated the area under a curve by subdividing the area into rectangles and then finding the sum of the rectangular areas. There are other numerical methods to approximate the area under a curve. For example, we can subdivide the area into trapezoids and then find the sum of these trapezoidal areas. The ***trapezoidal rule*** is based on the idea that the area under $f(x)$ from $x = a$ to $x = b$ is approximated by the area of the trapezoid whose boundaries are the secant line through $(a, f(a))$ and $(b, f(b))$, the x-axis, and the lines $x = a$ and $x = b$. In our exploration of the trapezoidal rule we will use the function $f(x) = 2x^3 - 3x + 3$ as an example. Since we can use the Fundamental Theorem to show that the value of the integral $\int_0^1 (2x^3 - 3x + 3)dx$ is 2, we can compare the trapezoidal approximations to the exact value of the integral.

trapezoidal rule

We begin by approximating the area under the curve with equation $f(x) = 2x^3 - 3x + 3$ using one trapezoid. The function $f(x) = 2x^3 - 3x + 3$ and a secant line segment from $(0, 3)$ to $(1, 2)$ are graphed in Figure 5.24. Notice that the bases of the trapezoid are the *vertical* line segments between the x-axis and the curve, and the height of the trapezoid is the distance between the vertical segments. Since the area of a trapezoid is given by

$$\frac{\text{(sum of bases)}}{2} \cdot \text{height,}$$

the approximation for $\int_0^1 (2x^3 - 3x + 3)dx$ illustrated in Figure 5.24 is given by

$$\frac{f(0) + f(1)}{2} \cdot 1 = \frac{3+2}{2} = 2.5.$$

We can see from the graph why this approximation overestimates the actual value of $\int_0^1 (2x^3 - 3x + 3)dx$.

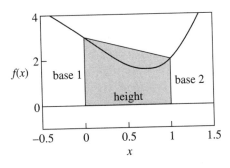

Figure 5.24 Using one trapezoid to approximate $\int_0^1 (2x^3 - 3x + 3)dx$

A better approximation can be found by using two trapezoids, as shown in Figure 5.25. This approximation yields

$$\frac{f(0) + f(0.5)}{2} \cdot 0.5 + \frac{f(0.5) + f(1)}{2} \cdot 0.5,$$

which has a value of

$$\frac{3 + 1.75}{2} \cdot 0.5 + \frac{1.75 + 2}{2} \cdot 0.5,$$

or 2.125. Figure 5.25 shows why a smaller value for Δx results in a better approximation for $\int_0^1 (2x^3 - 3x + 3)dx$.

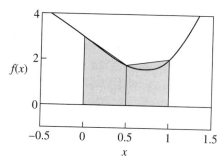

Figure 5.25 Using two trapezoids to approximate $\int_0^1 (2x^3 - 3x + 3)dx$

If four trapezoids are used, as shown in Figure 5.26, the value of the integral $\int_0^1 (2x^3 - 3x + 3)dx$ is approximated by

$$\frac{f(0)+f(0.25)}{2} \cdot 0.25 + \frac{f(0.25)+f(0.5)}{2} \cdot 0.25 + \frac{f(0.5)+f(0.75)}{2} \cdot 0.25 + \frac{f(0.75)+f(1)}{2} \cdot 0.25,$$

which has a value of approximately 2.031. As the number of trapezoids increases, the value of the approximation approaches the actual value of the integral. The approximation that comes from using 50 trapezoids is 2.0002.

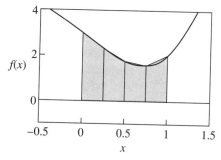

Figure 5.26 Using four trapezoids to approximate $\int_0^1 (2x^3 - 3x + 3)dx$

We can generalize the process used with our example to approximate the value of any definite integral using trapezoids, assuming the integrand is positive over the interval of integration. We know that the area of a trapezoid is the product of the height and the average of the bases. Each approximating trapezoid in Figure 5.26 has the same height, which we determined by dividing the length of the interval by the number of trapezoids. If we use n trapezoids over the interval from $x = a$ to $x = b$, then the height Δx of each trapezoid is

$$\Delta x = \frac{b-a}{n}.$$

The bases of each trapezoid are vertical segments between the x-axis and the curve. We let $x_0 = a$, $x_1 = x_0 + \Delta x$, $x_2 = x_1 + \Delta x$, and so on, so that in general $x_i = x_{i-1} + \Delta x$. The bases of the first trapezoid have lengths $f(x_0)$ and $f(x_1)$, the bases of the second trapezoid have lengths $f(x_1)$ and $f(x_2)$, and in general the ith trapezoid has bases with lengths $f(x_{i-1})$ and $f(x_i)$. The area of the ith trapezoid is the product of the average of the lengths of the bases and the height, which is

$$\frac{f(x_{i-1}) + f(x_i)}{2} \cdot \Delta x.$$

If n trapezoids are used to approximate $\int_a^b f(x)dx$, then the approximation is given by the sum of the areas of the n trapezoids, which is

$$\frac{f(x_0)+f(x_1)}{2} \cdot \Delta x + \frac{f(x_1)+f(x_2)}{2} \cdot \Delta x + \cdots + \frac{f(x_{n-2})+f(x_{n-1})}{2} \cdot \Delta x + \frac{f(x_{n-1})+f(x_n)}{2} \cdot \Delta x,$$

where $x_0 = a$ and $x_n = b$. Each term of this sum has a common factor $\frac{\Delta x}{2}$, so the sum can be written as

$$\left[f(x_0) + f(x_1) + f(x_1) + f(x_2) + \cdots + f(x_{n-2}) + f(x_{n-1}) + f(x_{n-1}) + f(x_n) \right] \cdot \frac{\Delta x}{2}.$$

Notice that each term in the sum appears twice except for $f(x_0)$ and $f(x_n)$, so we can simplify this sum by writing it as

$$\left[f(x_0) + 2f(x_1) + 2f(x_2) + \cdots + 2f(x_{n-1}) + f(x_n) \right] \cdot \frac{\Delta x}{2},$$

which in turn can be written as

$$\left[f(x_0) + f(x_n) + 2\sum_{i=1}^{n-1} f(x_i) \right] \frac{\Delta x}{2}.$$

We have shown that a definite integral can be approximated with trapezoids using the formula

$$\int_a^b f(x)dx \approx \left[f(x_0) + f(x_n) + 2\sum_{i=1}^{n-1} f(x_i) \right] \frac{\Delta x}{2},$$

trapezoidal rule

where $x_0 = a$, $x_n = b$, and $\Delta x = \frac{b-a}{n}$. This result is known as the **trapezoidal rule**.

Exercise Set 5.8

1. The examples in the previous section each involved estimating an integral in which the integrand is positive over the interval of integration. Why do the summation formulas for rectangles and trapezoids also work if the integrand has negative values over the interval of integration?

2. Use a calculator or computer to develop a numerical routine or spreadsheet to find the area under a curve $y = f(x)$ from $x = a$ to $x = b$ using the summation

$$\sum_{i=0}^{n-1} f(x_i) \cdot \Delta x.$$

Use your routine or spreadsheet to find the area under $y = \sin x$ between $x = 0$ and $x = \frac{\pi}{2}$. Compare your answer with the exact area found analytically.

3. When we estimate the area under the graph of $f(x)$ with the sum given in problem 2, we are using the function value at the left endpoint of each subinterval from x_{i-1} to x_i to determine the height of the ith rectangle. Modify your work in problem 2 to estimate area by calculating heights at the right endpoints of the rectangles. Modify your work again to calculate rectangle heights at midpoints. Compare your results for the integral $\int_1^5 x^3 dx$. Describe how the estimates with each method compare to each other as the number of rectangles increases.

4. Use a calculator or computer to develop a numerical routine or spreadsheet to estimate the area under a curve $y = f(x)$ from $x = a$ to $x = b$ using the trapezoidal rule. Use your routine or spreadsheet to estimate the area under $y = \sin x$ between $x = 0$ and $x = \frac{\pi}{2}$. Compare your answer with the exact area found analytically.

5. Use the trapezoidal rule to approximate the value of each integral to the nearest hundredth.

a. $\int_{-1}^{2} e^{-x^2} dx$ b. $\int_{1}^{5}(-x^2 + 7\ln x)dx$

c. $\int_{-1}^{5}(-3x^2 + 7x + 5)dx$ d. $\int_{0}^{\sqrt{\pi}} \sin(x^2) dx$

e. $\int_{0}^{300} 3xe^{-3x} dx$ f. $\int_{1}^{20} \frac{\ln x}{x} dx$

g. $\int_{0}^{1} \frac{1}{\sqrt{2\pi}} e^{-\frac{1}{2}x^2} dx$ h. $\int_{0}^{1} x \tan x \, dx$

6. Estimate the value of the integral of $f(x) = x^3$ from $x = 1$ to $x = 5$. Use the least number of rectangles with height calculated at the midpoint of each interval that are necessary for accuracy to the thousandths place. Also estimate the value of this integral using the least number of trapezoids necessary for the same accuracy. How does the number of rectangles compare with the number of trapezoids required to achieve three-place decimal accuracy for this integral?

5.9   Differential Equations and Data Analysis

A Model for a Projectile Traveling Along a Quadratic Path

The scatter plot and data listed in Figure 5.27 represent ordered pairs of height versus time for a projectile launched from ground level that experiences negligible air resistance.

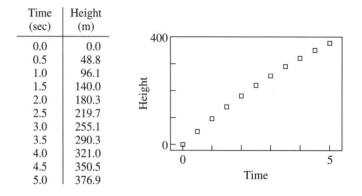

Time (sec)	Height (m)
0.0	0.0
0.5	48.8
1.0	96.1
1.5	140.0
2.0	180.3
2.5	219.7
3.0	255.1
3.5	290.3
4.0	321.0
4.5	350.5
5.0	376.9

Figure 5.27 Scatter plot and data for height of projectile versus time

If we ignore the effects of air resistance, then the only force acting on the projectile is the force of gravity. By Newton's second law of motion, force = mass × acceleration, and since the force of gravity is essentially constant near the surface of the earth, the projectile experiences constant acceleration. Based on our knowledge of derivatives, we know that constant acceleration implies that velocity is a linear function of time and that height is a quadratic function of time. The graph of the data seems consistent with a quadratic relationship. The quadratic function for height h in terms of time t can be expressed as

$$h = at^2 + bt + c. \tag{54}$$

where a is the acceleration due to gravity, b is the initial velocity, and c is the initial height. If the projectile's maximum height h_m occurs at time t_m, then equation (54) can be written in the form

$$h = a(t - t_m)^2 + h_m, \tag{55}$$

with the constant $a < 0$.

The values of a, b, and c in equation (54) can be found by using (a) quadratic least squares; (b) re-expression and linear least squares; or (c) calculus techniques. Most graphics calculators and data analysis software have an option to do quadratic least squares. If we wish to use linear least squares, we need to re-express the data to linearize it. Assuming that equation (55) is a good model for the data, we can linearize the data set by subtracting t_m from each t value, and then squaring the result. In other words, ordered pairs of the form $((t - t_m)^2, h)$ will be linear. A line fit to these ordered pairs will give us values for a and h_m in (55), and thus we can find a model for the original (time, height) data.

Our ability to carry out this planned re-expression depends, however, on knowing the value of t_m, or at least being able to guess its value. The data set in this example does not give a good indication of when the projectile reaches its maximum height. The techniques of calculus, however, do not require that we know the value of t_m. In this particular example, we can use quadratic least squares without linearizing the data; however, the techniques that follow are important for other problems where linearizing a data set is required.

The assumption that t and h are related by equation (55) implies that

$$\frac{dh}{dt} = 2a(t - t_m)$$
$$= 2at - 2at_m.$$

(56)

Differential equation (56) expresses a linear relationship between t and $\frac{dh}{dt}$. If we had values for the rate of change of h with respect to t, then we expect the ordered pairs $\left(t, \frac{dh}{dt}\right)$ to have a linear trend. The data set we have is the set of ordered pairs (t, h); nevertheless, we can use this data set to generate ordered pairs $\left(t, \frac{\Delta h}{\Delta t}\right)$. Values of the derivative $\frac{dh}{dt}$ are approximated by values of the difference quotients $\frac{\Delta h}{\Delta t}$, which are computed between consecutive data points by subtracting h's and t's and then dividing. Figure 5.28 gives the values of the difference quotients. Since each difference quotient approximates the derivative over an interval, we have calculated the midpoint of each time interval and have plotted the difference quotients versus these midpoint values. As we expected, this scatter plot has a linear trend.

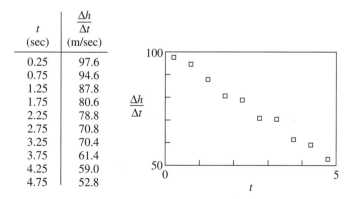

t (sec)	$\frac{\Delta h}{\Delta t}$ (m/sec)
0.25	97.6
0.75	94.6
1.25	87.8
1.75	80.6
2.25	78.8
2.75	70.8
3.25	70.4
3.75	61.4
4.25	59.0
4.75	52.8

Figure 5.28 Scatter plot and table of values for difference quotients versus time

The least squares line that fits the ordered pairs $\left(t, \frac{\Delta h}{\Delta t}\right)$ has the equation

$$\frac{\Delta h}{\Delta t} = -10.0t + 100.3.$$

Superimposing this line on the scatter plot from Figure 5.28 reveals that the residuals are relatively small and without a pattern, as shown in Figure 5.29. As a result, we know that the differential equation $\frac{dh}{dt} = -10.0t + 100.3$ models the ordered pairs $\left(t, \frac{\Delta h}{\Delta t}\right)$ obtained from the original data in Figure 5.27.

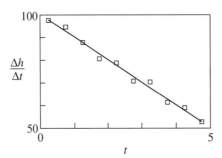 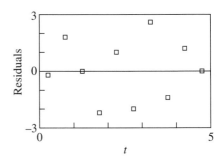

Figure 5.29 Least squares line and residual plot

The differential equation is solved for h in terms of t by integrating both sides of the equation with respect to t. This produces the equation

$$h = -5.0t^2 + 100.3t + C. \tag{57}$$

The constant of integration in equation (57) corresponds to the height of the projectile at time 0. Since the height is 0 at time 0, the value of C that best fits the data should be approximately 0. We can simply assign this value to C, or we can determine a best-fit value of C by some minimizing criterion, such as minimizing the sum of the absolute values of the residuals or minimizing the sum of the squares of the residuals. Using technology to experiment with various C values, we find with one decimal place accuracy that $C = 0.2$ minimizes the sum of the absolute values of the residuals. Using this criterion for our choice of C, we find that the quadratic function that best fits the data has equation

$$h = -5.0t^2 + 100.3t + 0.2.$$

Figure 5.30 shows this function graphed with a scatter plot of the original data and a residual plot. The solution presented for the height versus time data depends on the fact that a quadratic relationship between t and h implies a linear relationship between t and $\frac{dh}{dt}$.

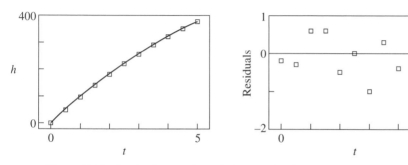

Figure 5.30 Graph of quadratic model with projectile data and residual plot

Using technology, a quadratic least squares curve can be fit to a data set, and the data set does not need to be linearized as it was in the previous example. Exponential and power least squares curves, on the other hand, require that we first shift the data into a standard position that depends on the type of curve. For example, a function of the form $y = ae^{kt}$ has a horizontal asymptote at $y = 0$; if the data we are modeling has an apparent asymptote $y = b (b \neq 0)$ then the data must be shifted vertically to give it an asymptote at $y = 0$. A least squares curve can then be fit to the shifted data by first linearizing the data, fitting a least squares line to the linearized data, and finally transforming the linear model into a model for the original data.

If our data is incomplete, so that we cannot shift the data into standard position, then we can use calculus techniques similar to those used in the preceding quadratic example. The following example introduces calculus techniques applied to an exponential model, and the lab concludes this example and also extends the techniques to a logistic model.

A Model for Cooling

Imagine that a cup of hot coffee is left in a room to cool. If we record data for the temperature and elapsed time as the coffee cools, we expect the data to have the shape of the curve in Figure 5.31. The coffee cools relatively quickly at first, then more slowly, and the coffee temperature eventually levels off at room temperature.

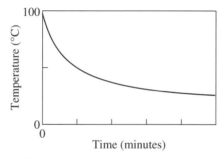

Figure 5.31 Shape of cooling curve

Data with the shape shown in Figure 5.31 can be modeled by an exponential function with a vertical shift. To determine the exponential curve that best fits the data, we first need to shift the data so it is asymptotic to the horizontal axis. An exponential least squares routine on a calculator or computer can then be used to fit an exponential least squares curve to the data. If such a routine is not available, the data can be re-expressed by taking the log of the temperature data. (We call this re-expression a *semi-log* transformation because we are taking the log of only one of the data coordinates.) This transformation will linearize the data. After fitting a least squares line to the re-expressed data, the linear model can be transformed to fit the original data.

How can we find a model if our data is incomplete, that is, if we do not know where the temperature levels off? If we do not know where the temperature levels off, then we do not know how much to shift the data. Many different shifts must be tried, and the shift that gives the best fit is used. This long and tedious process can be avoided, however, and in this section we outline a strategy that uses calculus to determine a model that fits the data.

Figure 5.31 suggests that the rate of cooling is most rapid when the difference between the temperature of the coffee and room temperature is greatest. As this difference decreases, so does the rate of cooling. As we noted in Section 4.2, the cooling behavior is modeled by the differential equation

$$\frac{dT}{dt} = k(T - A), \tag{58}$$

in which T is the temperature at time t, k is a negative constant, and A is the ambient temperature (room temperature). Differential equation (58) states that the rate of cooling is proportional to the difference between an object's temperature and the ambient temperature. This equation is a simplified form of Newton's law of cooling.

Differential equation (58) expresses $\frac{dT}{dt}$ as a linear function of T. We can compute the difference quotients $\frac{\Delta T}{\Delta t}$ from the original data and use these as approximations for $\frac{dT}{dt}$. Pairing $\frac{\Delta T}{\Delta t}$ with the dependent variable T from the original data set gives a set of ordered pairs that should appear linear when graphed. Using a least squares line fit to the ordered pairs $\left(T, \frac{\Delta T}{\Delta t}\right)$, we are able to determine a model for $\frac{dT}{dt}$, which is

$$\frac{dT}{dt} = mT + b. \tag{59}$$

We cannot solve differential equation (59) simply by integrating both sides of the equation because T is a function of t. However, we can use the technique of separation of variables to obtain an implicit model $T(t)$ for the temperature T as a function of time t. Solving this differential equation will introduce a constant of integration into the model $T(t)$. The value of the constant of integration is determined so that the model fits the original data as well as possible according to some criterion (such as minimizing the sum of the absolute values of the residuals).

Prior to this section, our data analysis techniques generally involved re-expression of the original data so that we could fit a least squares line to the transformed data. This linearization often required that we know some parameter of the model, such as the location of a vertex or an asymptote. This section has presented a technique for data analysis that we can use in situations where we do not have enough data to guess a reasonable value for one of these essential parameters.

We begin with a differential equation that we believe, in theory at least, governs the phenomenon for which we have data. We determine a linear model involving the derivative of the quantity we are measuring, use difference quotients to approximate the derivative, and thus linearize the data using difference quotients. The least squares line fit to the transformed data then yields values for the parameters in the differential equation model. After the differential equation is solved analytically, the constant of integration is determined to give the best fit of the model to the original data. In Lab 12 you will find models for two data sets using the techniques of this section.

Lab 12 *Difference Quotients and Data Analysis*

In this lab we will investigate Newton's model for cooling and the logistic model for population growth. The distinguishing feature of each part is that we have incomplete data. We can fit a curve to cooling data using semi-log re-expression if we know the ambient temperature. What if data are not recorded to the point where we can make a reasonable guess for the ambient temperature? Likewise, what if we do not know the limiting population for a logistic growth model? In both situations, modeling the derivative with difference quotients enables us to fit an appropriate curve to the data. We will fit a least squares line to difference quotients, and this will allow us to write a differential equation that represents the data. Then we can solve the differential equation to find a model for the original data set.

The first data set we will work with gives the temperature in degrees Celsius and elapsed time in seconds for a temperature probe that has been heated and then allowed to cool.

Time (t)	Temperature (T)
0	55.23
10	51.59
20	48.23
30	45.35
40	42.72
50	40.40
60	38.40
70	36.73
80	35.39

1. Enter the data on a spreadsheet or calculator and make a scatter plot of temperature versus elapsed time.

2. Using the difference quotients $\frac{\Delta T}{\Delta t}$ as approximations for the values of $\frac{dT}{dt}$, make a scatter plot of the difference quotients. This scatter plot should be linear based upon the differential equation for cooling, which is

$$\frac{dT}{dt} = k(T - A),$$

where T is the temperature at time t, A is the ambient temperature, and k is a constant of proportionality. Fit a least squares line to the scatter plot.

3. Use the slope and intercept from part 2 to write a differential equation for $\frac{dT}{dt}$.

4. Solve the differential equation in part 3 to find a model for the original data. Adjust the undetermined constant of integration so that the model fits the original data as closely as possible. Use some reasonable criterion to evaluate the fit, such as the sum of the absolute values of the residuals, or the sum of the squares of the residuals.

The second data set we will work with gives the number of cable television subscribers in the U.S. for certain years from 1970 to 1994. We assume that the data can be fit with a logistic growth curve that satisfies the differential equation

$$\frac{dP}{dt} = kP\left(1 - \frac{P}{M}\right),$$

where P is the number of cable television subscribers at time t, M represents the maximum attainable number of subscribers, and k is a constant of proportionality.

Year	Number of sub-scribers (millions)
1970	4.5
1975	9.8
1980	16.0
1984	29.0
1986	37.5
1988	44.0
1990	50.0
1992	53.0
1994	55.3

Reference
Television and Cable Fact Book No. 63, Series Volume p. I-76. Washington, D.C.: Warren Publishing, Inc. 1995.

The data we have does not allow us to predict easily a value for M. Since we do not know this value, we can use difference quotients to model the rate of change of population. The differential equation for logistic growth can be rewritten as

$$\frac{dP}{dt} = kP - \frac{kP^2}{M}$$

which means that the derivative $\frac{dP}{dt}$ is a quadratic function of P. There are several ways to linearize the ordered pairs $(P, \frac{dP}{dt})$. We will linearize the data based on the observation that $\frac{1}{P} \cdot \frac{dP}{dt}$ is linear in P since

$$\frac{1}{P} \cdot \frac{dP}{dt} = k\left(1 - \frac{P}{M}\right). \tag{60}$$

5. Enter the data on a spreadsheet or calculator and make a scatter plot of subscribers versus time, where t is number of years after 1970. That is, 1970 corresponds to $t = 0$.

6. Using the difference quotients $\frac{\Delta P}{\Delta t}$ as approximations for the values of $\frac{dP}{dt}$, make a scatter plot involving the difference quotients. This scatter plot should be linear based upon the re-expression suggested in (60). Fit a least squares line to the scatter plot.

7. Use your slope and intercept from part 6 to write a differential equation for $\frac{dP}{dt}$.

8. Solve the differential equation in part 7 to find a model for the original data. Adjust the undetermined constant of integration so that the model fits the original data as closely as possible.

Write a report summarizing your results and observations and evaluating your models.

Exercise Set 5.9

1. The following table gives the height (in feet) above the ground and elapsed time (in seconds) for a ball thrown off the top of a building. Use the techniques of this section to fit a least squares line to a transformation of this data set using difference quotients. Use this least squares line to fit a quadratic model to the original data.

Time	Height	Time	Height
0.2	96.1	2.7	90.4
0.6	99.4	3.2	82.5
0.9	100.8	3.6	73.3
1.4	101.2	4.0	63.6
2.1	97.4		

2. The following table gives the height (in meters) and elapsed time (in seconds) for the ascent phase of a small rocket. Use the techniques of this section to fit a least squares line to a transformation of this data set using difference quotients. Use this least squares line to fit a quadratic model to the original data.

Time	Height	Time	Height
0	0	2.5	219.7
0.5	48.8	3.0	255.1
1.0	96.1	3.5	290.3
1.5	140.0	4.0	321.0
2.0	180.3	4.5	350.5

3. Recall that in Lab 4, "Global Warming and the Chain Rule," we fit an exponential curve to the data for carbon dioxide concentration versus year. This curve includes a vertical shift to account for the ambient level of carbon dioxide in the atmosphere. Use the techniques of this section to fit an exponential curve to the data. What is the ambient level of carbon dioxide in the atmosphere?

4. Recall that in Lab 7, "Euler's Method," we used the differential equation for logistic growth with Euler's method to generate values that approximate the U.S. population from 1790-1910. Use the techniques of this section to fit a logistic curve to this data set. According to your model, what is the predicted upper limit for the U.S. population?

END OF CHAPTER EXERCISES

Review and Extensions

1. a. Evaluate $\int_0^3 x^2 \sqrt{x^3 + 9}\, dx$.

    b. Explain two ways of interpreting the integral in part a.

2. Two friends are driving across the country in an old automobile. Their odometer is broken and since they are short on funds, they decide not to get it fixed. Their speedometer still works, so while one is at the wheel the other records their speed periodically. They make the following observations.

time	10:00	10:30	10:50	11:30	12:00
speed	56 mph	63 mph	72 mph	40 mph	55 mph

    a. How far do you think they drove between 10:00 and 12:00? How much confidence do you have in your answer?

    b. What would you need to know to improve your answer to part a?

3. Use integrals to represent the area between $f(x) = \sin x + 1$ and $g(x) = \cos x$ on the interval $0 \le x \le 2\pi$.

4. Suppose a deer population grows exponentially at a rate of 10% per year. To keep the population size reasonable, the Park Service removes 20 deer each year.

    a. Write a differential equation for the rate of change of the size of the deer population with respect to time.

    b. Solve the differential equation.

5. Evaluate each integral. Use a table of integrals where appropriate.

    a. $\int \sin^6 x \cos x\, dx$

    b. $\int \sec x \tan x\, dx$

    c. $\int \dfrac{x^4 - 3x^2 + 8}{x}\, dx$

    d. $\int \dfrac{x-1}{x^2 - 7x + 10}\, dx$

    e. $\int \dfrac{\ln(2x)}{x}\, dx$

    f. $\int x \sin^3(x^2 + 7)\, dx$

    g. $\int x^2 \sqrt{9 - 4x^2}\, dx$

6. Solve the differential equation $\dfrac{dy}{dt} = \dfrac{(3 - y)\ln t}{5t}$.

7. The velocity of a particle moving along a straight line is given by $v(t) = 4\cos(2t)$ for $0 \le t \le \pi$. Find the displacement and the distance traveled by the particle on the interval $0 \le t \le \pi$.

8. a. Use 4 rectangles ($\Delta x = 0.5$) to approximate the area under $y = 2e^{-x^2}$ from 0 to 2. Use right endpoints to evaluate rectangle heights.

    b. Use 4 trapezoids ($\Delta x = 0.5$) to approximate the area under $y = 2e^{-x^2}$ from 0 to 2.

CHAPTER 6

Applications of the Definite Integral

In the previous chapter, we defined the definite integral of a function f over the interval from a to b as

$$\int_a^b f(x)dx = \lim_{\Delta x \to 0} \sum_{i=0}^{n-1} f(x_i) \cdot \Delta x$$

where $n = \frac{b-a}{\Delta x}$ and $x_i = a + i \cdot \Delta x$. In this chapter, we will use two important ideas in calculus that follow from this definition: the interpretation of definite integrals as areas and the relationship between definite integrals and infinite sums that represent quantities other than area. We will use these two ideas to develop a number of different applications. Through these applications, we will see how different situations can lead to similar integrals.

6.1 Geometric Probability

Suppose a dart is thrown at the target in Figure 6.1 so that it lands at a random point on the target—that is, the dart is equally likely to land on any point on the target. What is the probability that the dart lands in the shaded region that covers one-third of the target?

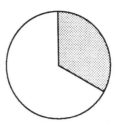

Figure 6.1 Partially shaded target

sample space
event space

probability of success

geometric probability

The region that represents all possible points where the dart can land is called the *sample space*. In this example, the sample space consists of the entire circular region of the target. The region that represents all successful outcomes is called the *event space*. In this example, the event space is the shaded region of the target. The *probability of success* is the ratio of the area of the event space to the area of the sample space. In this example, since one-third of the target is shaded, we say that the probability of hitting the shaded region on any one throw is $\frac{1}{3}$. This probability implies that if many darts land at random points on this target, we expect about one-third of them to land in the shaded region. Using ratios of areas to determine probabilities is what we mean by *geometric probability*.

Example 1

At a county fair, a game is played by tossing a coin onto a table with a grid of congruent squares superimposed on the top of the table, as pictured in Figure 6.2. If the coin lands entirely within some square and does not touch any edges, the player wins a prize. If the coin touches or crosses the edge of any square, the player wins nothing. Suppose the squares are S units on a side and the coin has a radius of R units. What is the probability of winning a prize on one toss?

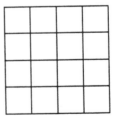

Figure 6.2 County fair game board

Solution   Before attempting to find a general solution, we will look at a few specific examples. Let the radius of the coin be 10 mm and the side of the square be 25 mm. What geometric shapes represent the sample and event spaces? The key to this problem is to think about where the center of the coin lands. If the distance from the center of the coin to a grid line is less than or equal to 10 mm, the coin will touch or cross a side of a square. So the center of the coin must be more than 10 mm from each side for the player to win a prize (see Figure 6.3).

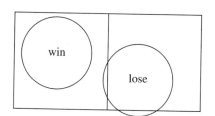 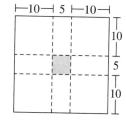

Figure 6.3 Winning and losing tosses and a geometric model for a winning toss

Throwing a coin at random corresponds to picking a point at random within a square, with the point representing the place where the center of the coin lands. A prize-winning throw corresponds to choosing that random point inside a 5×5 mm square in the center of the larger square. Speaking in terms of geometric probability, the sample space is the square with side 25 mm, and the event space is the square with side 5 mm, as shown in Figure 6.3. The ratio of these two areas is $\frac{25}{625} = \frac{1}{25}$. The probability of a prize-winning toss in this particular game is $\frac{1}{25}$.

If we change the side of the square to 20 mm while the radius of the coin remains 10 mm, what is the probability of winning? The diameter of the coin is equal to the length of a side of the squares, so every toss must touch or cross a grid line. The event space has an area of 0, so the probability of winning under these conditions is 0.

Now that we have looked at a couple of specific examples, we are ready to consider the general problem: a square of side S and a coin of radius R. To win, the center of the coin must be more than R units from the side. The area of the sample space is S^2, and the area of the event space is $(S-2R)^2$, as shown in Figure 6.4. The probability of a prize-winning toss is

$$P = \frac{(S-2R)^2}{S^2},$$

provided that R is less than $\frac{S}{2}$. If R is greater than or equal to $\frac{S}{2}$, then the diameter of the coin is greater than or equal to the length of the side of a square, and the probability of winning is 0. ■

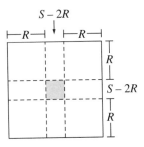

Figure 6.4 Sample space and event space

Example 2

Suppose we generate two random numbers, each between 0 and 1. What is the probability that their sum is less than one-half?

Solution First, we need to know the sample space. If we let x and y represent the two random numbers, then choosing two random numbers between zero and one is equivalent to choosing a random point in the unit square pictured in Figure 6.5. The region inside this square is the sample space. A one-to-one correspondence exists between each pair of numbers (x, y) and the points inside the square.

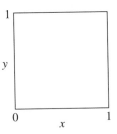

Figure 6.5 Sample space for Example 2

Second, we want to find a representation for the event space. We want to know the probability that the sum of the two numbers is less than one-half. In the language of algebra, we want to know the probability that $x + y < \frac{1}{2}$, which is equivalent to the inequality $y < -x + \frac{1}{2}$. The ordered pairs in the sample space that satisfy this inequality lie below the line $y = -x + \frac{1}{2}$. The event space, then, is the shaded region shown in Figure 6.6.

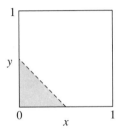

Figure 6.6 Event space for Example 2

The probability that the sum of two random numbers, each between zero and one, is less than one-half is the ratio of the area of the event space to the area of the sample space. The area of the sample space is 1, and the area of the event space is $\frac{1}{8}$, so the probability we seek is $\frac{1}{8}$. In practical terms, this means that if we generate 1000 pairs of random numbers between zero and one, we would expect about 125 of them to have a sum less than one-half. ■

Example 3

If we generate two random numbers between 0 and 1, what is the probability that their product is less than one-half?

Solution   Let x and y represent the two random numbers. The sample space is the same as in Example 2, a square with sides from 0 to 1 on the x- and y-axes. The event space is defined by $xy < \frac{1}{2}$, which is equivalent to $y < \frac{1}{2x}$. In the xy-plane, the event space is the region inside the sample space that is below or to the left of the curve $y = \frac{1}{2x}$. The sample space and event space are shown in Figure 6.7.

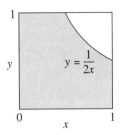

Figure 6.7 Sample space and event space for Example 3

The probability that the product of the two numbers is less than one-half is the ratio of the area of the event space to the area of the sample space. Since the area of the sample space is 1, the probability we seek is just the area of the event space. The area of the event space can be broken into two parts, the region beneath the portion of the curve $y = \frac{1}{2x}$ that shows in the event space and the rectangle to the left of but not beneath the curve. The shading in Figure 6.8 illustrates this division.

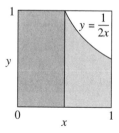

Figure 6.8 Divisions of event space for Example 3

The curve $y = \frac{1}{2x}$ intersects the boundary of the sample space where $x = 1$ and where $y = 1$; therefore, the two points of intersection of the curve with the sample space boundary are $(1, 0.5)$ and $(0.5, 1)$. The rectangle to the left of the curve has height 1 and width $\frac{1}{2}$, so its area is $\frac{1}{2}$. The curve within the sample space extends from $x = \frac{1}{2}$ to $x = 1$, so the area under the curve is equal to the value of the definite integral

$$\int_{1/2}^{1} \frac{1}{2x}\, dx.$$

The value of this integral is calculated as follows:

$$\int_{1/2}^{1} \frac{1}{2x}\, dx = \frac{1}{2} \int_{1/2}^{1} \frac{1}{x}\, dx$$
$$= \frac{1}{2} \ln|x| \Big|_{1/2}^{1}$$
$$= \frac{1}{2} \left[\ln(1) - \ln\left(\frac{1}{2}\right) \right]$$
$$\approx 0.347.$$

Adding the area under the curve to the area of the rectangle on the left yields a total area of about 0.847, which is the probability that the product of two randomly chosen numbers, each between 0 and 1, is less than $\frac{1}{2}$. If we generate 1000 pairs of random numbers between 0 and 1, we would expect about 847 of the pairs to have a product less than $\frac{1}{2}$. ∎

Exercise Set 6.1

1. In Example 1, suppose the square is 50 mm on a side. What radius of the coin provides the player with a 50% chance of winning?

2. In Example 1, under what conditions on S and R is the player guaranteed to lose the game? Under what conditions would the player be guaranteed to win?

3. Two real numbers are chosen at random between 0 and 10. What is the probability that their sum is less than 5? is more than 10?

4. Two real numbers, both between 0 and 2, are picked at random. What is the probability that their product is greater than 1?

5. Two real numbers between –2 and 2 are chosen at random. What is the probability that the sum of their squares is greater than 1?

6. If two numbers x and y are generated at random so that $0 < x < 1$ and $0 < y < 1$, determine the probability of each of the following events.

    a. $0.1 < xy < 0.2$
    b. $0.01 < xy < 0.02$
    c. $0.001 < xy < 0.002$
    d. the first nonzero digit of the product is 1
    e. $20 < \frac{x}{y} < 30$
    f. $2 < \frac{x}{y} < 3$
    g. $0.2 < \frac{x}{y} < 0.3$
    h. the first nonzero digit of the quotient is 2

7. Two real numbers between 0 and 4 are chosen at random. What is the probability that the sum of the two numbers exceeds the product of the two numbers?

8. One of the first examples of geometric probability is Buffon's needle problem of 1777, which can be stated as follows: "Suppose a flat surface is marked off with parallel lines spaced D units apart. If a needle of length L is dropped at a random place on the surface, with a random orientation, what is the probability that the needle will lie across one of the lines?" Buffon used the answer to this question to devise a method for approximating the value of π by dropping needles on the surface. We will investigate Buffon's method.

    As stated above, the distance between the lines is D. Let y represent the distance between the first line below the needle and the lower end of the needle and let θ represent the angle between the needle and the horizontal, as shown in Figure 6.9.

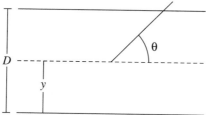

Figure 6.9 Buffon's needle problem

The needle will cross the first line above the needle if the vertical distance from one end of the needle to the other (which depends on θ) is greater than the distance from the lower end of the needle to the line above the needle (which depends on y and D). Use the concepts of geometric probability to solve this problem by first setting up a sample space consisting of all possible vertical positions and orientations of the needle (that is, $0 \le y < D$ and $0 \le \theta < \pi$) and then determining an appropriate event space.

6.2 Integrals and Probability Distribution Functions

The following scores are the results of a test given in a precalculus class: 66, 68, 72, 76, 79, 79, 82, 82, 82, 84, 87, 87, 89, 93, 93, 94, 95, and 98. To picture these test scores, we can create a histogram using intervals of 10 units (60–69, 70–79, 80–89, 90–99) on the horizontal axis, as shown in Figure 6.10. The height of each rectangle represents the number of students whose grades fall within the interval indicated on the horizontal axis for that particular rectangle.

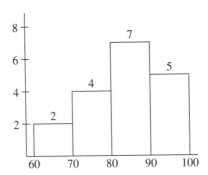

Figure 6.10 Histogram for test scores

If a student is selected at random from this class, what is the probability that the student's score is between 70 and 90? The list of scores indicates that 11 of the 18 students scored between 70 and 90, so the probability of a random student's score falling in that range is $\frac{11}{18}$. Referring to the histogram in Figure 6.10, we can also approach this question by thinking about area, as with geometric probability. The entire histogram can be thought of as the sample space, and the rectangles between 70 and 90 constitute the event space. The area of the sample space is the sum of the areas of all the rectangles. The four rectangles in the histogram, each with base 10, have areas 20, 40, 70, and 50; thus, the area of the sample space is 180. The area of the event space is the sum of the areas of the middle two rectangles, which is 110. The probability that a randomly chosen student has a score between 70 and 90 is the ratio of these two areas, $\frac{110}{180}$, or simply $\frac{11}{18}$.

Suppose the vertical axis of the histogram is rescaled so that the total area of the four rectangles is 1. This can be accomplished by dividing the height of each rectangle by the total area of all the rectangles, as in the rescaled histogram shown in Figure 6.11. This type of histogram is called a ***probability histogram.*** With a probability histogram, the ratio of the area of an event space to the area of the sample space is simply the area of the event space, since the area of the entire sample space is 1. The area of each rectangle in a probability histogram represents the probability that a randomly selected student has a test score that falls within the interval for that rectangle. If a student is selected at random, how likely is it that the individual's test score falls between 70 and 90? This probability is simply the total area represented by the rectangles of the probability histogram between 70 and 90, which is $\frac{40}{180} + \frac{70}{180}$, or $\frac{11}{18}$.

probability histogram

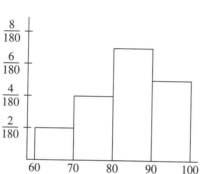

Figure 6.11 Probability histogram

A probability histogram often can be approximated by a continuous curve, which will allow us to apply the techniques of calculus to questions about probability. A function used to describe such a curve is called a ***probability density function*** or a ***probability distribution function***, each of which we abbreviate **PDF**. (One of the tasks in Lab 13 is to fit a probability density function to a probability histogram.) Probability density functions have two essential characteristics:

probability density function

probability distribution function

1. The function values for a PDF are non-negative. This implies that the value of a definite integral of a PDF can be interpreted as the area under a curve.

2. The area under the graph of a PDF over its entire domain is 1.

Suppose x is a number chosen randomly from the domain of a function f, a continuous probability density function. We wish to determine the probability that x lies between two values a and b. We represent this probability as $P(a < x < b)$. The region bounded by the graph of f and the x-axis can be thought of as a sample space. The event space is the area under the curve between $x = a$ and $x = b$, which can be found by evaluating a definite integral. Since f is a PDF, the area of the sample space is 1, and the ratio of the area of the event space to the area of the sample space is just the area of the event space. Therefore,

$$P(a < x < b) = \int_a^b f(x)\,dx.$$

Notice that we have used a continuous PDF to find the probability that an x value occurs in an interval. Since the x values are on a continuum, we cannot determine the probability of the occurrence of a specific x value from this infinity of x values. We can only determine the probability that an x value falls within a certain interval.

Example 1

Researchers in a curriculum project have gathered data for the number of minutes required by a sample of students to solve a particular calculus problem. From these data a histogram is constructed, with number of minutes on the horizontal axis and number of students requiring that number of minutes on the vertical axis. This histogram is then rescaled to become a probability histogram. The curve shown in Figure 6.12 is a continuous approximation for the probability histogram. The equation for this curve is

$$f(t) = \begin{cases} \dfrac{1}{4500}\left(30t - t^2\right) & \text{if } 0 < t < 30, \\ 0 & \text{elsewhere.} \end{cases}$$

Note in Figure 6.12 that t represents the time it takes a student to solve the problem and that all students can solve the problem within 30 minutes. Probabilities are represented by areas of regions under the curve.

Show that f in Figure 6.12 is a probability density function. Then use f to find the probability that the time required to solve the problem by a randomly chosen calculus student is between 12 and 25 minutes.

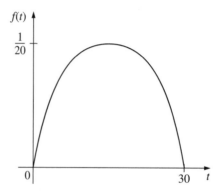

Figure 6.12 Continuous approximation for probability histogram

Solution   To show that f is a PDF, we must show that all values of f are non-negative and that the total area under the graph of f is 1. The graph in Figure 6.12 shows that f has non-negative values for $0 < t < 30$. Since $f(t) = 0$ for all other values of t, the first condition is satisfied. To show that the total area under the graph is 1, we must show that

$$\int_0^{30} \frac{30t - t^2}{4500}\, dt = 1.$$

The definite integral is calculated as follows:

$$\int_0^{30} \frac{30t - t^2}{4500}\, dt = \frac{1}{4500}\left(15t^2 - \frac{t^3}{3}\right)\Big|_0^{30}$$

$$= \frac{1}{4500}\left[15(30)^2 - \frac{(30)^3}{3}\right]$$

$$= 1.$$

To find the probability that the time required to solve the problem by a randomly chosen calculus student is between 12 and 25 minutes, we need to determine the area under the curve from $t = 12$ to $t = 25$, which is shown in Figure 6.13. This area equals the value of the definite integral

$$\int_{12}^{25} \frac{30t - t^2}{4500}\, dt. \tag{1}$$

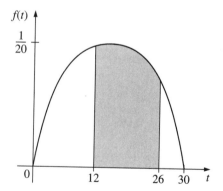

Figure 6.13 Area for probability calculation in Example 1

The value of the integral in (1) is calculated as follows:

$$\int_{12}^{25} \frac{30t - t^2}{4500}\, dt = \frac{1}{4500}\left(15t^2 - \frac{t^3}{3}\right)\Big|_{12}^{25}$$

$$= \frac{1}{4500}\left(\left[15(25)^2 - \frac{(25)^3}{3}\right] - \left[15(12)^2 - \frac{(12)^3}{3}\right]\right)$$

$$= \frac{1}{4500}\left(\frac{12500}{3} - 1584\right)$$

$$\approx 0.574.$$

The value of this integral is approximately 0.574, which is the probability that a randomly chosen calculus student takes between 12 and 25 minutes to solve the problem. In practical terms, if we assign this particular problem to 1000 randomly selected calculus students, we would expect about 574 of them to take between 12 and 25 minutes to solve the problem. ∎

Class Exercises

The following exercises provide an introduction to Lab 13. In the lab, we will use technology to generate a larger number of coin flips than we do here by hand, and then we will approximate the resulting probability histogram with a continuous function.

1. With a partner, flip a coin 30 times and record the number of heads. How many heads did you get? How many did you expect?

2. Tabulate the results for the entire class by making a histogram of the number of occurrences of each number of heads. In other words, the horizontal axis for the histogram is the number of heads in 30 coin flips, with one histogram rectangle for each number. The height of a rectangle is the number of occurrences of that number of heads.

3. What is the overall shape of the histogram? What other data do you think would have a similar shape?

Lab 13 *Simulation of Coin Tossing*

In this lab, we will use technology to simulate a large number of coin tosses. We will count the number of heads in 100 tosses, and we will repeat this experiment 100 times.

After displaying the results in the form of a probability histogram, we will fit a continuous PDF to the probability histogram. We will then compare predictions based on the continuous model with the simulation results.

1. Use the technology of your choice to simulate 100 coin tosses 100 times and record the number of heads in each set of 100 tosses. This can be done by generating a string of 100 randomly generated 0's and 1's, where 0 represents "tails" and 1 represents "heads." The sum of this sequence represents the number of heads in a set of 100 tosses. Display the results of 100 trials of 100 tosses as a histogram with the width of each rectangle equal to 1.

2. Use geometric probability and the histogram in part 1 to answer the following questions.

    a. Based on the histogram, what is the probability that the number of heads in 100 coin flips is less than or equal to 50? Is that what you expected?

    b. What is the probability that the number of heads is less than or equal to 46? What is the probability that the number of heads is between 48 and 52, inclusive?

3. Rescale the histogram to change the total area of the histogram to 1, which yields a probability histogram. What do the areas of the rectangles in the probability histogram represent?

4. Keeping the number of coin tosses in each trial constant at 100, vary the number of trials. What happens to the probability histogram as the number of trials is increased? Why is it better to compare the probability histograms than to compare the frequency histograms as we vary the number of trials?

5. Vary the number of coin tosses in the 100 trials. How does the probability histogram change as we increase the number of tosses in each trial?

We can smooth the probability histogram by approximating the rectangle heights with a continuous function. A function with equation

$$f(x) = e^{-x^2}$$

has the bell shape that we observe in the probability histogram from part 3.

6. Start with $f(x) = e^{-x^2}$ and shift, stretch, and compress this function to fit the probability histogram from part 3. Recall that shifting a graph to the right a units is accomplished by replacing x with $x - a$ on the right side of the function definition. Stretching a graph horizontally by a factor of h is accomplished by replacing x with $\frac{x}{h}$. To both shift and stretch a graph, replace x by $\frac{x-a}{h}$. The graph of a function $y = f(x)$ is compressed vertically by a factor v by multiplying $f(x)$ by v. (Hint: What is the maximum height of the graph of $y = e^{-x^2}$? What is the maximum height of the probability histogram? Use your answers to find an approximate value for v.)

7. For the function in part 6 to be a continuous probability density function, the total area under the graph of this function must equal 1. What is the area under the graph of the function you found in part 6 over the domain from 0 to 100? Modify this function, if needed, so that the area under the curve equals 1 and the curve is still a good fit for the probability histogram.

8. Use the function you found in part 6 (and possibly modified in part 7) and the ideas of geometric probability to answer the following questions.

    a. What is the probability that the number of heads in 100 coin flips is less than or equal to 50? less than or equal to 46? between 48 and 52 inclusive? How do your answers compare with the answers you found in part 2?

    b. What is the probability of obtaining at least 54 heads in 100 coin flips?

    c. What is the smallest value of H for which the probability of obtaining at least H heads in 100 coin flips is less than 0.05?

Write a report summarizing your observations for this lab. Be sure to explain your understanding of the various concepts introduced in this lab, such as simulating coin flips, generating a histogram of the results, rescaling a histogram, the effect of increasing the number of trials, the effect of increasing the number of coin tosses in each of the 100 trials, approximating the probability histogram with a continuous function, and the relationship between the continuous PDF and the probability histogram resulting from the simulation.

Exercise Set 6.2

---

1. For this exercise, use the function

$$f(x) = \begin{cases} kx^2 & \text{if } 0 \le x \le 10, \\ 0 & \text{elsewhere.} \end{cases}$$

    a. Find the value of k for which f will be a probability density function.

    b. Find $P(4 \le x \le 6)$ using the result from part a.

2. The milk dispensers in a school cafeteria are filled each morning with 35 gallons of milk. Records show that a probability density function f describing the daily milk consumption m (in gallons) is

$$f(m) = \begin{cases} 0.002m & \text{if } 0 \le m \le 20, \\ 0.04 & \text{if } 20 < m \le 35. \end{cases}$$

    a. Verify that f is a valid probability density function.

    b. What is the probability that students consume less than 15 gallons of milk on a particular day?

c. What is the probability that students consume between 15 and 25 gallons of milk on a particular day?

d. What is the probability that students consume between 21 and 22 gallons of milk on a particular day? between 25 and 26 gallons?

6.3 The Normal Probability Distribution

The most commonly used continuous probability distribution is the ***normal distribution function***, which is the PDF investigated in Lab 13. This distribution models many randomly distributed phenomena, from the height and weight of the students in a class to measurement errors in industrial processes. The normal probability density function is

normal distribution function

$$N(x) = \frac{1}{\sigma\sqrt{2\pi}} e^{-\frac{1}{2}\left(\frac{x-\mu}{\sigma}\right)^2}. \tag{2}$$

Note that $N(x)$ represents a family of functions with two parameters, traditionally represented by μ (the Greek letter *mu*) and σ (the Greek letter *sigma*).

The ***standard normal distribution*** has $\mu = 0$ and $\sigma = 1$, and its density function is defined as

standard normal distribution

$$f(x) = \frac{1}{\sqrt{2\pi}} e^{-\frac{x^2}{2}}.$$

A graph of this function is shown in Figure 6.14. We will revisit in Section 6.4 the question of how we know that the area under the standard normal curve is 1, which is one of the characteristics of a PDF. The maximum value of the standard normal distribution occurs where $x = 0$. The curve is symmetric about the line $x = 0$, so the area under the curve on either side of $x = 0$ is $\frac{1}{2}$.

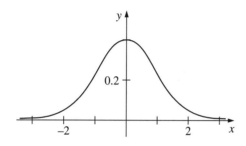

Figure 6.14 Standard normal distribution

mean

standard deviation

The graph of a normal distribution function is a bell-shaped curve. The parameter μ is called the ***mean***, and the parameter σ is known as the ***standard deviation***. Figure 6.15 shows the graph of a normal distribution with $\mu = 1$ and $\sigma = 3$. From this example and using our knowledge of graphical transformations, we can see that μ determines the center of the distribution and σ affects the spread of the distribution.

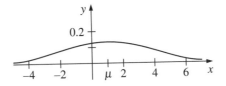

Figure 6.15 Normal distribution with $\mu = 1$ and $\sigma = 3$

Note that the parameter σ appears twice in the definition of $N(x)$ in (2). The σ that divides x in the exponent causes a horizontal stretch and thus affects the spread of the curve. The σ in the coefficient of the exponential causes a vertical compression and thus affects the height of the curve. For example, if σ is 3, then $N(x)$ has three times the spread and one-third the height of the standard normal curve.

Why are both appearances of σ in (2) necessary? Suppose we approximate the area under the standard normal curve with the area of n rectangles, each of which has a base of width Δx. The total area of the n rectangles is approximately 1. If we transform the standard normal curve

$$y = \frac{1}{\sqrt{2\pi}} e^{-\frac{x^2}{2}}$$

by dividing x by 3, resulting in the curve with equation

$$y = \frac{1}{\sqrt{2\pi}} e^{-\frac{1}{2}\left(\frac{x}{3}\right)^2},$$

then the new curve is stretched horizontally by a factor of 3. Each of the approximating rectangles is also stretched horizontally, so that each rectangle has a base of width $3 \cdot \Delta x$. The heights of the rectangles remain the same, however, so the total area of the rectangles is tripled by this transformation. For the stretched curve to remain a PDF, the area under it must be 1, so the curve must be compressed by a factor of 3 in the vertical direction. The curve with equation

$$y = \frac{1}{3\sqrt{2\pi}} e^{-\frac{1}{2}\left(\frac{x}{3}\right)^2}$$

has three times the spread and one-third the height of the standard normal curve, so the area under this curve is 1. In general, therefore, dividing x by σ in the exponent for the purpose of changing the spread of the standard normal curve must be accompanied by division by σ in the leading coefficient in order for the function to remain a PDF.

The area under a normal curve from $x = a$ to $x = b$ represents the probability of an occurrence in that interval of an event that is normally distributed, a property applied in the next two examples.

Example 1

Suppose the life of a certain brand of tire is normally distributed. This brand of tires averages 48,000 miles before wearing out ($\mu = 48,000$), with a standard deviation of 2,500 miles ($\sigma = 2,500$). What is the probability that a randomly selected tire of this brand lasts between 45,000 and 47,000 miles?

Solution   The probability that we seek corresponds to the area of the shaded region under the normal curve with $\mu = 48,000$ and $\sigma = 2,500$ shown in Figure 6.16.

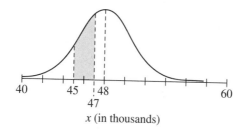

x (in thousands)

Figure 6.16 Area associated with $P(45,000 < x < 47,000)$, $\mu = 48,000$, $\sigma = 2500$

This area is given by the definite integral

$$\int_{45,000}^{47,000} \frac{1}{2500\sqrt{2\pi}} \cdot e^{-\frac{1}{2}\left(\frac{x-48,000}{2500}\right)^2} dx.$$

This integral cannot be evaluated analytically, but using a numerical technique we find that the integral has a value of approximately 0.23. This means that the probability of selecting a tire that lasts between 45 and 47 thousand miles is about 0.23, or you can expect approximately 23% of the tires to last between 45 and 47 thousand miles. ∎

Example 2

Suppose the tire production process for the brand in Example 1 is improved so that tires average 50,000 miles before wearing out. If the standard deviation remains 2500 miles, what is the probability of such a tire wearing out in the 45,000 to 47,000 mile interval? How does this probability change if the standard deviation is decreased to 1500 miles?

Solution   If $\mu = 50,000$ and $\sigma = 2500$, the probability that a tire lasts between 45,000 and 47,000 miles is given by

$$\int_{45,000}^{47,000} \frac{1}{2500\sqrt{2\pi}} \cdot e^{-\frac{1}{2}\left(\frac{x-50,000}{2500}\right)^2} dx. \tag{3}$$

The area corresponding to this probability is shown in Figure 6.17.

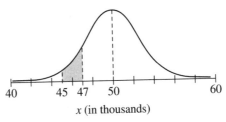

x (in thousands)

Figure 6.17 Area associated with $P(45,000 < x < 47,000)$, $\mu = 50,000$, and $\sigma = 2500$

Since the mean of the distribution is $\mu = 50,000$ and σ has not changed from Example 1, the region under the curve between $x = 45,000$ and $x = 47,000$ is farther from the mean and smaller than the region between the same x values in Figure 6.16. Using numerical techniques we find that the definite integral in (3) has a value of about 0.092.

The graph of the normal PDF with $\mu = 50,000$ and $\sigma = 1500$ is shown in Figure 6.18. It has a greater height at the mean and less spread than the graphs of the PDFs with $\sigma = 2500$ shown in Figures 6.16 and 6.17. The real-world significance of the smaller σ value is a reduction in the variability of tire lifetimes. Because the vast majority of tires last very close to 50,000 miles, it is now much less likely to get a tire that lasts between 45 and 47 thousand miles than it was when $\sigma = 2500$. This probability, which is given by the integral

$$\int_{45,000}^{47,000} \frac{1}{1500\sqrt{2\pi}} \cdot e^{-\frac{1}{2}\left(\frac{x-50,000}{1500}\right)^2} dx,$$

is only about 0.022. ■

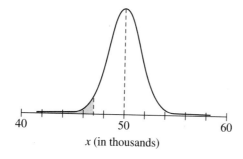

Figure 6.18 Area associated with $P(45,000 < x < 47,000)$, $\mu = 50,000$, and $\sigma = 1500$

Exercise Set 6.3

1. Parts a–d refer to a normal probability density function.

   a. For a standard normal distribution, what is the probability of getting a value between –1 and 1, that is, within 1 standard deviation of the mean? within 2 standard deviations of the mean? within 3? within 4?

   b. Suppose $\mu = 0$ and $\sigma = 2$. What is the probability of getting a value between –2 and 2, that is, within 1 standard deviation of the mean? within 2 standard deviations of the mean? within 3? within 4?

   c. Suppose $\mu = 0$ and $\sigma = 0.5$. What is the probability of getting a value between –0.5 and 0.5, that is, within 1 standard deviation of the mean? within 2 standard deviations of the mean? within 3? within 4?

   d. Based on your results from parts a through c, for a normal distribution with mean μ and standard deviation σ, what do you think is the probability of getting a value within 1 standard deviation of the mean? within 2? within 3? within 4?

2. Over what interval, symmetric about 0 and specified to the nearest tenth, should the definite integral for the standard normal distribution function be calculated to achieve an area of 1 accurate to 5 decimal places? In other words, find a so that the following inequality is satisfied:

$$0.99999 < \int_{-a}^{a} N(x)\,dx < 1.00001, \quad \text{where } N(x) = \frac{1}{\sqrt{2\pi}}\,e^{-\frac{x^2}{2}}.$$

Areas for any normal distribution can be calculated using a standard normal distribution by transforming values into standard deviation units. For instance, in Example 2, we found the fraction of tires that wear out between 45,000 and 47,000 miles using a *normal distribution with mean 50,000 and standard deviation 2500. Our answer would be the same if we used the standard normal distribution and calculated the fraction of values between –2.00 (which is $\frac{45,000-50,000}{2500}$) and –1.20 (which is $\frac{45,000-50,000}{2500}$). In other words, since 45,000 is 2.00 standard deviations below the mean and 47,000 is 1.20 standard deviations below the mean, our answer is the same as the area between –2.00 and –1.20 under the standard normal curve.*

*To convert a number s for the normal distribution with mean μ and standard deviation σ to an equivalent number for the standard normal distribution, subtract μ and divide by σ. The converted number $\frac{s-\mu}{\sigma}$ is called a **z-score**. The reason for converting to z-scores and using the standard normal is that values of probabilities for the standard normal are easily looked up in tables. Such a table is given in the appendix. This table gives the area to the left of each z value from –3.40 to 3.49 in increments of 0.01. To answer the first question from Example 2, we need to find –2.00 and –1.20 in the table. The area under the standard normal curve to the left of –2.00 is about 0.0228, and the area to the left of –1.20 is about 0.1151. The area between –2.00 and –1.20 is the difference between these two values, 0.0923, which is close to the answer found in the solution to Example 2.*

z-score

3. Intelligence Quotient (IQ) scores are distributed normally with mean 100 and standard deviation 15. Answer questions a–c using the table of z-scores provided in the appendix.

    a. Within how many standard deviations from the mean do 68% of the scores lie? To what interval of IQ scores, symmetric about 100, does this correspond?

    b. What percentage of IQ scores are between 100 and 110? between 100 and 120? between 100 and 130?

    c. How unusual is an IQ score above 140? above 150?

4. Suppose the number of peanuts found in a snack package is normally distributed with a mean of 120 and a standard deviation of 7. What is the probability of getting a bag with between 110 and 120 peanuts? What is the probability of getting more than 100 peanuts in a bag? Less than 130?

5. The amount of cereal in a box is determined by weight, not volume, and actual weights are normally distributed about a target weight. Suppose a box of cereal is to weigh 16 ounces. If the machine filling the boxes produces a standard deviation of 0.6 ounce and the target weight is 16 ounces, how likely is the occurrence of a box weighing less than 15 ounces? Between 15.9 and 16.1 ounces? Why do you think companies might set the target weight higher than the weight indicated on a box of cereal?

6. Suppose a particular packing machine can be set to any target value, but the standard deviation is 15 grams. If a weight of 350 grams will be printed on the box and state law requires that at most 5% of the boxes can be under the weight printed on the box, at what target value should you set the packing machine? Assume that the weights are normally distributed about the target weight.

7. A bottle-filling machine can be adjusted so that it dispenses a target amount of liquid ranging between 16 and 32 ounces. Whatever the setting, the actual amounts dispensed are normally distributed with a standard deviation of 0.3 ounces. Suppose the machine is used to fill 28-ounce juice bottles. What is the highest setting that should be used so that the juice overflows the bottle no more than 2% of the time?

8. A company needs to buy several cars for its motor pool. Managers have specified that the new cars should get at least 30 miles per gallon (mpg) on the highway. Two models are being considered, and the gas mileage for both is normally distributed. For one model the mean gasoline consumption is 35 mpg with a standard deviation of 3 mpg. The other model averages 34 mpg with a standard deviation of 1.5 mpg. Which model should the managers choose? Support your answer.

9. Suppose a random sample of data consists of n observations, which we symbolize by x_i, where i has values from 1 to n. The mean $\bar{x}$ of the data set is the average of the x_i's, which is given by the formula

$$\bar{x} = \frac{1}{n} \cdot \sum_{i=1}^{n} x_i.$$

The standard deviation is given by

$$s.d. = \sqrt{\frac{1}{n-1} \cdot \sum_{i=1}^{n} (x_i - \bar{x})^2}.$$

Calculate the mean and standard deviation of the data generated by the coin-tossing simulation in Lab 13. Use these values to find the equation of a normal curve to fit the data. How does this function compare with the one you determined in Lab 13?

10. The following data set gives the heights reported by a random sample of 70 females at the start of their junior year in high school. Use the formulas given in problem 9 to fit a normal curve to the data. How well does the normal curve fit the data? (Hint: Construct a histogram of the data using intervals of one inch.)

Heights (in inches) of 70 females:
61, 65, 63.5, 65, 66, 64, 67, 64, 65, 66.5, 64, 71, 63, 62, 65, 57, 60, 63, 59, 58, 61, 65, 65, 68, 62, 63, 63, 65, 67, 62, 63, 66, 63, 66, 58, 65, 58.5, 65, 69, 68, 64, 67.5, 64.5, 66, 68, 67.5, 70, 65, 67, 66, 62.5, 65, 70.5, 63, 66, 65, 66, 64.5, 66, 60, 66, 62, 65, 66, 68, 65, 67.5, 62, 64, 72

11. Use the first and second derivatives of the normal distribution function

$$N(x) = \frac{1}{\sigma\sqrt{2\pi}} e^{-\frac{1}{2}\left(\frac{x-\mu}{\sigma}\right)^2}$$

to show that the maximum value of $N(x)$ is found at $x = \mu$ and that the points of inflection are found at $x = \mu \pm \sigma$.

12. Find the quadratic approximation about $x = 0$ for the standard normal curve. Determine the area under the quadratic approximation between -0.5 and 0.5, between -1 and 1, and between -2 and 2. How does the area under the quadratic approximation compare with the area under the standard normal curve on these intervals?

6.4   Improper Integrals and Infinite Series

When we defined the standard normal distribution with equation

$$N(x) = \frac{1}{\sqrt{2\pi}} \cdot e^{-\frac{x^2}{2}}$$

in the preceding section, we stated that the area under the entire graph of the standard normal PDF is 1. How can we verify this statement? Our first step is to write a definite integral that represents the area, then we need to evaluate this integral. Since the domain of $N(x)$ is all real numbers and we want to find the area between this curve and the x-axis, we use infinite limits of integration on the definite integral. Thus, the definite integral that represents the total area under the standard normal curve is

$$\int_{-\infty}^{\infty} \frac{1}{\sqrt{2\pi}} \cdot e^{-\frac{x^2}{2}} dx.$$

improper integral   A definite integral with one or two infinite limits of integration is called an *improper integral*.

We will see applications of improper integrals in this section and the exercises that follow, but first we will discuss in more depth how to evaluate and interpret such integrals. Recall that in Chapter 5 we defined a definite integral as the limit of a Riemann sum:

$$\int_a^b f(x)dx = \lim_{\Delta x \to 0} \sum_{i=0}^{n-1} f(x_i)\Delta x.$$

The values of n and Δx in this Riemann sum are closely related. If we assume that Δx is a constant equal to $\frac{b-a}{n}$, letting n increase without bound has the same effect as letting Δx approach 0. If $f(x) \geq 0$ for $a \leq x \leq b$, and if the terms in the summation are interpreted as areas of rectangles, then increasing n creates more rectangles in the finite interval from a to b. Increasing the number of rectangles decreases the width of the rectangles, thus giving a better approximation for the actual area under the curve.

If we place infinite limits on a definite integral, such as in $\int_{-\infty}^{\infty} f(x)dx$, then we are dealing with infinity in two separate forms. The rectangles that approximate this integral have widths that approach zero in the limit, and we must have infinitely many such rectangles. Furthermore, in contrast to a definite integral with finite limits of integration, the rectangles must cover an interval that is infinite in extent. In other words, to define an improper integral in terms of a Riemann sum requires an infinite number of exceedingly thin rectangles that cover an infinitely long interval.

To help us try to unravel this tangle of infinities, we will examine infinite sums of real numbers. We know that $\frac{1}{3}$ can be written as the repeating decimal $0.333\overline{3}$. This repeating decimal can be written as the infinite sum

$$S = \frac{3}{10} + \frac{3}{100} + \frac{3}{1000} + \frac{3}{10000} + \frac{3}{100000} + \frac{3}{1000000} + \cdots = \sum_{i=1}^{\infty} \frac{3}{10^i}. \quad (4)$$

To show that the sum in (4) is really equal to $\frac{1}{3}$, we can let

$$S_n = \frac{3}{10} + \frac{3}{100} + \frac{3}{1000} + \frac{3}{10000} + \frac{3}{100000} + \frac{3}{1000000} + \cdots + \frac{3}{10^n}. \quad (5)$$

Multiplying both sides of (5) by $\frac{1}{10}$ gives

$$\frac{S_n}{10} = \frac{3}{100} + \frac{3}{1000} + \frac{3}{10000} + \frac{3}{100000} + \frac{3}{1000000} + \cdots + \frac{3}{10^{n+1}}. \quad (6)$$

Subtracting equation (6) from equation (5) we see that

$$S_n - \frac{S_n}{10} = \frac{3}{10} - \frac{3}{10^{n+1}}.$$

Solving for S_n, we find that

$$S_n = \frac{1}{3} - \frac{1}{3 \cdot 10^n}. \quad (7)$$

Since S_n represents the sum of the first n terms of S given in (4), then as n increases S_n gets closer to S. In fact, we can write $\lim_{n \to \infty} S_n = S$. Now as $n \to \infty$, the term $\frac{1}{3 \cdot 10^n}$ on the right side of equation (7) approaches zero, so $\lim_{n \to \infty} S_n = \frac{1}{3}$. Since $\lim_{n \to \infty} S_n = S$, we conclude that $S = \frac{1}{3}$.

There is a significant connection between infinite sums, which we usually call *infinite series*, and improper integrals. Eventually we will use an improper integral to help determine if an infinite series has a finite sum, but first we will use the infinite series in (4), which we know has a finite sum, to help us understand that an improper integral can have a finite value.

infinite series

Example 1

Evaluate the improper integral $\int_1^\infty \frac{3}{10^x}\,dx$.

Solution   The infinite sum in (4), which has a value of $\frac{1}{3}$, can be represented geo-metrically as the total area of the rectangles shown in Figure 6.19. Although the rectangles get small quickly, the top of each rectangle is above the curve. Each rect-angle has width 1 and height $\frac{3}{10^n}$, where n is a positive integer.

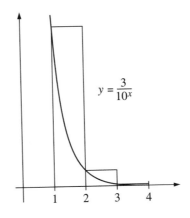

$$y = \frac{3}{10^x}$$

Figure 6.19 Rectangles with heights $\frac{3}{10^n}$

The value of the improper integral $\int_1^\infty \frac{3}{10^x}\,dx$ represents the area to the right of $x = 1$ that lies below the curve $y = \frac{3}{10^x}$ shown in Figure 6.19. The region defined by this integral has a domain with corresponding positive y values that is infinite in extent along the x-axis. Can a region with positive y values over an infinitely long interval have a finite area? The answer is yes, just as the sum of an infinite number of terms can have a finite value. Since the region that has an area represented by the integral $\int_1^\infty \frac{3}{10^x}\,dx$ is entirely contained within the rectangles that have a total area of $\frac{1}{3}$, the value of this improper integral must be less than $\frac{1}{3}$.

We know that the improper integral $\int_1^\infty \frac{3}{10^x}\,dx$ has a finite value, but how do we find this value? If we replace ∞ with some large, finite number k, then our calcula-tions would involve finding the value of a definite integral with finite limits of integration. This we can do analytically using a table of integrals and the Fundamen-tal Theorem of Calculus as follows:

$$\int_1^k \frac{3}{10^x}\,dx = 3\int_1^k 10^{-x}\,dx$$

$$= -\frac{3}{\ln 10}\left(10^{-x}\right)\Big|_1^k$$

$$= -\frac{3}{\ln 10}\left(10^{-k} - 10^{-1}\right).$$

The larger k is, the closer the value of the integral is to $\frac{3}{10\ln 10}$. In fact, we can make the value of this integral as close to $\frac{3}{10\ln 10}$ as we like by making k sufficiently large. This last statement fits our intuition about limits, so we find the value of $\int_1^\infty \frac{3}{10^x}\,dx$ with the statement

$$\int_1^\infty \frac{3}{10^x}\,dx = \lim_{k\to\infty} \int_1^k \frac{3}{10^x}\,dx,$$

and we have shown that the value of this limit is $\frac{3}{10\ln 10}$. ∎

In general, improper integrals are defined as follows, where a and b are finite constants.

$$\int_a^\infty f(x)dx = \lim_{k\to\infty}\int_a^k f(x)dx,$$

$$\int_{-\infty}^b f(x)dx = \lim_{m\to-\infty}\int_m^b f(x)dx,$$

and

$$\int_{-\infty}^\infty f(x)dx = \int_{-\infty}^a f(x)dx + \int_a^\infty f(x)dx$$

$$= \lim_{m\to-\infty}\int_m^a f(x)dx + \lim_{k\to\infty}\int_a^k f(x)dx.$$

If an improper integral has a finite value, that is, if each limit in the definition of the improper integral exists, then we say that the integral *converges*; otherwise we say it *diverges*. Deciding whether an improper integral converges or diverges is sometimes straightforward. For example, the integral $\int_0^\infty x\,dx$ represents the area under the function $f(x) = x$, which increases without bound as the upper limit of integration increases without bound, so the integral diverges. The divergence or convergence of other improper integrals is not always so easy to determine, and actually may conflict with our intuition, as in the following example.

converges

diverges

Example 2

Determine whether $\int_1^\infty \frac{1}{x}\,dx$ converges or diverges.

Solution   The integral $\int_1^\infty \frac{1}{x}\,dx$ represents the area under the graph of $y = \frac{1}{x}$ to the right of $x = 1$. Since this improper integral is defined by

$$\int_1^\infty \frac{1}{x}\,dx = \lim_{k\to\infty}\int_1^k \frac{1}{x}\,dx,$$

the value of $\int_1^\infty \frac{1}{x} dx$ is the limiting value of the area under $y = \frac{1}{x}$ between $x = 1$ and $x = k$ as k gets infinitely large. The boundaries of this area are $y = \frac{1}{x}$, the x-axis, $x = 1$, and $x = k$. As $k \to \infty$, the right boundary at $x = k$ moves to the right. The graph of $y = \frac{1}{x}$ is asymptotic to the positive x-axis, so as the right boundary moves to the right, the added area becomes less and less. This line of thinking might lead us to conclude that the integral converges; however, the work that follows shows that this is not the case. This integral is evaluated as follows:

$$\int_1^\infty \frac{1}{x} dx = \lim_{k \to \infty} \int_1^k \frac{1}{x} dx$$

$$= \lim_{k \to \infty} \left. \ln|x| \right|_1^k$$

$$= \lim_{k \to \infty} (\ln k - \ln 1)$$

$$= \lim_{k \to \infty} (\ln k).$$

The convergence or divergence of $\int_1^\infty \frac{1}{x} dx$ is determined by the behavior of $\ln k$ as k increases without bound. As $k \to \infty$, the natural logarithm of k also increases without bound, so $\lim_{k \to \infty} (\ln k)$ does not exist. Thus we conclude that $\int_1^\infty \frac{1}{x} dx$ diverges. We can enclose as much area as we like under the curve $y = \frac{1}{x}$ by letting k be large enough. If we wanted to enclose an area of 1000, for example, then we would let $k = e^{1000}$ which gives us $\int_1^k \frac{1}{x} dx = 1000$. ∎

converges

diverges

We conclude this section by returning to the subject of infinite series. If an infinite series has a finite value, that is, if the limit of the sum of the terms has a finite value as the number of terms increases without bound, then we say that the infinite series **converges**; otherwise we say it **diverges**. Now that we know that the improper integral $\int_1^\infty \frac{1}{x} dx$ diverges, we will be able to use this knowledge to help us investigate an infinite series. The improper integral $\int_1^\infty \frac{1}{x} dx$ is related to the infinite series

$$1 + \frac{1}{2} + \frac{1}{3} + \frac{1}{4} + \cdots = \sum_{n=1}^{\infty} \frac{1}{n}, \tag{8}$$

as shown in Figure 6.20. The height of each rectangle in Figure 6.20 is determined by the curve with equation $y = \frac{1}{x}$, and the width of each rectangle is 1. The areas of these rectangles correspond to the terms in the infinite series in (8). At first inspection, this infinite series may appear to converge, since each term is smaller than the term before it, and the values of the terms approach zero. After 100 terms the sum is only a little greater than 5, and after 1000 terms the sum is still less than 7.5.

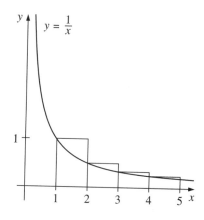

Figure 6.20 Rectangles with heights determined by $y = \frac{1}{x}$

Each rectangle in Figure 6.20 has some part of its interior above the curve $y = \frac{1}{x}$, so the total area of these rectangles is greater than the area under the curve to the right of $x = 1$. Therefore, we can write

$$\int_1^\infty \frac{1}{x}\, dx \le 1 + \frac{1}{2} + \frac{1}{3} + \frac{1}{4} + \cdots.$$

Analyzing this inequality and knowing that the improper integral $\int_1^\infty \frac{1}{x} dx$ diverges, we conclude that the series in (8) also must diverge. The infinite series $1 + \frac{1}{2} + \frac{1}{3} + \frac{1}{4} + \cdots$ is called the **harmonic series**. We have shown that this series diverges by comparing the area it represents to an area that we can analyze with an improper integral.

harmonic series

The two series considered in this section are important because they give us insight regarding the nature of infinite sums. We studied the harmonic series $1 + \frac{1}{2} + \frac{1}{3} + \frac{1}{4} + \cdots$, and we saw that it diverges. We studied the series

$$\sum_{n=1}^\infty \frac{3}{10^n} = \frac{3}{10} + \frac{3}{100} + \frac{3}{1000} + \frac{3}{10000} + \frac{3}{100000} + \frac{3}{1000000} + \cdots, \qquad (9)$$

and we saw that it converges to $\frac{1}{3}$. The series in (9) is an example of an **infinite geometric series**. In a geometric series each term (after the initial term) is generated by multiplying the preceding term by a constant factor. This factor is called the **common ratio** of the series. In the series in (9), the common ratio is one-tenth. The terms of this series approach 0, and they approach 0 quickly enough that the sum never exceeds $\frac{1}{3}$. If we add a finite number of the terms of this series, the sum will always be less than $\frac{1}{3}$; however, we can get as close as we like to $\frac{1}{3}$ by adding enough terms. The infinite sum therefore converges to $\frac{1}{3}$.

infinite geometric series

common ratio

The sum of the first $n+1$ terms of a geometric series has the general form

$$S_n = a + ar + ar^2 + ar^3 + \cdots + ar^n, \tag{10}$$

where a is the first term of the series and r is the common ratio. As we did previously to find the sum of the series in (9), we can find a simpler form for S_n. First, multiply both sides of equation (10) by r, which yields

$$rS_n = ar + ar^2 + ar^3 + ar^4 \cdots + ar^n + ar^{n+1}. \tag{11}$$

Subtracting equation (11) from equation (10) gives us

$$S_n - rS_n = a - ar^{n+1}. \tag{12}$$

Factoring the left side of (12) yields

$$(1-r)S_n = a - ar^{n+1}$$

and solving for S_n gives

$$S_n = \frac{a - ar^{n+1}}{1-r} \tag{13}$$

if $r \neq 1$. Equation (13) is an equation for the sum of the geometric series (10).

In general, the sum S of an infinite geometric series, if this sum exists, is equal to the limit of S_n as n increases without bound. We can find an expression for the sum of an infinite geometric series by taking the limit as $n \to \infty$ of each side of equation (13), which yields

$$S = \lim_{n\to\infty} S_n = \lim_{n\to\infty} \frac{a - ar^{n+1}}{1-r}, \tag{14}$$

if $r \neq 1$. If $r = 1$, then the series equals $a + a + a + \ldots$, which diverges. If $r = -1$, then the series equals $a - a + a - a + \ldots$, which oscillates between a and 0 as the number of terms increases, so the series diverges. If $|r| > 1$, then the term ar^{n+1} in (14) increases without bound as $n \to \infty$, and the series diverges. If $|r| < 1$, then the term ar^{n+1} in (14) approaches 0 as $n \to \infty$, and the series converges to $\frac{a}{1-r}$.

We can now say the following about an infinite geometric series and its sum S:

$$S = \frac{a}{1-r} \qquad \text{if } |r| < 1;$$
$$\text{the series diverges} \qquad \text{if } |r| \geq 1.$$

In contrast to an infinite geometric series with $|r| < 1$, the harmonic series

$$\sum_{n=1}^{\infty} \frac{1}{n} = 1 + \frac{1}{2} + \frac{1}{3} + \frac{1}{4} + \frac{1}{5} + \cdots$$

does not converge. Even though the terms approach 0, the sum nevertheless increases without bound. Comparing the harmonic series to a geometric series, we observe that what determines whether an infinite series converges is not *whether* the terms approach 0, but *how quickly* the terms approach 0.

Exercise Set 6.4

1. Evaluate the following infinite sums.

    a. $\displaystyle\sum_{n=0}^{\infty} \frac{7}{10^n}$

    b. $\displaystyle\sum_{n=1}^{\infty} 5^{-n}$

    c. $\displaystyle\sum_{n=1}^{\infty} \frac{31}{100^n}$

    d. $\displaystyle\sum_{n=0}^{\infty} 1.2^n$

    e. $\displaystyle\sum_{n=0}^{\infty} 0.5^{2n}$

    f. $\displaystyle\sum_{n=0}^{\infty} 0.5^{2n+1}$

2. Evaluate the following improper integrals:

    a. $\displaystyle\int_1^{\infty} \frac{dx}{x^2}$

    b. $\displaystyle\int_1^{\infty} \frac{dx}{\sqrt{x}}$

    c. $\displaystyle\int_1^{\infty} \frac{dx}{x^p}$ for $p = 1.1,\ 1.01,\ 1.001$

    d. $\displaystyle\int_1^{\infty} \frac{dx}{x^p}$ for $p = 0.9,\ 0.99,\ 0.999$

    e. $\displaystyle\int_1^{\infty} \frac{dx}{x^p}$ for $p > 1$

    f. $\displaystyle\int_1^{\infty} \frac{dx}{x^p}$ for $p < 1$

    g. $\displaystyle\int_0^{\infty} x \cdot e^{-x^2}\, dx$

    h. $\displaystyle\int_{-\infty}^0 x \cdot e^{-x^2}\, dx$

    i. $\displaystyle\int_{-\infty}^{\infty} x \cdot e^{-x^2}\, dx$

    j. $\displaystyle\int_1^{\infty} \frac{\ln x}{x}\, dx$

    k. $\displaystyle\int_0^{\infty} \frac{x}{x^2+1}\, dx$

    l. $\displaystyle\int_2^{\infty} \frac{dx}{x^2-1}$

    m. $\displaystyle\int_2^{\infty} \frac{dx}{x \ln x}$

    n. $\displaystyle\int_{-\infty}^{\infty} \frac{dx}{x^2+1}$

3. Given a probability density function (see section 6.2), we define the **median** to be   *median*
    the number m such that m divides the area under the PDF in half.

    a. Find the median of the probability density function $f(t) = \frac{1}{10} e^{-\frac{t}{10}}$, $0 \le t < \infty$.

b. Find the median of the probability density function $f(x) = \frac{1}{36} x(6 - x)$, $0 \le x \le 6$.

c. Find the median of the standard normal probability density function.

interquartile range

4. Suppose that $f(x)$ is a PDF with domain $a \le x \le b$. If $\int_a^{k_1} f(x)dx = 0.25$ and $\int_a^{k_2} f(x)dx = 0.75$, then $(k_2 - k_1)$ is called the ***interquartile range.*** Describe what information the interquartile range gives you.

5. The number of hours taken by students competing in a 1.5 hour mathematics contest is recorded as they hand in their contest papers. Any student not finished at the end of the 1.5 hours is recorded as taking 1.5 hours. The time in hours is then found to have a probability density function given by

$$f(x) = \begin{cases} cx^2 + x & \text{if } 0 \le x \le 1.5, \\ 0 & \text{elsewhere.} \end{cases}$$

a. Find c so that f is a PDF.

b. Use $f(x)$ to find the probability that a student takes less than one hour.

c. What is the probability that a student finishes within one-half hour?

d. What is the median time taken by the students?

6.5   The Integral Test for Convergence of an Infinite Series

In Section 6.4 we saw that an infinite geometric series converges when the common ratio is between -1 and 1. We also saw that the harmonic series does not converge. One way to think about the divergence of the harmonic series is that the terms being added do not approach zero fast enough. In problem 2 of Exercise Set 6.4, you may have seen that improper integrals of the form $\int_1^\infty \frac{1}{x^p} dx$ converge if $p > 1$ and diverge if $0 < p < 1$. The formal justification for these conclusions about $\int_1^\infty \frac{1}{x^p} dx$ follows.

Assuming $p > 0$ and $p \neq 1$, we can evaluate $\int_1^\infty \frac{1}{x^p} dx$ using the following sequence of calculations:

$$\int_1^\infty \frac{1}{x^p} dx = \lim_{b \to \infty} \left(\frac{1}{-p+1} x^{-p+1} \right) \Big|_1^b$$

$$= \lim_{b \to \infty} \left[\frac{1}{-p+1} \left(b^{-p+1} - 1 \right) \right]$$

$$= \lim_{b \to \infty} \left[\frac{1}{-p+1} \left(\frac{1}{b^{p-1}} - 1 \right) \right].$$

If $p > 1$, then $p - 1 > 0$. This means that b^{p-1} approaches infinity as $b \to \infty$, so $\frac{1}{b^{p-1}}$ approaches zero as $b \to \infty$. This in turn means that $\frac{1}{b^{p-1}} - 1$ approaches -1, so the expression $\left(\frac{1}{-p+1}\right)\left(\frac{1}{b^{p-1}} - 1\right)$ approaches $\left(\frac{1}{-p+1}\right)(-1) = \frac{1}{p-1}$ as $b \to \infty$. Thus, the integral $\int_1^\infty \frac{1}{x^p} \, dx$ converges to $\frac{1}{p-1}$ if $p > 1$.

If $0 < p < 1$, then $-1 < p - 1 < 0$. This means that b^{p-1} approaches zero as $b \to \infty$, so $\frac{1}{b^{p-1}}$ approaches infinity as $b \to \infty$, and the entire expression $\left(\frac{1}{-p+1}\right)\left(\frac{1}{b^{p-1}} - 1\right)$ approaches infinity. Therefore, the integral $\int_1^\infty \frac{1}{x^p} \, dx$ diverges if $0 < p < 1$.

If $p = 1$, then

$$\int_1^\infty \frac{1}{x^p} \, dx = \int_1^\infty \frac{1}{x} \, dx$$

$$= \lim_{b \to \infty} \ln|x| \Big|_1^b$$

$$= \lim_{b \to \infty} (\ln b - \ln 1)$$

$$= \lim_{b \to \infty} \ln b.$$

Since $\ln b$ increases without bound as b increases without bound, this limit does not exist. Therefore, if $p = 1$, we know that the integral $\int_1^\infty \frac{1}{x^p} \, dx$ diverges.

We have now shown that $\int_1^\infty \frac{1}{x^p} \, dx$ converges if $p > 1$ and diverges if $p \le 1$. Using this information, we know, for example, that $\int_1^\infty \frac{1}{\sqrt{x}} \, dx$ and $\int_1^\infty \frac{1}{x^{0.8}} \, dx$ both diverge, while $\int_1^\infty \frac{1}{x^3} \, dx$ and $\int_1^\infty \frac{1}{x^{1.4}} \, dx$ both converge.

We can use integrals of the form $\int_1^\infty \frac{1}{x^p} \, dx$ to investigate another type of infinite series, called the *p-series*, which is

p-series

$$\sum_{n=1}^\infty \frac{1}{n^p} = 1 + \frac{1}{2^p} + \frac{1}{3^p} + \frac{1}{4^p} + \cdots.$$

We will compare series of the form $\sum \frac{1}{n^p}$ to improper integrals of the form $\int_1^\infty \frac{1}{x^p} \, dx$. We will use the same strategy we used in Section 6.4, which is to place rectangles whose areas represent the terms of the series either above or below the graph of the function $f(x) = \frac{1}{x^p}$ and use the convergence or divergence of the improper integral to determine the convergence or divergence of the associated series.

First we investigate the integral with $p = \frac{1}{2}$. Since p is between 0 and 1, we know that the integral $\int_1^\infty \frac{1}{\sqrt{x}} \, dx = \int_1^\infty \frac{1}{x^{1/2}} \, dx$ diverges, so we suspect that the associated series $\sum \frac{1}{\sqrt{n}}$ also diverges. Since our initial thinking points to divergence, we will begin by testing for divergence. We will show that the series is greater than or equal to the integral, and we will use this fact to conclude that the series must diverge.

We first place the rectangles associated with the series so that the top of each rectangle is above the graph of the function $f(x) = \frac{1}{\sqrt{x}}$, as shown in Figure 6.21. We draw a vertical line at each integer value on the x-axis to the corresponding point on the curve and then draw the top of each rectangle by moving one unit to the right. Each rectangle has an area equal to a term of the series, and each rectangle has part of its interior above the curve.

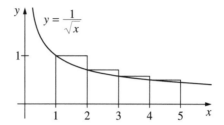

Figure 6.21 Series of rectangles with tops above the curve

The sum of the areas of the rectangles is

$$\frac{1}{\sqrt{1}} \cdot 1 + \frac{1}{\sqrt{2}} \cdot 1 + \frac{1}{\sqrt{3}} \cdot 1 + \frac{1}{\sqrt{4}} \cdot 1 + \cdots = \sum_{n=1}^{\infty} \frac{1}{\sqrt{n}},$$

and the area under the curve to the right of $x = 1$ is $\int_1^{\infty} \frac{1}{\sqrt{x}} dx$. Since the area of the rectangles is greater than the area under the curve, we know that

$$\int_1^{\infty} \frac{1}{\sqrt{x}} dx < \sum_{n=1}^{\infty} \frac{1}{\sqrt{n}}.$$

The improper integral diverges because the area under the curve $y = \frac{1}{\sqrt{x}}$ to the right of $x = 1$ is infinitely large. The sum $\sum \frac{1}{\sqrt{n}}$ is greater than this area, so the infinite series diverges.

Now consider the series $\sum \frac{1}{n^2}$ and the related integral $\int_1^{\infty} \frac{1}{x^2} dx$. Since the improper integral converges ($p > 1$), we suspect that the associated series also converges. Since we think the series converges, we will draw rectangles that remain *under* the function $f(x) = \frac{1}{x^2}$, as shown in Figure 6.22. Again, we will draw vertical lines from the x-axis to the curve at each positive integer value. This time, however, we draw the top of each rectangle to the left to ensure that the interior of each rectangle is entirely below the curve.

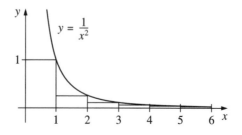

Figure 6.22 Series of rectangles below the curve

Beginning from the y-axis and moving to the right, the first rectangle has an area of 1, and the rest of the rectangles are below the part of the curve to the right of $x = 1$. The sum of the areas of the rectangles is

$$\frac{1}{1^2} + \frac{1}{2^2} + \frac{1}{3^2} + \frac{1}{4^2} + \cdots = \sum_{n=1}^{\infty} \frac{1}{n^2}.$$

All of the rectangles except for the first are below the graph of $y = \frac{1}{x^2}$ for $x > 1$. The sum of the areas of the rectangles excluding the first is $\sum_{n=2}^{\infty} \frac{1}{n^2}$, so

$$\sum_{n=2}^{\infty} \frac{1}{n^2} < \int_1^{\infty} \frac{1}{x^2}\, dx.$$

We know that the integral $\int_1^{\infty} \frac{1}{x^2}\, dx$ converges, so the area under the curve with equation $y = \frac{1}{x^2}$ has a finite value. Since the sum $\sum_{n=2}^{\infty} \frac{1}{n^2}$ is less than this area, the sum converges. Since $\sum_{n=1}^{\infty} \frac{1}{n^2} = 1 + \sum_{n=2}^{\infty} \frac{1}{n^2}$, the infinite series $\sum_{n=1}^{\infty} \frac{1}{n^2}$ also converges.

The preceding discussion shows how to use improper integrals to determine if an infinite series converges or diverges. This process of comparing a series to an improper integral can be generalized. The result is called the **integral test**, which is stated as: *integral test*

Let f be a decreasing continuous function that has positive values $f(x)$ for all $x \geq 1$. The infinite series $\sum_{n=1}^{\infty} f(n)$ converges if the improper integral $\int_1^{\infty} f(x)\, dx$ converges, and the infinite series $\sum_{n=1}^{\infty} f(n)$ diverges if the improper integral $\int_1^{\infty} f(x)\, dx$ diverges.

Drawing rectangles to approximate the area under a curve provides a way for us to see that the series $\sum_{n=1}^{\infty} f(n)$ and the improper integral $\int_1^{\infty} f(x)\, dx$ either both converge or both diverge. Although it has not been proven here, the generalization holds for any decreasing, continuous function f. In practice, we can simply consider the integral, and there is no need to draw the rectangles that represent the areas of the terms of the series.

Based on the integral test, we know that a p-series $\sum \frac{1}{n^p}$ converges if $p > 1$ (since $\int_1^{\infty} \frac{1}{x^p}\, dx$ converges if $p > 1$) and the p-series diverges if $p \leq 1$ (since $\int_1^{\infty} \frac{1}{x^p}\, dx$ diverges if $p \leq 1$).

Exercise Set 6.5

Use the integral test to determine whether each infinite series converges or diverges.

1. $\displaystyle\sum_{n=1}^{\infty} \frac{\ln n}{n}$

2. $\displaystyle\sum_{n=2}^{\infty} \frac{1}{n \ln n}$

3. $\displaystyle\sum_{n=1}^{\infty} n \cdot e^{-n^2}$

4. $\displaystyle\sum_{n=2}^{\infty} \frac{n}{n^2 - 1}$

5. $\displaystyle\sum_{n=2}^{\infty} \frac{2}{n^2 - 2}$

6. $\displaystyle\sum_{n=1}^{\infty} \frac{1}{2n - 1}$

6.6 The Exponential Probability Distribution

A certain brand of light bulb is advertised as having an average life of 1000 hours. The values in Table 6.23 represent the number of hours until failure (burnout) for a sample of 100 of these light bulbs.

440	301	515	813	1064	1259	1117	1901	127	933
2227	83	609	1906	549	853	27	618	542	814
2392	1637	1604	372	683	1580	427	488	627	415
365	1948	1146	419	122	417	85	2720	70	413
4265	1109	2647	465	5	49	1521	329	374	480
272	102	2386	2029	942	2063	8	2572	1909	530
1488	338	59	3375	1233	393	749	450	1046	366
294	768	3680	1151	25	3622	158	636	276	156
165	2929	2922	379	412	367	701	614	2179	581
877	55	398	714	2843	2517	626	593	2067	613

Table 6.23 Time to failure for a sample of 100 light bulbs

The histogram shown in Figure 6.24 shows the distribution of the sample using an interval length of 0.2 thousand hours on the horizontal axis. We can see that some of the light bulbs fail in a short time, while some last several thousand hours. The average lifetime for this sample is 1015 hours, so the advertised lifetime of 1000 hours is a reasonably accurate statement. The median lifetime, however, is 616 hours. (The median of a data set is the number such that half of the values are greater than it and half of the values are less than it.) So even though the average lifetime of a bulb is about 1000 hours as advertised, most of the bulbs fail in less time.

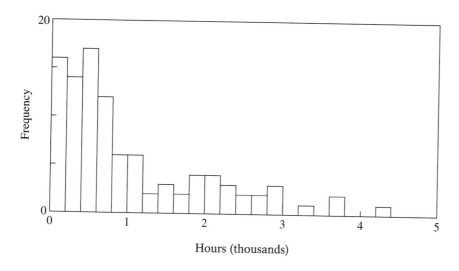

Figure 6.24 Histogram for light bulb lifetimes

What probability distribution function describes the time until failure for these light bulbs? In other words, what function can we use to estimate the probability that a light bulb will fail within q hours, where q might be 500, 1000, or 2000?

We can transform the histogram in Figure 6.24 into a probability histogram, which means the total area equals 1, if we divide the heights of the rectangles by the total area under the histogram. The original histogram consists of 22 rectangles over the domain from 0 to 4.4, each of which has width 0.2 thousand hours. Some of the rectangles have height 0. The area of each rectangle is equal to the height multiplied by the width. Numbering the rectangles from $i = 1$ to $i = 22$, the sum of the areas of the rectangles is

$$\sum_{i=1}^{22} 0.2 \cdot f_i,$$

with the height of the ith rectangle given by the associated frequency f_i. This summation can be rewritten as

$$0.2 \cdot \sum_{i=1}^{22} f_i.$$

Since there are 100 light bulbs in our sample, the sum of the frequencies is 100, so the total area of the frequency histogram is $0.2 \cdot 100$, or 20. Dividing the height of each rectangle by 20 gives the rescaled histogram shown in Figure 6.25. Each rectangle in this histogram has an area that represents the proportion of the light bulbs in the sample that fail in the time interval covered by the rectangle. This proportion is an estimate for the probability that any randomly selected light bulb will fail during that time interval.

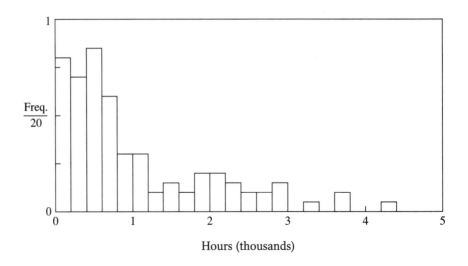

Figure 6.25 Probability histogram

We now wish to fit a continuous probability density function to the probability histogram in Figure 6.25. The main feature of this histogram is that it starts high and gradually drops off to zero. Two elementary functions have these characterisitcs, a reciprocal power function and a decreasing exponential function. A reciprocal power function has a vertical asymptote, which is not a characteristic of our data; therefore, we will investigate the fit of an exponential function.

We can construct a data set to represent the histogram by using the midpoint of each interval of the histogram as the value of the independent variable and the height of each rectangle as the corresponding value of the dependent variable. This yields the data set and scatter plot shown in Figure 6.26.

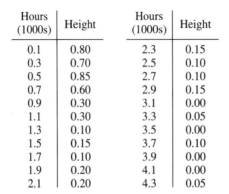

Hours (1000s)	Height	Hours (1000s)	Height
0.1	0.80	2.3	0.15
0.3	0.70	2.5	0.10
0.5	0.85	2.7	0.10
0.7	0.60	2.9	0.15
0.9	0.30	3.1	0.00
1.1	0.30	3.3	0.05
1.3	0.10	3.5	0.00
1.5	0.15	3.7	0.10
1.7	0.10	3.9	0.00
1.9	0.20	4.1	0.00
2.1	0.20	4.3	0.05

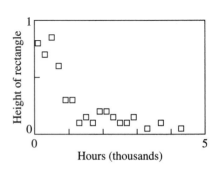

Figure 6.26 Data set and scatter plot corresponding to probability histogram

To determine whether the data set in Figure 6.26 is exponential, we re-express the data by taking the natural logarithm of the non-zero values of the dependent variable. This re-expression yields the data set with the scatter plot shown in Figure 6.27. The re-expressed data has a linear trend with negative slope, so we are reasonably confident that a decreasing exponential function fits the shape of the original data. To find the equation of the exponential function, we could use the least squares line for the re-expressed data, or we could fit an exponential least squares curve to the data set in Figure 6.26. Neither of these options is appropriate, however, because we are not simply fitting an arbitrary exponential curve to the data. We are trying to fit a probability density function, and a PDF must enclose an area of 1 between its graph and the horizontal axis.

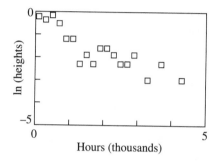

Figure 6.27 Scatter plot of re-expressed data

The equation of a decreasing exponential function p that has the horizontal axis as an asymptote must have the form

$$p(x) = a \cdot e^{-bx},$$

where a and b are positive parameters. If p is a PDF for the time until failure for light bulbs, then its domain is all non-negative real numbers. Furthermore, the area under the graph of p over its domain must be 1, so

$$\int_0^\infty a \cdot e^{-bx} dx = 1.$$

We can determine the restrictions that the area requirement places on a and b by evaluating this improper integral.

$$\int_0^\infty a \cdot e^{-bx}\,dx = \lim_{k\to\infty} \int_0^k a \cdot e^{-bx}\,dx$$

$$= \lim_{k\to\infty} \frac{a}{-b} \cdot e^{-bx}\Big|_0^k$$

$$= \lim_{k\to\infty} -\frac{a}{b}\left(e^{-bk} - e^0\right)$$

$$= -\frac{a}{b}(0-1) = \frac{a}{b}.$$

Since $\frac{a}{b}$ must equal 1, a must equal b, and the equation for p has the form

$$p(x) = a \cdot e^{-ax}.$$

We seek the value of a for which the graph of p fits the scatter plot in Figure 6.26. Using a computer or calculator, we can experiment with the value of a with the goal of minimizing the residuals. The value of a that minimizes the sum of the squares of the residuals is approximately 0.979; therefore, a PDF for the time until failure for this particular brand of light bulb is

$$p(x) = 0.979 \cdot e^{-0.979x}. \tag{15}$$

A graph of $p(x)$ superimposed on the scatter plot of the data from the heights of the histogram is shown in Figure 6.28.

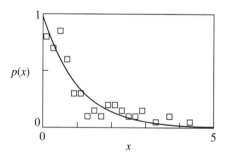

Figure 6.28 The PDF $p(x)$ fit to the data from the histogram

Example 1

Based on the exponential density function p in (15), what is the probability that a randomly selected light bulb fails within the first 1000 hours of use? How does this probability compare with the proportion of failures from the data set given in Table 6.23?

Solution   The probability of failure within the first 1000 hours equals the area under the graph of $p(x)$ over the interval from $x = 0$ to $x = 1$, since the units of x are thousands of hours. This area is represented by a definite integral, which is evaluated analytically as follows:

$$
\begin{aligned}
P(0 \leq x \leq 1) &= \int_0^1 0.979 \cdot e^{-0.979x} \, dx \\
&= -e^{-0.979x} \Big|_0^1 \\
&= -e^{-0.979} + e^0 \\
&\approx 0.624.
\end{aligned}
$$

In the sample, 65 of the 100 light bulbs failed within 1000 hours, a proportion that compares favorably with the theoretical probability of 0.624. ■

The PDF in (15) for the time until failure for light bulbs is an example of an ***exponential probability distribution function***. The equation of an exponential PDF usually is written as $f(x) = \lambda \cdot e^{-\lambda x}$, where λ (the Greek letter lambda) is a positive parameter that depends upon the phenomenon being modeled. Exponential probability density functions are used to model events such as the distance between successive cars on a highway, the time between calls at a telephone hot-line, and the life expectancy of electronic components. The domain of an exponential PDF is the set of non-negative real numbers.

exponential probability distribution function

Example 2

Suppose you are the luncheon host at a restaurant and you know that the time between the arrivals of successive customers, measured in minutes, is exponentially distributed with $\lambda = 1$. The distribution of the time between arrivals tells you how long you are likely to wait between successive customers. After a customer arrives, how likely is it that you will wait more than 2 minutes before your next customer arrives?

Solution   The PDF for an exponential distribution with $\lambda = 1$ is $p(x) = e^{-x}$, so the probability of waiting more than 2 minutes for the next customer is given by

$$
P(x \geq 2) = \int_2^\infty e^{-x} \, dx,
$$

which is equivalent to

$$
\lim_{b \to \infty} \int_2^b e^{-x} \, dx.
$$

This expression is evaluated as follows:

$$\lim_{b \to \infty} \int_2^b e^{-x} dx = \lim_{b \to \infty} -e^{-x} \Big|_2^b$$

$$= \lim_{b \to \infty} -e^{-b} + e^{-2}$$

$$= e^{-2}$$

$$\approx 0.135.$$

This means that approximately 13.5% of the time you will wait more than two minutes before your next customer arrives. ■

In problem 11 of the following exercise set, you are asked to show that an exponential distribution is "memoryless," meaning that the probability of a product failing in the next k years given the product has lasted a certain number of years is the same as the probability of a product failing in the first k years. For example, if the life span of a type of electronic component is distributed exponentially, then the probability that one component fails during the first two years of service is the same as the probability that the component fails between years 22 and 24, provided the component has already lasted 22 years. Furthermore, the memoryless property implies that the rate of failure for products with an exponential distribution is constant. A constant failure rate means that the number of products that will fail in the next year, expressed as a percentage of the products that have not yet failed, is a constant and does not depend on the time that the products have been in service. The constant failure rate for exponential distributions is analogous to the constant percentage growth rate property for exponential functions.

Exercise Set 6.6

1. Sketch the graphs of the exponential probability density functions with $\lambda = 1$, 0.5, 0.1, and 0.05. Describe what information each graph gives you about the distribution. How do exponential distributions compare to normal distributions?

2. The lengths of phone calls to a business, measured in minutes, have an exponential distribution. List some reasons that might cause the value of λ for one business to be different from the λ value for another.

3. Show that the area under the graph of an exponential density function is exactly 1 by showing that $\int_0^\infty \lambda \cdot e^{-\lambda x} dx = 1$.

4. In problem 3 you showed that $\int_0^\infty \lambda \cdot e^{-\lambda x} dx = 1$. If we split the domain from 0 to ∞ at some point $x = k$, then the left side of this equation can be split into two integrals, so that

$$\int_0^k \lambda \cdot e^{-\lambda x} dx + \int_k^\infty \lambda \cdot e^{-\lambda x} dx = 1.$$

We can rewrite this equation in the form

$$\int_0^k \lambda \cdot e^{-\lambda x} dx = 1 - \int_k^\infty \lambda \cdot e^{-\lambda x} dx.$$

Since each of the integrals represents a probability, explain what this equation means in terms of the probability of time to failure of a piece of equipment.

5. The PDF for the time t to failure (t is in months) of a particular engine ignition component is given by $f(t) = \frac{1}{40} e^{-0.025t}$. Find the probability that a randomly selected component will last

   a. at most 20 months;

   b. at least 40 months; and

   c. between 40 and 50 months.

6. The length of time, measured in minutes, that customers must wait for service at a fast food restaurant has an exponential density function $f(t) = \frac{1}{3} e^{-t/3}, t > 0$.

   a. Find the probability that a customer will have to wait more than 3 minutes for service.

   b. Suppose the restaurant advertises that they will give a free order of fries to anyone who has to wait more than a certain number of minutes for service. If the restaurant manager does not want to give away fries to more than 10% of the customers, what is the least number of minutes they can advertise?

7. The probability that a radio transistor fails after a months and before b months is $\int_a^b \lambda \cdot e^{-\lambda t} dt$ for some constant λ. The probability that it will fail in the first year is 0.2. What is the probability that the transistor will fail during the second year?

8. Let x represent the life span in hours of an electrical component with the following probability distribution function:

$$f(x) = \begin{cases} \dfrac{1}{50} e^{-\frac{x}{50}} & \text{if } x > 0, \\ 0 & \text{elsewhere.} \end{cases}$$

   a. Determine the probability that the life span of a randomly selected component is less than 50 hours.

   b. Determine the probability that the life span of a randomly selected component will exceed 70 hours.

   c. What life span will half of the components exceed?

9. What is the median time between customer arrivals at a bank if the interarrival times are exponentially distributed with $\lambda = 0.45$?

10. What is the median time to failure for light bulbs that have an exponential probability distribution for time to failure, with $\lambda = 0.979$? How does this compare with the median for the sample of 100 light bulbs in Table 6.23?

11. Suppose a product has a life span that is distributed exponentially with x measured in years.

    a. Find the probability that the product will fail in the first year by calculating the value of $\int_0^1 \lambda \cdot e^{-\lambda x} dx$.

    b. Suppose the product has already lasted for five years. What is the probability that it will fail within the next year? Since the product has already lasted five years, the domain of the distribution is now $5 < x < \infty$. The probability desired is the ratio of the area under the graph of the PDF between $x = 5$ and $x = 6$ to the area of the sample space, so we compare the area for $5 < x < 6$ to the area for $5 < x < \infty$. The probability is, therefore,

    $$\frac{\int_5^6 \lambda \cdot e^{-\lambda x} dx}{\int_5^\infty \lambda \cdot e^{-\lambda x} dx}.$$

    Compute this ratio and compare it with the answer to part a.

    c. Show that the probability of the product failing in the first k years is the same as the probability of the product failing in k *additional* years, given that it has already lasted n years.

    d. Many electrical components for automobiles have a life span that is distributed exponentially. What do you think this result has to do with the typical policy of auto shops, which is not to allow returns on electrical components?

    e. Show that the probability of the product lasting at least k years is the same as the probability of the product lasting at least k *additional* years, given that it has already lasted n years. This shows that an exponential distribution is "memoryless."

6.7 Expected Value

Suppose you are planning a fund raising game night for your school. One student proposes a dice game that costs $1.50 to play. Players get one roll of a die with the following payoffs:

– if an odd number is rolled, the player wins nothing;

– if a 2 or 4 is rolled, the player wins a small prize that costs the school $1.95;

– if a 6 is rolled, the player wins a large prize that costs the school $4.50.

If 100 people play this game, how much profit do you expect to make for the school?

To help answer the question, consider the following chart which gives the school's profit for each of the six possible rolls of the die:

Number	1	2	3	4	5	6
Profit	$1.50	–$0.45	$1.50	–$0.45	$1.50	–$3.00

The sum of the six numbers in the profit column is $0.60. If each of the numbers 1 through 6 appears once in six plays, then the school receives 60 cents in profit, for an average of 10 cents profit per play. Though it is unlikely that all six numbers will occur within six plays, in the long run we do expect that each of the six numbers will occur about the same number of times. In the long run, therefore, we expect this game to yield an average of 10 cents profit per play, an amount that is called the ***expected value*** of this game. Based on this expected value, the school should make about $10.00 if the game is played 100 times.

expected value

For this simple game, the expected value is the sum of the profits divided by 6, which is

$$\frac{1.50 - 0.45 + 1.50 - 0.45 + 1.50 - 3.00}{6}.$$

Instead of adding the six numbers individually, we gain an important insight about the concept of expected value if we group the numbers as in the expression

$$\frac{3(1.50) + 2(-0.45) + 1(-3.00)}{6}.$$

Rewriting this expression yields

$$\frac{3(1.50)}{6} + \frac{2(-0.45)}{6} + \frac{1(-3.00)}{6},$$

which is equivalent to

$$\frac{3}{6}(1.50) + \frac{2}{6}(-0.45) + \frac{1}{6}(-3.00). \tag{16}$$

Notice that $\frac{3}{6}$ is the probability of rolling an odd number, and $1.50 is the profit for the school when a player rolls an odd number; $\frac{2}{6}$ is the probability of rolling a 2 or 4, each of which gives the school a loss of 45 cents; and $\frac{1}{6}$ is the probability of rolling a 6, in which case the school loses $3.00. Grouping the numbers as in (16) leads us to believe that the expected value can be calculated by multiplying each possible outcome and its probability and then adding the products.

In general, for any random experiment with discrete outcomes $x_1, x_2, \ldots, x_n$, each with a probability of occurrence $p(x_i)$, the expected value E of this experiment is

$$E = \sum_{i=1}^{n} x_i \cdot p(x_i).$$

mean

If the experiment is repeated a large number of times, we expect the average of the outcomes to be about E. The expected value is also called the **mean** of the outcomes.

Now we wish to extend the concept of expected value to situations in which the outcomes can take on any value along a continuum and are distributed according to a continuous probability density function. In these situations, we cannot list the possible outcomes. We will investigate the continuous case in the context of the following scenario.

A wholesale supplier knows that the number of gallons of a particular soft drink that a typical customer will order in any week is distributed according to the PDF

$$s(x) = \begin{cases} x^3 & \text{if } 0 \le x \le 1; \\ 1 & \text{if } 1 < x \le 1.75; \text{ and} \\ 0 & \text{elsewhere,} \end{cases}$$

where x is in units of 100 gallons. A graph of $s(x)$ is shown in Figure 6.29. This PDF fits a probability histogram constructed from data for the number of gallons of soft drink per week that customers order. How can we use this PDF to determine the expected value of the weekly order for a typical customer?

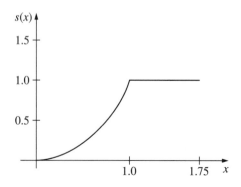

Figure 6.29 PDF for soft drink orders

Since x can take on any value in the interval from 0 to 1.75, we cannot determine the expected value by calculating the sum of all the products of the x values and their probabilities. Suppose instead that we divide the interval from 0 to 1.75 into n equal length subintervals, where n is large. We label the left endpoints of the subintervals x_0, x_1, x_2, and so on, through x_{n-1}, with values given by the recursive equations

$$x_0 = 0,$$
$$x_i = x_{i-1} + \Delta x.$$

Since s is a PDF, the probability $P(i)$ of choosing an x value in the subinterval with endpoints x_i and x_{i+1} is equal to the area under $s(x)$ between x_i and x_{i+1}. This area is approximately equal to the area of a rectangle of width Δx and height given by $s(x_i)$. In Figure 6.30, $P(i)$ is the area of the shaded region, and the rectangular approximation to this region is also shown. We can write this approximation as

$$P(i) \approx s(x_i) \cdot \Delta x. \tag{17}$$

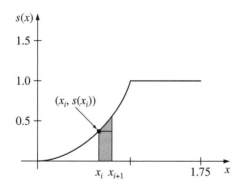

Figure 6.30 Approximation of $P(i)$ with the area of a rectangle

The product $s(x_i) \cdot \Delta x$ in (17) is an approximation for the probability of choosing an x value between x_i and x_{i+1}. The sum of the products of x_i and $s(x_i) \cdot \Delta x$ is an approximation for the expected value E of the set of x values with PDF s with one x value chosen in each subinterval. We write this approximation as

$$E \approx \sum_{i=0}^{n-1} x_i \cdot s(x_i) \cdot \Delta x. \tag{18}$$

As the number of subintervals increases without bound, or as Δx goes to 0, the approximation for the expected value in (18) has a limit equal to the expected value of the continuum of x values. This limit has the form

$$\lim_{\Delta x \to 0} \sum_{i=0}^{n-1} x_i \cdot s(x_i) \cdot \Delta x. \tag{19}$$

The expression in (19) is the limit of a Riemann sum, which by definition is a definite integral. Therefore, the expected value E of the continuous set of x values representing soft drink orders is

$$E = \int_0^{1.75} x \cdot s(x)\, dx.$$

Since s is a piecewise-defined function, we split the interval between the limits of integration according to the two non-zero pieces of s, so that

$$E = \int_0^1 x \cdot x^3\, dx + \int_1^{1.75} x \cdot 1 \cdot dx.$$

Evaluating these integrals, we have

$$E = \left.\frac{x^5}{5}\right|_0^1 + \left.\frac{x^2}{2}\right|_1^{1.75}$$

which simplifies to

$$E = \left(\frac{1}{5} - 0\right) + \left(\frac{1.75^2}{2} - \frac{1}{2}\right) \approx 1.23.$$

The expected value of the demand for soft drinks is about 123 gallons a week since x is in units of 100 gallons. This means that if a large sample of the retail stores that are customers of this supplier is chosen at random, we would expect the average of their orders to be about 123 gallons.

expected value E of a random event

In general, the **expected value E of a random event** distributed according to a continuous PDF is

$$E = \int_a^b x \cdot f(x)\, dx,$$

where f is the PDF for x, and the interval $a \le x \le b$ is the domain of f.

Example

What is the expected value of the standard normal distribution (see section 6.3)?

Solution   The expected value E of the standard normal distribution is

$$E = \int_{-\infty}^{\infty} x \cdot N(x)\, dx$$

$$= \int_{-\infty}^{\infty} \frac{1}{\sqrt{2\pi}}\, x e^{-\frac{1}{2}x^2}\, dx.$$

This integral is evaluated as follows:

$$E = \int_{-\infty}^{\infty} \frac{1}{\sqrt{2\pi}} x e^{-\frac{1}{2}x^2} dx$$

$$= \lim_{a \to -\infty} \int_a^0 \frac{1}{\sqrt{2\pi}} x e^{-\frac{1}{2}x^2} dx + \lim_{b \to \infty} \int_0^b \frac{1}{\sqrt{2\pi}} x e^{-\frac{1}{2}x^2} dx$$

$$= \lim_{a \to -\infty} \frac{-1}{\sqrt{2\pi}} e^{-\frac{1}{2}x^2} \Big|_a^0 + \lim_{b \to \infty} \frac{-1}{\sqrt{2\pi}} e^{-\frac{1}{2}x^2} \Big|_0^b$$

$$= \lim_{a \to -\infty} \left[-\frac{1}{\sqrt{2\pi}} e^0 + \frac{1}{\sqrt{2\pi}} e^{-\frac{1}{2}a^2} \right] + \lim_{b \to \infty} \left[-\frac{1}{\sqrt{2\pi}} e^{-\frac{1}{2}b^2} + \frac{1}{\sqrt{2\pi}} e^0 \right]$$

$$= -\frac{1}{\sqrt{2\pi}} + \frac{1}{\sqrt{2\pi}}$$

$$= 0.$$

The expected value being 0 means that if random selections are made from a set of numbers with a standard normal distribution, then we expect the average of these selections to be about 0. This is consistent with the fact that the mean μ of the standard normal distribution is 0. ∎

Exercise Set 6.7

1. Let x represent the life span in hundreds of hours of a particular engine part. The probability density function is given by $f(x) = \frac{3}{x^4}$, $x \geq 1$.

    a. Sketch a graph of $f(x)$.

    b. Verify that f satisfies the definition of a PDF.

    c. Find the expected value of the life span of this engine part.

    d. Find the median value for the life span of this engine part.

    e. What characteristic of f makes the expected value and median unequal?

2. The expected value of the exponential distribution $f(x) = \lambda e^{-\lambda x}$ is given by $\int_0^\infty \lambda x e^{-\lambda x} dx$. Though we have not yet discussed an analytical technique for evaluating this integral, we can use numerical methods to evaluate it. Estimate the expected value of $f(x) = \lambda e^{-\lambda x}$ by evaluating the integral numerically for various values of λ, such as $\lambda = 0.05$, 0.1, 0.25, 1, 2, 5. Based on the results, what do you think is the expected value of an exponential distribution in terms of the parameter λ?

3. The length of time, measured in minutes, that customers must wait for service at a fast food restaurant has an exponential density function $f(t) = \frac{1}{3} e^{-t/3}$, $t > 0$.

    a. Find the median length of time that customers must wait for service.

    b. Estimate the expected length of time that customers must wait for service.

6.8   Integration by Parts

Problem 2 of the previous exercise set asked you to find the expected value of an exponential distribution by finding the value of the integral $\int_0^\infty \lambda x\, e^{-\lambda x} dx$. To calculate the value of this integral, you must have used a numerical method because you did not know an antiderivative of the integrand. Now we will examine a technique for finding such an antiderivative.

The technique of substitution of variables is used to integrate some products. For example, the integrand in $\int x \cdot \sin x^2 dx$ is the product of x and $\sin x^2$. This indefinite integral can be evaluated by using substitution of variables with $u = x^2$ and $du = 2x\, dx$, so that

$$\int x \cdot \sin x^2 dx = \int \sin u \cdot \frac{1}{2}\, du = -\frac{1}{2}\cos u + C,$$

which equals $-\frac{1}{2}\cos x^2 + C$. This technique is based on the chain rule for derivatives.

Often, however, we encounter an integrand that is a product of functions that cannot be connected through the chain rule. In these situations, substitution of variables does not help in finding an antiderivative. For example, substitution of variables does not help us deal with the product in $\int x \cdot \sin x\, dx$. A technique known as ***integration by parts*** will enable us to find this antiderivative.

integration by parts

To develop the integration by parts technique, we start with the product rule for derivatives, which is

$$\frac{d}{dx}[f(x)g(x)] = f(x) \cdot \frac{d}{dx} g(x) + g(x) \cdot \frac{d}{dx} f(x). \tag{20}$$

If we integrate both sides of equation (20) with respect to x, we have

$$\int \left(\frac{d}{dx}[f(x)g(x)] \right) dx = \int \left[f(x) \cdot \frac{d}{dx} g(x) \right] dx + \int \left[g(x) \cdot \frac{d}{dx} f(x) \right] dx. \tag{21}$$

The left side of equation (21) is the antiderivative of a derivative, so by simplifying and rearranging terms, we have

$$\int \left[f(x) \cdot \frac{d}{dx} g(x) \right] dx = f(x) \cdot g(x) - \int \left[g(x) \cdot \frac{d}{dx} f(x) \right] dx. \tag{22}$$

We can write (22) as

$$\int \left(f \cdot \frac{dg}{dx} \right) dx = f \cdot g - \int \left(g \cdot \frac{df}{dx} \right) dx,$$

which, from our work with differentials in Section 5.7, we know is equivalent to

$$\int f\, dg = f \cdot g - \int g\, df. \tag{23}$$

From equation (23), we see that an integral of the form $\int f\,dg$ can be rewritten as the difference between the product $f \cdot g$ and another integral $\int g\,df$. What is the advantage of turning $\int f\,dg$ into $f \cdot g - \int g\,df$? Such an exchange is helpful only if the second integral is simpler than the first. In other words, equation (23) allows us to trade integrals, which is useful only if we can trade a difficult integral for an easier one.

As an example, consider the integral $\int x\,e^{-x}dx$. The integrand is a product, so we want to think of $x \cdot e^{-x}dx$ as $f\,dg$. The key question with integration by parts always is: what is f and what is dg? In this case, we have two choices, $f = x$ and $dg = e^{-x}dx$, or $f = e^{-x}$ and $dg = x\,dx$. In either case we need to find df and g. That is, we need to differentiate f and integrate dg. The consequences of each of these two choices are outlined below:

1. If $f = x$ and $dg = e^{-x}dx$, then $df = dx$ and $g = -e^{-x}$, so $\int xe^{-x}dx$ becomes $-x \cdot e^{-x} - \int -e^{-x}dx$. Note that $\int -e^{-x}dx$ is an integral that we can evaluate analytically.

2. If $f = e^{-x}$ and $dg = x\,dx$, then $df = -e^{-x}dx$ and $g = \frac{1}{2}x^2$, so $\int xe^{-x}dx$ becomes $\frac{1}{2}(x^2 \cdot e^{-x}) - \int \left(\frac{1}{2}x^2\right) \cdot \left(-e^{-x}\right)dx$. Note that $\int \left(\frac{1}{2}x^2\right) \cdot \left(-e^{-x}\right)dx$ is a more difficult integral than the original one.

Clearly the first choice is a better one, as it allows us to trade an integral that we cannot evaluate for one that we can evaluate.

Completing the integration in the first choice, we have

$$\int x \cdot e^{-x}dx = -xe^{-x} - \int -e^{-x}dx$$

$$= -xe^{-x} - e^{-x} + C.$$

We can check this result by verifying that the derivative of our answer is equal to the integrand, that is, $\frac{d}{dx}\left(-xe^{-x} - e^{-x} + C\right) = xe^{-x}$:

$$\frac{d}{dx}\left(-xe^{-x} - e^{-x} + C\right) = (-x)\left(-e^{-x}\right) + \left(e^{-x}\right)(-1) - \left(-e^{-x}\right)$$

$$= xe^{-x} - e^{-x} + e^{-x}$$

$$= xe^{-x}.$$

Example 1

Find the expected value of the exponential PDF $\lambda \cdot e^{-\lambda x}$.

Solution   Since the domain of the exponential PDF is the set of nonnegative real numbers, the expected value is given by the integral

$$\int_0^\infty x \cdot \lambda e^{-\lambda x}dx.$$

Applying the technique of integration by parts to the integrand, we let $f = x$ and $dg = \lambda e^{-\lambda x} dx$, so that $df = dx$ and $g = -e^{-\lambda x}$. Applying equation (23) to the integral $\int x \cdot \lambda e^{-\lambda x} dx$, we have

$$\int x \cdot \lambda e^{-\lambda x} dx = x\left(-e^{-\lambda x}\right) - \int \left(-e^{-\lambda x}\right) dx$$

$$= -xe^{-\lambda x} - \frac{1}{\lambda} e^{-\lambda x} + C.$$

The expected value of the exponential distribution is given by

$$\int_0^\infty x \cdot \lambda e^{-\lambda x} dx = \lim_{b \to \infty} \int_0^b x \cdot \lambda e^{-\lambda x} dx$$

$$= \lim_{b \to \infty} \left(-xe^{-\lambda x} - \frac{1}{\lambda} e^{-\lambda x}\right)\Big|_0^b$$

$$= \lim_{b \to \infty} \left[-be^{-\lambda b} - \frac{1}{\lambda} e^{-\lambda b} - \left(0 - \frac{1}{\lambda}\right)\right].$$

We know that $\lim_{b \to \infty} \frac{1}{\lambda} e^{-\lambda b} = 0$, and we know that $\lim_{b \to \infty} \frac{1}{\lambda} = \frac{1}{\lambda}$. However, evaluating $\lim_{b \to \infty} \left(-be^{-\lambda b}\right)$ presents a problem. As b increases without bound, the expression $be^{-\lambda b}$ is the product of a quantity that increases without bound and a quantity that goes to 0. Does the limit of this product increase without bound, does it approach 0, or does it approach some other value? This situation is similar to the dilemma posed by limits of the form $\frac{0}{0}$ that we encountered at the end of Chapter 2. Recall that l'Hôpital's rule states that if $\lim_{x \to a} f(x) = 0$ and similarly $\lim_{x \to a} g(x) = 0$, then $\lim_{x \to a} \frac{f(x)}{g(x)} = \lim_{x \to a} \frac{f'(x)}{g'(x)}$.

We can rewrite $\left(-be^{-\lambda b}\right)$ as $\left(\frac{-b}{e^{\lambda b}}\right)$, which is of the form $\frac{\infty}{\infty}$, or the ratio of two quantities increasing without bound. An extension of l'Hôpital's rule states that if $\lim_{x \to \infty} f(x) = \infty$ and $\lim_{x \to \infty} g(x) = \infty$, then $\lim_{x \to \infty} \frac{f(x)}{g(x)} = \lim_{x \to \infty} \frac{f'(x)}{g'(x)}$. Applying this extension of l'Hôpital's rule yields

$$\lim_{b \to \infty} \left(\frac{-b}{e^{\lambda b}}\right) = \lim_{b \to \infty} \left(\frac{-1}{\lambda e^{\lambda b}}\right) = 0.$$

Returning to the problem of finding the expected value of an exponential distribution, we have

$$\int_0^\infty x \cdot \lambda e^{-\lambda x} dx = \lim_{b \to \infty} \left(-be^{-\lambda b} - \frac{1}{\lambda} e^{-\lambda b} + \frac{1}{\lambda}\right) = \frac{1}{\lambda}.$$

The expected value of an exponential distribution with parameter λ is $\frac{1}{\lambda}$. ∎

Example 2

Evaluate $\int x \cos x \, dx$.

Solution   If we choose $f = x$ and $dg = \cos x \, dx$, then $df = dx$ and $g = \sin x$. By substitution, we have

$$\int x \cos x \, dx = x \cdot \sin x - \int \sin x \, dx,$$

which simplifies to

$$\int x \cos x \, dx = x \cdot \sin x + \cos x + C. \qquad \blacksquare$$

When using integration by parts, knowing how to choose f and dg may seem difficult at first, but skill can be acquired through practice. As we apply the technique of integration by parts, two important guidelines will be helpful. First, try not to pick an expression for dg that we cannot integrate. Second, try to pick an expression for f that simplifies when differentiated. Using these two rules, we can try various possibilities and see what happens. In addition, the new integral that results from an application of integration by parts may require another application of integration by parts, a substitution, partial fractions, or some other technique of integration.

Exercise Set 6.8

1. Evaluate the following integrals by the method of integration by parts.

    a. $\int x \cdot \ln x \, dx$
    b. $\int x \cdot \sin x \, dx$

    c. $\int x^2 \cdot \sin x \, dx$
    d. $\int x \cdot e^x \, dx$

    e. $\int_1^e \ln x \, dx$
    f. $\int \sec^3 x \, dx$

    g. $\int_0^1 \tan^{-1} x \, dx$
    h. $\int x^2 \cdot e^x \, dx$

    i. $\int \sin^2 x \, dx$
    j. $\int e^x \cdot \sin x \, dx$

    k. $\int \ln(x^2 - 1) \, dx$
    l. $\int_{-1}^1 (6x + 5) \cdot e^x \, dx$

    m. $\int_0^\pi x \cdot \cos(3x) \, dx$
    n. $\int x^n \cdot \ln x \, dx$

2. a. If you have a data set that is distributed exponentially, how would you find the appropriate value to use for λ?

    b. The time between arrivals of phone calls at a mail-order warehouse is exponentially distributed. The average time between arrivals of calls is 1.25 minutes. What is the probability that the next call will arrive within the next minute?

c. The number of people waiting in line at a fast food store is exponentially distributed. The mean number in line is 1. How likely is it to find more than 3 people in line at this store? How likely would it be if the median number in line were 1?

3. One of the duties of the switchboard operator at a mail-order house is to answer incoming calls and direct them to the proper department. During the operator's first morning on the job, the times (in minutes) between calls arriving at the switchboard are:

7.1	3.8	10.9	3.7	9.0	14.8	5.9	0.3	4.5
0.7	1.3	2.4	11.3	1.2	3.3	2.5	2.0	0.2
2.1	7.7	0.9	5.0	1.7	2.5	1.4		

a. Construct a histogram of the data. Does it appear to be exponentially distributed?

b. Assuming the times between calls are exponentially distributed, what is an estimate of λ?

c. Assume that times between calls are exponentially distributed. The times between calls vary from 0.2 minutes to 14.8 minutes. Within what interval do you expect the shortest 20% of the times to fall? That is, find k so that $P(0 \le t \le k) = 0.20$. Repeat for 40%, 60%, 80%, and 100% of the times.

d. Consider the shortest 20% of the actual times between calls. What is the actual interval for these times? Repeat for 40%, 60%, 80%, and 100% of the times.

e. Based on the results of parts c and d, how justified is the assumption that this data is exponentially distributed? Explain your reasoning.

4. A sales representative working for the company in problem 3 records the length of each call the representative receives. The following are sample data for one day.

3.5	5.6	9.0	0.5	7.3	6.1	3.9	3.3	12.0
2.7	6.2	7.1	3.2	4.7	5.2	8.8	10.5	1.3
8.1	6.9	2.4	13.3	5.1	4.4	7.4	3.8	

Are the lengths of the calls distributed exponentially? Are they distributed normally? Justify your answer.

5. The distribution of maximum values for physical events such as wind velocities, flood levels, size of earthquakes, and so on, often is described by the *Gumbel distribution*, which has the probability density function

$$G(x) = \frac{1}{\delta} \cdot e^{\frac{x-\lambda}{\delta}} \cdot e^{-e^{\frac{x-\lambda}{\delta}}}$$

The parameters of the distribution are λ, which is the mode, or most likely value, and δ, which is determined by the standard deviation and the number of values in the data set. Suppose you were given the maximum wind velocities for the past 50 years on Grandfather Mountain in North Carolina as well as the values of δ and λ. Explain how you could determine whether the Gumbel distribution is a reasonable model for these maximum wind velocities. Be specific.

Reference
Lawless, J.F. *Statistical Models and Methods for Lifetime Data.* New York: John Wiley and Sons, 1982.

6.9   Time Spent Waiting in Line

All of us have found ourselves at the end of a long line, either at the bank, the grocery store, or a drive-in window, only to find that once we are served, no one else is left in line. If we had come ten minutes later, we would not have had to wait at all. The ebb and flow of lines have been the subject of much study, and we can explain many of our personal experiences by looking at a mathematical model of this process, which is the subject of Lab 14.

Suppose you have a business, the Quick Change oil change station, and in this business you serve customers one at a time. If one customer arrives while you are busy with another, the newcomer must wait until you are finished with your present customer. How long will a customer generally expect to wait before being served? Will there often be a line? If so, how long will it be on average? Our work with the exponential distribution gives us a tool to answer all of these questions. Recall that the exponential density function is $P(x) = \lambda e^{-\lambda x}$, $0 < x < \infty$, and that $\frac{1}{\lambda}$ is the expected value, or mean, of the distribution. The parameter λ is also called the ***rate of the distribution***.

rate of the distribution

We assume that the inter-arrival time of customers is exponentially distributed with an average rate of λ and the service time for each customer is also exponentially distributed with an average rate of μ. In your business, suppose you spend, on average, 10 minutes serving each customer, and customers arrive on average 12 minutes apart. Converting these numbers to rates, we have that $\lambda = 5$ customers per hour and $\mu = 6$ customers per hour.

From our work with the exponential PDF in Section 6.6, we know that this distribution is memoryless. This means that the probability that a customer will arrive within Δt units of time after time t is the same at any time t; therefore, this probability equals the probability that a customer will arrive in the interval of time between $t = 0$ and $t = \Delta t$. If the arrival rate is λ, then the probability that a customer will arrive within the next Δt units of time is given by the definite integral

$$\int_0^{\Delta t} \lambda e^{-\lambda t} dt.$$

The value of this integral can be approximated by a rectangle with height equal to the value of the integrand at $t = 0$ and width equal to the width of the interval of integration. The height of this rectangle is λ, and the width is Δt. Thus, we write

$$\int_0^{\Delta t} \lambda e^{-\lambda t} dt \approx \lambda \, \Delta t.$$

We assume that Δt is sufficiently small that there is only one arrival or one departure in any given time interval of duration Δt.

With this information, we are ready to begin building the model. Let S_j be the state of having j customers in the system. The state of the system is characterized by the number of customers waiting in line plus the customer being served at the Quick Change station. If $j = 3$, then one customer is being served while two are waiting, and the system is in state S_3. Let $P_j(t)$ be the probability of being in state S_j at time t.

The crucial step in building our model comes from considering the question, "How can the system be in state S_j at time $t + \Delta t$?" We answer this question relative to the state of the system at time t. For example, if we let $\Delta t = 1$, how can the system be in state S_3 (3 people) at time $t = 10$? We examine all possible configurations at time $t = 10 - \Delta t = 9$ that could result in 3 people in the system at time $t = 10$. One way to have 3 people in the system at time 10 is to have 2 people in the system at time 9, and have one person come in and no one leave during the interval from $t = 9$ to $t = 10$. Here are all the ways to be in S_3 at time $t = 10$:

- have 2 people at time 9 and one enters and no one leaves;
- have 4 people at time 9 and one leaves (after service) and no one enters;
- have 3 people at time 9 and no one enters and no one leaves;
- have 3 people at time 9 and one person enters and one leaves.

There are four possible situations at time 9 that would lead to 3 people at time 10. (Remember, Δt is assumed to be sufficiently small so that there is at most one arrival or one departure in any given time interval.) In each case the probability that one person arrives is approximately $\lambda \, \Delta t$ and that one person leaves (after being served) is approximately $\mu \, \Delta t$. The probabilities that no one arrives and no one leaves are approximately $1 - \lambda \, \Delta t$ and $1 - \mu \, \Delta t$, respectively. In general, there are four ways of being in state S_j at time $t + \Delta t$.

- Be in state S_{j-1} at time t with 1 arriving and 0 leaving. This happens with probability $P_{j-1}(t) \cdot (\lambda \, \Delta t) \cdot (1 - \mu \, \Delta t)$.

- Be in state S_{j+1} at time t with 0 arriving and 1 leaving. This happens with probability $P_{j+1}(t) \cdot (1 - \lambda \, \Delta t) \cdot (\mu \, \Delta t)$.

–   Be in state S_j at time t with 0 arriving and 0 leaving. This happens with probability $P_j(t) \cdot (1 - \lambda\,\Delta t) \cdot (1 - \mu\,\Delta t)$.

–   Be in state S_j at time t with 1 arriving and 1 leaving. This happens with probability $P_j(t) \cdot (\lambda\,\Delta t) \cdot (\mu\,\Delta t)$.

Notice that each of the four ways of being in state S_j at time $t + \Delta t$ have probabilities that include a factor $P_{j-1}(t)$, $P_j(t)$, or $P_{j+1}(t)$, each of which is also a probability.

The four ways of being in state S_j at time $t + \Delta t$ are mutually exclusive. Therefore, the probability $P_j(t + \Delta t)$ of being in state S_j at time $t + \Delta t$ is given by the sum of the probabilities of the four ways of being in state S_j at time $t + \Delta t$, which is

$$P_j(t + \Delta t) = P_{j-1}(t)(\lambda\Delta t)(1 - \mu\Delta t) + P_{j+1}(t)(1 - \lambda\Delta t)(\mu\Delta t)$$
$$+ P_j(t)(1 - \lambda\Delta t)(1 - \mu\Delta t) + P_j(t)(\lambda\Delta t)(\mu\Delta t). \tag{24}$$

Multiplying all the terms on the right side of equation (24) and combining like terms gives

$$P_j(t + \Delta t) = P_{j-1}(t)\lambda\Delta t - P_{j-1}(t)\lambda\mu\Delta t^2 + P_{j+1}(t)\mu\Delta t - P_{j+1}(t)\lambda\mu\Delta t^2 +$$
$$P_j(t) - P_j(t)\lambda\Delta t - P_j(t)\mu\Delta t + 2P_j(t)\lambda\mu\Delta t^2. \tag{25}$$

Ultimately, we would like to have a function for $P_j(t)$. Analytically, this is impossible to find. If we knew the derivative of $P_j(t)$, we would be able to calculate values of $P_j(t)$. We can easily turn equation (25) into a difference quotient. We subtract the term $P_j(t)$ from both sides of equation (25) and then divide both sides by Δt, which yields

$$\frac{P_j(t + \Delta t) - P_j(t)}{\Delta t} = P_{j-1}(t)\lambda - P_{j-1}(t)\lambda\mu\Delta t + P_{j+1}(t)\mu -$$
$$P_{j+1}(t)\lambda\mu\Delta t - (\mu + \lambda)P_j(t) + 2P_j(t)\lambda\mu\Delta t. \tag{26}$$

The left side of equation (26) is a difference quotient $\frac{\Delta P_j}{\Delta t}$. We take the limit of both sides of the equation as $\Delta t \to 0$, so that

$$\lim_{\Delta t \to 0} \frac{P_j(t + \Delta t) - P_j(t)}{\Delta t} = \lim_{\Delta t \to 0} [P_{j-1}(t)\lambda - P_{j-1}(t)\lambda\mu\Delta t + P_{j+1}(t)\mu - P_{j+1}(t)\lambda\mu\Delta t -$$
$$(\mu + \lambda)P_j(t) + 2P_j(t)\lambda\mu\Delta t],$$

which is equivalent to the differential equation

$$\frac{dP_j(t)}{dt} = \lambda P_{j-1}(t) + \mu P_{j+1}(t) - (\mu + \lambda)P_j(t). \tag{27}$$

Equation (27) is valid for $j = 1, 2, 3$, and so on, but not for $j = 0$. It is not valid for $j = 0$ because the right side includes the term $P_{j-1}(t)$, which is not defined for $j = 0$.

Separately, we have to determine the initial equation involving P_0. In the lab, you will be asked to show that

$$P_0(t + \Delta t) = P_1(t) \cdot (1 - \lambda \Delta t) \cdot (\mu \Delta t) + P_0(t) \cdot (1 - \lambda \Delta t) \cdot 1. \qquad (28)$$

Simplifying equation (28) by expanding the last term and subtracting $P_0(t)$ from both sides gives

$$P_0(t + \Delta t) - P_0(t) = P_1(t) \cdot (1 - \lambda \Delta t) \cdot (\mu \Delta t) - P_0(t) \cdot \lambda \Delta t,$$

and dividing by Δt gives

$$\frac{P_0(t + \Delta t) - P_0(t)}{\Delta t} = P_1(t) \cdot (1 - \lambda \Delta t) \cdot \mu - P_0(t) \cdot \lambda. \qquad (29)$$

The limit as Δt approaches 0 of the left side of equation (29) is the derivative $\frac{dP_0(t)}{dt}$, so taking the limit as Δt approaches 0 of both sides of equation (29) yields

$$\frac{dP_0(t)}{dt} = \mu P_1(t) - \lambda P_0(t). \qquad (30)$$

Thus we have the system of differential equations defined by (27) and (30).

Lab 14 *Modeling the Time Spent Waiting in Line*

We are interested in the long-term behavior of the model for time spent waiting in line given by the system of differential equations

$$\frac{dP_0(t)}{dt} = \mu P_1(t) - \lambda P_0(t),$$
$$\frac{dP_j(t)}{dt} = \lambda P_{j-1}(t) + \mu P_{j+1}(t) - (\mu + \lambda) P_j(t). \qquad (31)$$

We do not care very much about the initial behavior of the system, since people might be lined up outside when the doors first open. We care more about the behavior during periods in which things have settled down, which is known as the ***steady state*** of the system. All of our analysis is related to what occurs after the system has reached a steady state.

steady state

A steady state implies that the behavior of the system is not changing, at which time the differential equations in (31) are both equal to 0. If $\frac{dP_0(t)}{dt} = 0$, then from the differential equation in (31), we have $\mu P_1(t) - \lambda P_0(t) = 0$, or

$$P_1(t) = \frac{\lambda}{\mu} P_0(t).$$

If $\dfrac{dP_j(t)}{dt} = 0$, then from the differential equation in (31), we have in steady state

$$\lambda P_{j-1}(t) + \mu P_{j+1}(t) - (\mu + \lambda)P_j(t) = 0,$$

or

$$P_{j+1}(t) = \left(\frac{\mu + \lambda}{\mu}\right)P_j(t) - \frac{\lambda}{\mu}P_{j-1}(t). \tag{32}$$

1. By considering the various ways of being in state S_0 at time $t + \Delta t$, derive the equation for $P_0(t + \Delta t)$ in (28).

2. From this point on we will write P_j instead of $P_j(t)$ to make our algebraic work less cluttered.

    a. By substitution into equation (32), determine values of P_2, P_3, and P_4 in terms of the initial value P_0 and the parameters μ and λ. Determine a general equation for P_j in terms of P_0 and the parameters μ and λ.

    b. Use the fact that the sum of all the probabilities for a fixed t, $\sum_{j=0}^{\infty} P_j$, equals 1 to find P_0 in terms of λ, μ, and j.

3. Use the information from part 2 to evaluate P_j for $j = 1$ to 10 for the Quick Change scenario given at the beginning of Section 6.9, which has $\lambda = 5$ and $\mu = 6$. Based on the evaluation of P_j for $j = 1$ to 10, approximate the probability that there will be more than 3 people in the system after it reaches a steady state. Based on the evaluation of P_j for $j = 1$ to 10, approximately what percentage of the time will a customer have to wait for service?

4. Using the values of P_j from part 3, approximate the expected value of the number of people in the system assuming $\lambda = 5$ and $\mu = 6$.

5. To determine a general expression for the average number of customers in the system, we must evaluate the expected value,

$$E = \sum_{j=0}^{\infty} j \cdot P_j.$$

    To simplify the algebraic manipulations, let $\rho = \frac{\lambda}{\mu}$ and rewrite your expressions from part 2. Rewrite the summand for E in terms of ρ and j only.

    a. In the summation for E, ρ is a constant, so any factors that depend only on ρ can be factored outside the summation. If you factor $\rho(1-\rho)$ from each term, what remains inside the summation is an expression for a derivative with respect to ρ. The expression remaining inside the summation is the derivative of what function?

    b. Recall that the derivative of a sum is equal to the sum of the derivatives. Write the summation as a derivative and evaluate the sum first, then differentiate.

    c. Show that the average number of customers in the system is $\frac{\rho}{1-\rho}$.

6. Assume that a customer arrives at the same moment that another customer is leaving, so that at that moment all the customers are in line. On average, how long will it take the new customer to get through the system assuming $\lambda = 5$ and $\mu = 6$?

7. **Extension:** To simulate an exponential distribution with rate λ using a random number generator, let $x = -\frac{1}{\lambda} \ln r$ where r is a random number between 0 and 1. Create a table of inter-arrival and service times for 100 customers with $\lambda = 5$ and $\mu = 6$. Use these values to determine the arrival and exit times from the business. How does the average length of time in the system in the simulation compare to the theoretical value found in part 5?

Write a report that includes your work on parts 1–6 and summarizes your observations for this lab.

6.10 Using Integrals to Add Rates

At the beginning of this chapter, we stated that the concepts of this chapter are based on the definition of the definite integral and on two of the most important ideas in calculus, the interpretation of a definite integral as area, and the relationship between definite integrals and infinite sums that represent quantities other than area. So far, the material in this chapter has been based mainly on area; the rest of the chapter depends primarily on interpreting a definite integral as an infinite sum. Our goals for the remainder of the chapter are to learn some specific applications of the definite integral, and also to learn how similar the integrals are in different situations. Recognizing these similarities will enable us to apply these ideas and techniques to new situations.

Example

The rate of consumption of oil in the United States since 1974 can be modeled by the function $C(t) = 21.3 \cdot e^{0.04t}$, where t is in years since 1974 and $C(t)$ is in billions of barrels of oil per year. Based on this model, how much oil was used in the United States in the decade starting in 1980 and ending at the end of 1989?

The discrete solution The function C measures the *rate* at which oil is being consumed, so a particular value of $C(t)$ tells us the rate at which oil is being consumed, in billions of barrels per year, at a particular time. Though the function C does not measure the amount of oil consumed, we can use the values of $C(t)$ to compute the amount of oil consumed in the decade of the 1980s.

At the beginning of 1980, oil was consumed at a rate given by the value of $C(t)$ at $t = 6$, which is approximately 27.1 billion barrels per year. If we assume that the rate of consumption remained constant for the whole year, then the amount of oil consumed in 1980 was about 27.1 billion barrels. At the beginning of 1981, we find that oil was consumed at a rate of approximately 28.2 billion barrels per year. Assuming that this rate remains constant for the year gives us an oil consumption of 28.2 billion barrels for 1981. This is simply the product of the consumption rate $C(7)$ and the length of time, Δt, which is 1. We can make a similar approximation for the amount of oil consumed during each year of the 1980s, then add these values to find an approximation for the total oil consumption during the decade. We can write this sum in symbols as

$$\sum_{t=6}^{15} C(t) \cdot \Delta t,$$

which is equivalent to

$$\sum_{t=6}^{15} 21.3e^{0.04t}, \tag{33}$$

since $\Delta t = 1$. This sum is approximately 326.3 billion barrels.

The continuous solution The discrete approximation of 326.3 billion barrels can be improved by decreasing the length of the interval over which each approximation is made. For example, we can find the rate of consumption at half-year intervals and use that rate to compute consumption for 6 month time periods. Using this strategy, the approximate consumption for the first half of 1980 would be $C(6)$ billion barrels per year times one-half year, which is about 13.5 billion barrels. The consumption for the second half of 1980 would be $C(6.5)$ times 0.5, or approximately 13.8 billion barrels. Continuing with half-year approximations for the entire decade gives the summation

$$\sum_{i=0}^{19} 21.3e^{0.04t_i} \cdot 0.5, \tag{34}$$

where $t_0 = 6$, $t_1 = 6.5$, $t_2 = 7$ and so on through $t_{19} = 15.5$. The value of this summation is approximately 329.6 billion barrels.

We can use the form of the summation in (34) to estimate the total oil consumption using any interval size, an estimate equal to

$$\sum_{i=0}^{n-1} 21.3e^{0.04t_i} \cdot \Delta t, \tag{35}$$

where $t_0 = 6$, $t_1 = 6 + \Delta t$, $t_2 = 6 + 2\Delta t$, and so on through $t_{n-1} = 16 - \Delta t$. The values of Δt and n are related by the equation $\Delta t = \frac{16-6}{n}$. The approximation in (35) improves as Δt decreases (or n increases), which in the limit as Δt approaches 0 (or n approaches infinity) becomes a definite integral, so that

$$\lim_{\Delta t \to 0} \sum_{i=0}^{n-1} 21.3e^{0.04t_i} \cdot \Delta t = \int_6^{16} 21.3e^{0.04t} dt.$$

The value of this integral is found by the Fundamental Theorem:

$$\int_6^{16} 21.3e^{0.04t} dt = \frac{21.3}{0.04} e^{0.04t} \Big|_6^{16}$$

$$= \frac{21.3}{0.04}\left(e^{0.04 \cdot 16} - e^{0.04 \cdot 6}\right)$$

$$\approx 332.9.$$

We estimate that the total oil consumption in the U.S. in the 1980s was about 332.9 billion barrels.

The sum we computed in (33) is identical in form to the summations we have used previously to approximate areas under curves. Referring to the graph of $C(t)$ over the domain from $t = 6$ to $t = 16$ shown in Figure 6.31, each term in the summation represents the area of a rectangle with width 1 and height given by $C(t)$ for an integer value of t. Because C is increasing, all of the rectangles lie completely below the curve, so our discrete approximation is lower than the actual consumption. The actual consumption has a value equal to the area under the curve that represents the rate of consumption. ∎

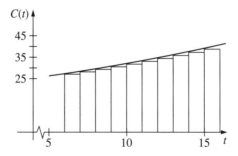

Figure 6.31 Area under $C(t)$ and approximation for oil consumption

The oil consumption example is a specific formulation of the following more general question: If we know a rate-of-change function, how can we analyze the accumulation of that rate function? We have two strategies for accumulating total change over an interval when we know a rate-of-change function. We can use a discrete summation, or we can use the definite integral. If we can find an antiderivative of the rate function, then we can evaluate the definite integral by the Fundamental Theorem. If we cannot find an antiderivative of the rate function, then the strategy of using a definite integral is not helpful. If the limits are known values rather than unknown constants, we can use a discrete summation, which is a numerical approximation for the definite integral.

We spent the first half of this text investigating problems formulated in terms of rates of change, which involve observing phenomena over small intervals. As we observed things over these small pieces, we observed the local linearity of differentiable functions. The behavior of a function at a point was described in terms of the derivative. In this segment of the text, we are reversing the process. We are now taking little pieces, which are products involving rates of change over small intervals, and adding them up to reconstruct the original phenomenon. The definite integral is the primary concept we use to sum up the little pieces.

In the examples in this section we have accumulated change over a fixed interval. In the exercise set we will investigate what happens when we accumulate change over a variable interval.

Exercise Set 6.10

1. The rate of consumption of oil prior to the price increase in 1974 can be modeled by the function $C(t) = 16.1e^{0.07t}$, where t is in years since 1970. How much oil was saved between 1980 and 1989 by the decrease in the rate of consumption that occurred in 1974? Use the rate function $C(t) = 21.3 \cdot e^{0.04t}$ for post-1974 as given in the example.

2. The removal of pollutants from the exhaust at a chemical plant is controlled by filters in the exhaust system. The filters are changed every 2 months, and d days after the filter is changed, the rate at which pollutants are released into the air is $3.27\sqrt{\frac{d}{8.2}}$ kg/day. How many kilograms of pollutants enter the air during each 2 month period between changings of filters?

3. The rate of production for a new calculator is $100\left(\frac{t}{t+12}\right)^2$ units per week t weeks after initial production. How many calculators are made in the first 10 weeks of production?

*Suppose $C(t)$ represents the rate of oil consumption in year t. We can accumulate this rate from year 0 until year x, and we can represent this amount with the integral $\int_0^x C(t)dt$. Since the total amount of oil consumed from year 0 until year x depends on the value of x, $\int_0^x C(t)dt$ is a function of x, and we can write $F(x) = \int_0^x C(t)dt$. How can we interpret the derivative $\frac{dF}{dx}$? Since $F(x)$ represents the total oil consumption up to year x, $\frac{dF}{dx}$ is the rate of change of oil consumption with respect to time x. The value of $\frac{dF}{dx}$ at any particular x value gives the rate of change of oil consumption with respect to time in year x. The function C evaluated at year x gives exactly the same information, so $\frac{dF}{dx} = C(x)$. Thus we can conclude that $\frac{d}{dx}(\int_0^x C(t)dt) = C(x)$, provided C is a continuous function. This result is called the **Second Fundamental Theorem of Calculus**. In general, it does not matter where we start the accumulation. If $F(x) = \int_a^x C(t)dt$ and $G(x) = \int_b^x C(t)dt$, F and G will differ only by a constant, so the derivatives will be identical.*

Second Fundamental Theorem of Calculus

4. Find $\frac{dF}{dx}$.

    a. $F(x) = \int_0^x \sin t \, dt$ b. $F(x) = \int_{-2}^x \sin t^2 \, dt$

    c. $F(x) = \int_x^4 \ln(t^2 + 1) \, dt$

5. Find the zero of each function F in problem 4.

6. Define a function F so that $\frac{dF}{dx} = \tan x^2$ and $F(0) = \frac{\pi}{4}$. What is the domain of F?

7. Explain the difference between what $\int_a^x f(t) \, dt$ represents and what $\int f(x) \, dx$ represents.

8. Use the chain rule to find $\frac{dF}{dx}$ for $F(x) = \int_0^{x^2} e^{\cos(t)} \, dt$.

6.11 Volumes

The Great Pyramid of Cheops at Al Giza, Egypt, has the shape of a regular pyramid with a square base. The base is 230 meters on each side, and the height of the pyramid is 147 meters. The pyramid is constructed from blocks that are each 1.5 meters high. The blocks are arranged in 98 levels of squares. Each of the 98 levels is a solid with a height of 1.5 meters and a square base. What is the total volume of the pyramid?

    The total volume of the pyramid is the sum of the volumes of the levels of the pyramid. Suppose we number the levels from 0 to 97, with 0 at the bottom and 97 at the top. Let v_i represent the volume of the ith level, and let s_i represent the length of a side of the square base of the ith level. Since $v_i = (s_i)^2(1.5)$, we need an expression for s_i to determine the volume. The value of s_i depends on where the level is located within the pyramid; that is, s_i is larger near the base and smaller near the peak.

    Using the fact that each level is 1.5 meters high, we see that the bottom of the ith level is $1.5i$ meters above the base of the pyramid. (See Figure 6.32.) For example, the base with side s_0 is at height 0, the base with side s_1 is at height 1.5, and the base with side s_2 is at height $(1.5)(2) = 3$.

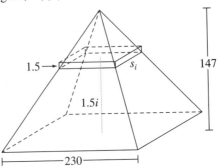

Figure 6.32 Pyramid showing the ith level at height $1.5i$

We can use similar right triangles to find an expression for s_i. The right triangle with legs 147 and 115 (ΔBGC) in Figure 6.33 is similar to the smaller right triangle with legs $147 - 1.5i$ and $\frac{s_i}{2}$ (ΔBDF). The lengths of corresponding sides of these triangles are proportional, which implies that

$$\frac{147}{147 - 1.5i} = \frac{115}{\frac{s_i}{2}},$$

or

$$s_i = \frac{2 \cdot 115(147 - 1.5i)}{147}.$$

The volume of each level is simply

$$v_i = \left(s_i\right)^2 \cdot 1.5$$

$$= \left[\frac{230(147 - 1.5i)}{147}\right]^2 \cdot 1.5.$$

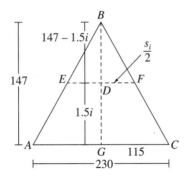

Figure 6.33 Similar triangles for side view of pyramid

To find the volume of the pyramid, we must find the sum V of the volumes of all of the levels, which is

$$V = \sum_{i=0}^{97} \left[\frac{230(147 - 1.5i)}{147}\right]^2 \cdot 1.5.$$

This sum yields a volume of approximately $2,631,910 \text{ m}^3$.

What would the volume be if the pyramid were a geometrically perfect pyramid made without steps and with smooth sides? We can answer this question if we modify the technique we just used. We divide the pyramid into n levels, numbered 0 to $n-1$, with 0 at the base of the pyramid and $n-1$ at the top. Let h_i represent the distance between the base of the pyramid and the bottom of the ith level, and let Δh represent the thickness of each level. (See Figure 6.34.)

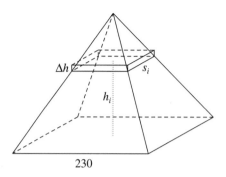

Figure 6.34 Smooth pyramid with cross-section

The volume of the ith level is given by $v_i = (s_i)^2 \cdot \Delta h$, where s_i is the length of a side of the ith level. Using similar triangles as before, we find that

$$s_i = \frac{230(147 - h_i)}{147},$$

so the volume of each level is given by

$$v_i = \left[\frac{230(147 - h_i)}{147} \right]^2 \cdot \Delta h.$$

The volume of the entire pyramid is approximated by the sum of the volumes of all the levels, which is

$$V_{\text{pyramid}} \approx \sum_{i=0}^{n-1} \left[\frac{230}{147}(147 - h_i) \right]^2 \cdot \Delta h. \tag{36}$$

This approximation improves as we increase n, which decreases Δh, so the volume of the smooth pyramid is the limit of the value of the summation in (36) as Δh approaches 0.

$$V_{\text{pyramid}} = \lim_{\Delta h \to 0} \sum_{i=0}^{n-1} \left[\frac{230}{147}(147 - h_i) \right]^2 \cdot \Delta h. \tag{37}$$

The right side of equation (37) is the limit of a Riemann sum, which is equivalent to the definite integral

$$\int_0^{147} \left[\frac{230}{147}(147 - h) \right]^2 dh.$$

We can evaluate this integral analytically as shown below:

$$\int_0^{147} \left[\frac{230}{147}(147 - h) \right]^2 dh = \int_0^{147} \left(230 - \frac{230h}{147} \right)^2 dh$$

$$= -\frac{147}{230} \frac{\left(230 - \frac{230h}{147} \right)^3}{3} \Bigg|_0^{147}$$

$$= -\frac{147}{230} \left[\frac{(230 - 230)^3}{3} - \frac{(230 - 0)^3}{3} \right]$$

$$= -\frac{147}{230} \left[0 - \frac{12167000}{3} \right]$$

$$= 2592100.$$

The volume of the pyramid with smooth sides is $2,592,100$ m^3. If the smooth pyramid is to have the same height and base as originally given, we have to remove stone to make the sides of the pyramid smooth. This is why the volume of the smooth pyramid is less than the volume of the pyramid with levels 1.5 meters high.

The formula from solid geometry for the volume of a pyramid is $\frac{1}{3}Bh$, where B is the area of the base and h is the height, which gives the same result as the calculus answer for the smooth pyramid.

While the volume of a pyramid may be found fairly quickly using the formula $V = \frac{1}{3}Bh$, the procedure we used illustrates a technique that can be used to calculate the volumes of many different solids. We find an approximation for the volume as a sum of thin slices that are perpendicular to an axis of the solid. As the thickness of the slices approaches zero, the approximation improves, and the limit of the sum represents the value of a definite integral. If the definite integral involves a known antiderivative, then the volume can be determined exactly by analytic techniques; otherwise, the value of the integral can be computed numerically.

Example

The formula from solid geometry for the volume of a right circular cone with radius r and height h is $V = \frac{\pi r^2 h}{3}$. Verify this formula using calculus.

Solution   A cone can be generated by revolving the region in Figure 6.35 about the y-axis, which produces what is known as a *solid of revolution*. The revolved region is bounded by the x-axis, the y-axis, and the line containing the points $(0, h)$ and $(r, 0)$. The equation of this line is $y = -\frac{h}{r}x + h$.

solid of revolution

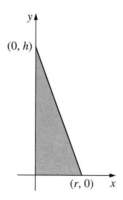

Figure 6.35 Region to rotate about the y-axis to form a cone

The volume of the cone can be approximated by the sum of the volumes of slices perpendicular to the axis of symmetry of the cone. Each slice is a cylinder with height Δy. Assume we have n such slices. The slices all have the same thickness, but the radius of a slice depends on the location of the slice within the cone. Figure 6.36 shows a representative slice with radius x_i and thickness Δy.

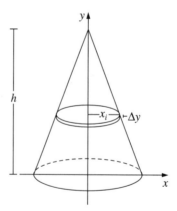

Figure 6.36 Cone with representative slice

The volume of the ith slice is the area of the circular base times the height, $\pi x_i^2 \Delta y$, and the volume of the entire cone is approximated by the sum of the volumes of the n slices, which is

$$\sum_{i=0}^{n-1} \pi x_i^2 \Delta y.$$

The limit of this sum as the thickness approaches 0 is the actual volume of the cone:

$$V_{\text{cone}} = \lim_{\Delta y \to 0} \sum_{i=0}^{n-1} \pi x_i^2 \Delta y. \tag{38}$$

The expression on the right side of equation (38) resembles the definition of a definite integral, except for a slight discrepancy. A Riemann sum must have the form

$$\sum_{i=0}^{n-1} f(y_i) \cdot \Delta y.$$

The sum on the right side of (38) does not have this form, so we need to rewrite x_i in terms of y_i, the height of the ith slice. Since the boundary of the region in Figure 6.35 has equation $y = -\frac{h}{r} x + h$, we know that $x = (y - h)\left(-\frac{r}{h}\right)$ or $x = -\frac{r}{h} y + r$. Thus (38) can be rewritten as

$$V_{\text{cone}} = \lim_{\Delta y \to 0} \sum_{i=0}^{n-1} \pi \left(-\frac{r}{h} y_i + r\right)^2 \Delta y. \tag{39}$$

Equation (39) expresses the volume as the limit of a Riemann sum with $y_0 = 0$ and $y_n = h$, so the volume of the cone equals the definite integral

$$\int_0^h \pi \left(-\frac{r}{h} y + r\right)^2 dy.$$

Recalling that π, r, and h are all constants, we can evaluate this integral analytically:

$$\int_0^h \pi \left(-\frac{r}{h} y + r\right)^2 dy = \frac{\pi}{3} \left(-\frac{h}{r}\right)\left(-\frac{r}{h} y + r\right)^3 \Big|_0^h$$

$$= \frac{\pi}{3} \left(-\frac{h}{r}\right)\left(0^3 - r^3\right)$$

$$= \frac{\pi}{3} r^2 h. \qquad \blacksquare$$

Exercise Set 6.11

1. Use calculus to verify that the volume of a sphere with radius r is $V = \frac{4}{3}\pi r^3$.

2. The base of a certain solid is the unit circle in the xy-plane. Find the volume of the solid if all cross sections perpendicular to the x-axis are:

   a. squares;

   b. isosceles right triangles with a leg in the xy-plane; or

   c. equilateral triangles.

3. The region bounded by the curve $y = \sqrt{x}$, the x-axis, and the line $x = 4$ is re-
    volved around the x-axis. Use a summation of cross sections perpendicular to the
    x-axis to find a definite integral for the volume of this solid. Show that the vol-
    ume of this solid is 8π.

4. The region bounded by the curve $y = \sqrt{x}$, the x-axis, and the line $x = 4$ is re-
    volved around the y-axis. Use a summation of cross sections perpendicular to the
    y-axis to find a definite integral for the volume of this solid. What is the volume
    of this solid?

5. The region between the curves $y = x^2$ and $y = \sqrt{x}$ is rotated about the y-axis.
    Find the volume of the solid generated.

6. a. Find the volume of the solid formed by rotating $y = \frac{1}{x}$ about the x-axis over
        the interval
        i.   from $x = 1$ to $x = 2$;
        ii. from $x = 1$ to $x = 10$; and
        iii. from $x = 1$ to $x = k$.

    b. What happens to the volume from $x = 1$ to $x = k$ as $k \to \infty$?

6.12 Average Value of a Function

We often hear the average temperature reported for a certain city. This could be an
average for the entire year, for a month, or for a day. Cities in the warmer parts of the
country use this figure to attract visitors from colder parts, while farmers use this
figure to help them determine what crops they can grow and when to plant them. To
find an average value for a finite set of discrete numbers, we add up all the numbers
and divide by the number of values in the set. In actuality, temperatures over a year or
even over just one day do not consist of a finite number of values. If we want the exact
average rather than an approximation, we cannot simply add up values and divide by
the number of values because the values are defined on a continuum and do not form
a discrete set. If we have a continuous function that models the temperature, then we
can extend the concept of average value from the discrete case to the continuous case.

Suppose that the temperature for a particular day in Durham, North Carolina, is
modeled with the function

$$T(t) = -15.3 \cdot \cos\left(\frac{\pi}{8.1}t\right) + 44.6,$$

where t represents hours after sunrise, T is in degrees Fahrenheit, and the domain of T
is the interval $0 \le t \le 12$. (On this day there are twelve hours of daylight in Durham.

Sunrise is at 6:23 A.M. and sunset is at 6:23 P.M.) What was the average temperature on this day during the twelve daylight hours?

How do we find the average value of a continuous function over an interval? We can approximate the average temperature by averaging a certain number of equally spaced values from throughout the day. (This is how the average temperature in the weather reports is calculated.) In this case, we will use temperatures at one hour intervals starting at sunrise, as shown in Table 6.37. The average temperature using these twelve values is $\frac{1}{12} \sum_{i=0}^{11} T_i$, which is approximately 47.2 degrees.

Time	6:23	7:23	8:23	9:23	10:23	11:23	12:23	1:23	2:23	3:23	4:23	5:23
t_i	0	1	2	3	4	5	6	7	8	9	10	11
T_i	29.3	30.4	33.7	38.5	44.3	50.1	55.1	58.5	59.9	59.0	55.9	51.2

Table 6.37 Time and temperature data

We can obtain a better approximation by recording the temperature every half hour, or better yet, every quarter hour. If we take n equally spaced temperature measurements at times $t_0, t_1, \cdots, t_{n-1}$, our approximation for the average temperature is

$$\frac{1}{n} \sum_{i=0}^{n-1} T(t_i).$$

The approximation improves as we use larger values of n, and the actual average value of the temperature is the limit as n increases without bound, which is written as

$$\lim_{n \to \infty} \frac{1}{n} \sum_{i=0}^{n-1} T(t_i). \tag{40}$$

The summation in (40) is similar to a Riemann sum, but it does not contain Δt. Since the t_i's are equally spaced over the twelve-hour interval, n and Δt are related by the equation $\Delta t = \frac{12}{n}$, which is equivalent to $\frac{1}{n} = \frac{\Delta t}{12}$. As n increases without bound, Δt approaches 0, so the summation in (40) can be rewritten as

$$\lim_{\Delta t \to 0} \frac{1}{12} \sum_{i=0}^{n-1} T(t_i) \cdot \Delta t . \tag{41}$$

The summation in (41) is the limit of a Riemann sum, and it is equal to the definite integral

$$\frac{1}{12} \int_0^{12} T(t) \, dt.$$

The value of this integral is determined below:

$$
\begin{aligned}
\frac{1}{12}\int_0^{12} T(t)\, dt &= \frac{1}{12}\int_0^{12}\left[-15.3\cdot\cos\left(\frac{\pi}{8.1}t\right)+44.6\right]dt \\
&= \frac{1}{12}\left[-15.3\left(\frac{8.1}{\pi}\right)\sin\left(\frac{\pi}{8.1}t\right)+44.6t\right]_0^{12} \\
&= \frac{1}{12}\left[-15.3\left(\frac{8.1}{\pi}\right)\sin\left(\frac{\pi}{8.1}\cdot 12\right)+44.6(12)\right] \\
&\approx 47.9.
\end{aligned}
$$

The average temperature in Durham during the daylight hours of the given day was 47.9° F.

We can apply the technique used to calculate an average temperature to find the average value of any function f that we can integrate over an interval from $x = a$ to $x = b$. To see this, consider an approximation of the average value of $f(x)$ over the interval $a \le x \le b$ using n equally spaced values $x_0, x_1, \cdots, x_{n-1}$, which is

$$
\frac{1}{n}\sum_{i=0}^{n-1} f(x_i). \tag{42}
$$

The average value of f over the interval is the limit of the sum in (42) as n increases without bound, which is written as

$$
\lim_{n\to\infty}\frac{1}{n}\sum_{i=0}^{n-1} f(x_i). \tag{43}
$$

Since the x_i's are equally spaced from a to b, Δx and n are related by the equation $\Delta x = \frac{b-a}{n}$ which is equivalent to $\frac{1}{n} = \frac{\Delta x}{b-a}$. As n increases without bound, Δx approaches 0, so the summation in (43) can be rewritten as

$$
\lim_{\Delta x\to 0}\frac{1}{b-a}\sum_{i=0}^{n-1} f(x_i)\cdot\Delta x. \tag{44}
$$

average value

The expression in (44) contains the limit of a Riemann sum, and thus it contains a definite integral. Therefore, we define the **average value**, f_{ave}, of the function f over the interval $a \le x \le b$ as

$$
f_{ave} = \frac{1}{b-a}\int_a^b f(x)\, dx.
$$

A graphical interpretation of average value is based upon rewriting the equation for f_{ave} in the equivalent form

$$
(b-a)\cdot f_{ave} = \int_a^b f(x)\, dx. \tag{45}
$$

If we assume that $f(x)$ is positive for $a \le x \le b$, then the right side of equation (45) is the area under the curve $y = f(x)$. The left side of equation (45) is the area of a rectangle with base along the x-axis from $x = a$ to $x = b$ and height f_{ave}. These two areas are equal; therefore, f_{ave} is the unique value such that the rectangle with height f_{ave} and length $(b - a)$ has the same area as the region below $y = f(x)$ and above the x-axis. Figure 6.38 illustrates this graphical interpretation.

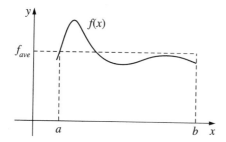

Figure 6.38 Average value of a function

Exercise Set 6.12

1. Let the function $T(t) = -15.3 \cdot \cos\left(\frac{\pi}{8.1}t\right) + 44.6$ represent the temperature in Durham, NC, on the day used in the example for this section. Find the average value of the temperature for the first six hours after sunrise, and for the last six hours before sunset. Recall that t is the number of hours after sunrise and there are exactly twelve hours of daylight. How do these two values compare to the average temperature of the entire day, from sunrise to sunset? Why is this result reasonable?

2. Suppose the temperature of a cup of coffee cooling on a desk is modeled by $T(t) = 122e^{-0.027t} + 71$, where t is measured in minutes since we placed the cup on the desk and T is measured in degrees Fahrenheit. Find the average temperature for the first 30 minutes of cooling. What is the average temperature for the first hour? Can you use these two average temperatures to find the average temperature for the second half hour? Check your answer.

3. A ball is thrown into the air, and the height is given by $H(t) = -4.9t^2 + 15t + 2$, where t is time in seconds, $H(t)$ is in meters, and $H(t) = 0$ is ground level. What is the average height for the first 2 seconds of the flight? For the remainder of the flight? Can you use these two averages to find the average height over the course of the flight? Explain why you cannot simply average the averages.

4. Suppose that, in an effort to study a method of cutting down on tooth decay, a candy company is studying the acid levels in the mouth after an individual eats candy. The pH level measures acidity or alkalinity of liquids. On a scale from 0 to 14, 0 represents the highest level of acidity and 14 the most alkaline. Neutral is designated as a pH of 7. A mathematical model for the pH level of the saliva in the mouth t minutes after eating a piece of candy is found experimentally to be

$$A(t) = 6.5 - \frac{20t}{t^2 + 36}.$$

   a. What is the pH of the saliva before the candy is eaten?

   b. How long after eating the candy does it take for the saliva to return to within 0.1 of normal?

   c. At what rate is the pH changing after 3 minutes?

   d. At what time does the pH begin to return to normal?

   e. What is the average pH of the mouth during the first ten minutes after eating the candy? during the first hour after eating the candy?

5. a. Find the average value of $y = e^{-x}$ from
       i. $x = 0$ to $x = 1$;
       ii. $x = 0$ to $x = 10$; and
       iii. $x = 0$ to $x = k$.

   b. What happens to the average value from $x = 0$ to $x = k$ as $k \to \infty$?

6. Let $f(x) = x^2$. The average value of f over the interval $1 \le x \le 3$ is m.

   a. Determine the value of m.

   b. Compare the graphs of $y = f(x)$ and $y = m$. How do the areas under the graphs over the interval $1 \le x \le 3$ compare? Use this comparison to give a geometric interpretation of the average value of f over this interval.

7. Let $h(x) = e^{\sin x}$.

   a. Sketch a graph of $y = h(x)$ from $x = -\pi$ to $x = \pi$.

   b. Let m be the average value of h over this interval. Approximate the value of m numerically.

   c. How is the graph of $y = m$ related to the graph of $y = h(x)$?

   d. What is the value of $\int_{-\pi}^{\pi} [h(x) - m]\, dx$? Interpret your answer.

8. a. Let f be an odd function and g be an even function. Find
       i. the average value of f over $[-a, a]$;
       ii. the average value of g over $[-a, a]$ minus the average value of g over $[0, a]$.

   b. Explain your results graphically and algebraically.

9. The height in meters of a falling object is given by the function

$$h(t) = \begin{cases} -4.9t^2 + 10t + 1000 & \text{if } t \le 12, \\ -107.6t + 1705.6 & \text{if } t > 12. \end{cases}$$

a. What is the average value of the object's velocity function $v(t) = \frac{dh}{dt}$ during the first 12 seconds of its fall ($t = 0$ is when the object first began to fall)?

b. What is the average velocity of the object during the first 12 seconds?

c. What, if any, is the distinction between the answers to parts a and b?

d. What is the average value of the object's velocity function $v(t) = \frac{dh}{dt}$ during the first 20 seconds of its fall?

e. What is the average velocity of the object during the first 20 seconds?

f. What, if any, is the distinction between the answers to parts d and e?

10. Give a graphical interpretation similar to the illustration in Figure 6.38 for the average value of a function f over the interval from $x = a$ to $x = b$, where the values of f are not positive over this entire interval.

11. Compare and contrast the concepts of expected value and average value. How are they similar? How are they different?

6.13 Length of Path

Length of Path for Equations in Cartesian Form

A projectile is fired across a level plain and follows a path given by the equation $y = 0.1x(10 - x)$, where the ordered pair (x, y) gives the horizontal and vertical coordinates (in kilometers) of the projectile. The projectile is fired from the origin of our coordinate system. We can determine several characteristics of the flight of the projectile based on our knowledge of the graph of $y = 0.1x(10 - x)$. Finding where the projectile lands is equivalent to determining where y equals 0. The value of y is 0 when $x = 0$ and $x = 10$, so the projectile lands 10 km from its launch point. Since the path is parabolic, the maximum height occurs at the point with $x = 5$, where the height is 2.5 km. We know that the projectile rises 2.5 km and falls 2.5 km, and it travels 10 km across the plain. What is the length of the actual path that the projectile follows? In other words, if a piece of rope was strung through each point that the projectile passes through, how much rope would we need? Finding the length of the path of the projectile involves finding the length of the curve $y = 0.1x(10 - x)$ from $(0,0)$ to $(10,0)$.

The length of the line segment from (0, 0) to (10, 0) is 10 km, so the length of the path of the projectile is greater than 10 km. By the Pythagorean theorem, the length of the line segment from (0, 0) to the highest point of the path (5, 2.5) is approximately 5.6 km. Therefore, a path composed of line segments from the origin to the point of maximum height and then to the landing point (10, 0) has a length of about 11.2 km. We can refine our estimate by choosing points on the parabola at $x = 2.5$, 5, and 7.5 km, and then finding the length of a path composed of line segments starting at the origin, going through those points, and ending at (10, 0). By continuing the process of using more and more points along the parabola, we obtain better and better estimates of the length of the actual path.

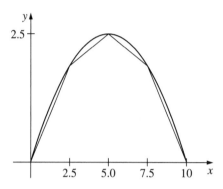

Figure 6.39 Approximations for the length of the projectile's path

Determining the length of a broken-line path involves adding together the lengths of secant line segments, as illustrated in Figure 6.39. By approximating a curve $y = f(x)$ with secant line segments, we obtain an estimate for the length of the actual path, which is given by the summation

$$\sum_{i=1}^{n} \Delta s_i \, ,$$

where Δs_i is the length of the ith secant segment. Each secant line segment is actually the hypotenuse of a right triangle as shown in Figure 6.40. Using the Pythagorean theorem transforms the summation to

$$\sum_{i=1}^{n} \sqrt{\left(\Delta x_i\right)^2 + \left(\Delta y_i\right)^2}$$

where Δx_i and Δy_i are the lengths of the legs of the triangle shown in Figure 6.40.

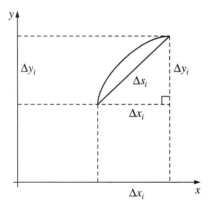

Figure 6.40 Length of secant segments

The secant segments are constructed through a sequence of points along the curve. We assume that Δx, the difference between successive x values for the endpoints of the secant segments, is constant. The changes in the y values are not constant and are determined by the x values at the endpoints of the secant segments, specifically, $\Delta y_i = f(x_i) - f(x_{i-1})$ where x_i is the horizontal coordinate of the right endpoint of the segment. An approximation for the length of the path is therefore

$$\sum_{i=1}^{n} \sqrt{(\Delta x)^2 + (\Delta y_i)^2} . \tag{46}$$

If we factor Δx^2 from each term inside the radical in expression (46), we obtain the expression

$$\sum_{i=1}^{n} \sqrt{\left(1 + \left(\frac{\Delta y_i}{\Delta x}\right)^2\right) \cdot \Delta x^2} ,$$

or

$$\sum_{i=1}^{n} \Delta x \sqrt{1 + \left(\frac{\Delta y_i}{\Delta x}\right)^2} . \tag{47}$$

We obtain a better approximation for the length of the path by shrinking the size of Δx in the summation in (47). The limit of this expression as Δx approaches zero is the exact value of the length of the path. This value can also be represented by a definite integral for the length of the path from $x = a$ to $x = b$, which is

$$\int_{a}^{b} \sqrt{1 + \left(\frac{dy}{dx}\right)^2}\, dx = \lim_{\Delta x \to 0} \sum_{i=1}^{n} \Delta x \sqrt{1 + \left(\frac{\Delta y_i}{\Delta x}\right)^2} , \tag{48}$$

where a and b are the x values at the beginning and end of the path.

For the path of our projectile, since $y = 0.1x(10 - x)$, we know that $\frac{dy}{dx} = 1 - 0.2x$. Substituting into the left side of equation (48), the length of the curve followed by the projectile from its initial point at $x = 0$ to its final point at $x = 10$ is

$$\int_0^{10} \sqrt{1 + (1 - 0.2x)^2}\, dx . \tag{49}$$

We can evaluate the integral in (49) by consulting the table of integrals in the appendix, in which formula 25 is

$$\int \sqrt{u^2 \pm a^2}\, du = \frac{u}{2} \sqrt{u^2 \pm a^2} \pm \frac{a^2}{2} \ln\left| u + \sqrt{u^2 \pm a^2} \right| + C . \tag{50}$$

This formula can be applied to expression (49) by letting $a = 1$ and $u = 1 - 0.2x$, which means that $du = -0.2dx$, or $dx = -5du$, which yields

$$\int \sqrt{1 + (1 - 0.2x)^2}\, dx = -5 \int \sqrt{u^2 + a^2}\, du . \tag{51}$$

Combining (51) with (50) yields

$$\int \sqrt{1 + (1 - 0.2x)^2}\, dx =$$
$$-5\left(\frac{1 - 0.2x}{2} \sqrt{1 + (1 - 0.2x)^2} + \frac{1}{2} \ln\left| 1 - 0.2x + \sqrt{1 + (1 - 0.2x)^2} \right| \right) + C . \tag{52}$$

The value of the definite integral in expression (49) is found by evaluating the right side of equation (52) at 10 and at 0, and then finding the difference of these two values. Performing these calculations gives

$$\int_0^{10} \sqrt{1 + (1 - 0.2x)^2}\, dx \approx 11.48 ,$$

so the length of the path of the projectile is approximately 11.48 km.

Although we now have an analytic expression for length of path,

$$\int_a^b \sqrt{1 + \left(\frac{dy}{dx} \right)^2}\, dx,$$

this expression often is not an integral to which we can apply the Fundamental Theorem of Calculus. The square root expression in the integrand frequently does not possess an antiderivative that we can find analytically. So in practice, we generally calculate the length of path using a numerical method to evaluate the definite integral.

Length of Path for Equations in Parametric Form

In many situations we will not know the equation of the path followed by a projectile. Instead, we may know only that a 10 kg projectile is fired at an initial angle θ to the horizontal with an initial velocity of v_0 m/sec. In addition, suppose the projectile travels across a flat plain and comes to rest a certain distance from the starting position. What is the length of the path followed by the projectile?

Let us begin our analysis of this problem by ignoring the effects of air resistance. The location of the projectile at any time is given by the coordinates $(x(t), y(t))$, where x is the horizontal distance from the starting position expressed as a function of time and y is the height above the flat plain as a function of time. We can determine equations for $x(t)$ and $y(t)$ by beginning with parametric equations for the acceleration of the projectile, and subsequently determining parametric equations for the velocity and the position of the projectile.

The acceleration can be separated into a horizontal component $a_x(t)$ and a vertical component $a_y(t)$. There are no forces acting in the horizontal direction (we are ignoring air resistance), so the horizontal component of acceleration is zero, or

$$a_x(t) = 0. \tag{53}$$

The only force in the vertical direction is the downward force due to gravity, and we assign negative values to the downward direction, so

$$a_y(t) = -g. \tag{54}$$

The value of g, the acceleration due to gravity, is approximately 9.8 m/sec^2.

The horizontal and vertical components of the velocity, $v_x(t)$ and $v_y(t)$, are antiderivatives of the acceleration components. Equation (53) implies that $v_x(t)$ is constant for all t, which means that $v_x(t)$ equals the horizontal component of the initial velocity. The horizontal component $v_x(0)$ and the vertical component $v_y(0)$ of the initial velocity are the lengths of the sides of a right triangle with angle θ (measured in radians) and hypotenuse of length v_0 (the initial velocity), as shown in Figure 6.41. The length $v_x(0)$ of the side adjacent to angle θ divided by the length v_0 of the hypotenuse is equal to $\cos\theta$, so

$$v_x(0) = v_0 \cos\theta.$$

Therefore, the horizontal component of the velocity is

$$v_x(t) = v_0 \cos\theta,$$

which is constant for all t.

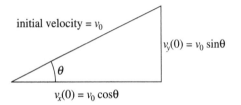

Figure 6.41 Components of the initial velocity

The vertical component of the velocity, found by integrating both sides of equation (54) with respect to t, is the linear function

$$v_y(t) = -gt + C.$$ (55)

If we substitute $t = 0$ into equation (55), we see that the constant of integration C is equal to $v_y(0)$, the vertical component of the initial velocity, so that

$$v_y(t) = -gt + v_y(0).$$ (56)

Referring to Figure 6.41, we see that the length $v_y(0)$ of the side opposite angle θ divided by the length v_0 of the hypotenuse is equal to $\sin\theta$, so

$$v_y(0) = v_0 \sin\theta,$$

which can be substituted into equation (56) to give

$$v_y(t) = -gt + v_0 \sin\theta.$$

The horizontal component $x(t)$ and the vertical component $y(t)$ of the position are determined by integrating the expressions for the velocity components with respect to t, which yields

$$x(t) = \int \left(v_0 \cos\theta\right) dt$$ (57)

and

$$y(t) = \int \left(-gt + v_0 \sin\theta\right) dt.$$ (58)

Evaluating the integrals on the right sides of equations (57) and (58) gives

$$x(t) = \left(v_0 \cos\theta\right) t + C_1$$

and

$$y(t) = -\frac{1}{2} gt^2 + \left(v_0 \sin\theta\right) t + C_2.$$

We assume that the initial position of the projectile has coordinates $(0,0)$, so C_1 and C_2 are both 0. Thus the position of the projectile over time is given by the parametric equations

$$x(t) = \left(v_0 \cos \theta\right)t,$$

$$y(t) = -\frac{1}{2} gt^2 + \left(v_0 \sin \theta\right)t. \qquad (59)$$

How can we find the length of the path for the projectile with parametric equations given in (59)? Recall that the integral for length of path in Cartesian form,

$$\int_a^b \sqrt{1 + \left(\frac{dy}{dx}\right)^2} \, dx,$$

was developed by summing up the lengths of secant segments along a curve. This resulted in the summation

$$\sum_{i=1}^n \Delta s_i \, ,$$

where Δs_i is the length of the ith secant segment. We changed this sum using the Pythagorean Theorem to

$$\sum_{i=1}^n \sqrt{\left(\Delta x_i\right)^2 + \left(\Delta y_i\right)^2} \, , \qquad (60)$$

where Δx_i and Δy_i are the horizontal and vertical displacements along the ith secant segment. When we used a Cartesian equation of the form $y = f(x)$, the value of Δx_i determined the value of Δy_i. With parametric equations, specifying Δx_i does not necessarily determine the value of Δy_i. Instead, the change in the parameter t determines the changes in x and y along a secant segment. If we use a constant value for Δt, the change in t, and then factor out Δt from the terms in the summation in (60), then an approximation for the length of the path of the projectile is

$$\sum_{i=1}^n \sqrt{\left(\frac{\Delta x_i}{\Delta t}\right)^2 + \left(\frac{\Delta y_i}{\Delta t}\right)^2} \cdot \Delta t. \qquad (61)$$

The approximation in (61) is improved by shrinking Δt, which corresponds to using more secant segments to approximate the total length of path. The limit of the summation as Δt approaches 0 gives the exact expression for the length of path,

$$\lim_{\Delta t \to 0} \sum_{i=1}^n \sqrt{\left(\frac{\Delta x_i}{\Delta t}\right)^2 + \left(\frac{\Delta y_i}{\Delta t}\right)^2} \cdot \Delta t,$$

which is the limit of a Riemann sum. In the limit as $\Delta t \to 0$, $\frac{\Delta x_i}{\Delta t}$ approaches $\frac{dx}{dt}$ and $\frac{\Delta y_i}{\Delta t}$ approaches $\frac{dy}{dt}$; therefore, the length of path specified with parametric equations is given by the definite integral

$$\int_a^b \sqrt{\left(\frac{dx}{dt}\right)^2 + \left(\frac{dy}{dt}\right)^2}\ dt,, \tag{62}$$

where a and b are t values at the beginning and end of the path. Equation (62) is important because not all paths can be described in the Cartesian form $y = f(x)$. If a path can be described in the parametric form $x = g(t)$ and $y = h(t)$, then equation (62) allows us to find the length of the path.

Example

Use parametric equations to find the length of the path of a projectile fired across a flat plain with an initial velocity of 100 m/sec at an angle of $\theta = \frac{\pi}{6}$ radians to the horizontal.

Solution   The coordinates of the projectile are given by the parametric equations (59). Substituting $g = 9.8$, $v_0 = 100$, and $\theta = \frac{\pi}{6}$ radians gives

$$x(t) \approx 86.6t,$$

$$y(t) = -4.9t^2 + 50t.$$

The projectile travels through the air over an interval of time from launch, which is when $t = 0$, until it returns to the ground, which is when $y(t) = 0$. The solutions to $y(t) = 0$ are $t = 0$ and $t \approx 10.2$, so the projectile lands after about 10.2 seconds.

Now we are prepared to find the length of the path using expression (62). The derivatives of x and y with respect to time are

$$\frac{dx}{dt} \approx 86.6,$$

$$\frac{dy}{dt} = -9.8t + 50.$$

Substituting these expressions into (62) yields a length of path approximately equal to

$$\int_0^{10.2} \sqrt{(86.6)^2 + (-9.8t + 50)^2}\ dt.$$

This integral can be evaluated with a table of integrals. It can also be evaluated numerically, which yields a value of about 930 meters. ∎

Exercise Set 6.13

1. Find the lengths of the curves described in parts a–d by calculating values of the appropriate integrals analytically.

   a. $y = x^{3/2}$ from $x = 0$ to $x = 4$

   b. $y = \frac{1}{3}x^{3/2} - x^{1/2}$ from $x = 0$ to $x = 4$

   c. $y = \dfrac{\left(e^x + e^{-x}\right)}{2}$ from $x = 0$ to $x = \ln 3$

   d. $y = x^2$ from $x = 0$ to $x = 2$ (You may need to consult a table of integrals.)

2. Find the lengths of the curves described in parts a–d by calculating values of the appropriate integrals numerically.

   a. $y = \sin x$ from $x = 0$ to $x = \pi$

   b. $y = x^2$ from $x = 0$ to $x = 2$

   c. $y = e^x$ from $x = -1$ to $x = 1$

   d. $y = \ln x$ from $x = \frac{1}{2}$ to $x = 2$

3. Write a definite integral in terms of the parameter t that gives the length of the path for a projectile's motion described by the parametric equations (59). Be sure to specify the limits of integration.

4. Use the integral from problem 3 to find the length of the path for a projectile that is fired across a flat plain with an initial velocity of 100 m/sec at an angle of $\theta = \frac{\pi}{4}$ radians above the horizontal. Ignore the effects of air resistance.

5. Find the length of the path of a projectile that is fired across a flat plain with an initial velocity of 100 m/sec at an angle of $\theta = \frac{\pi}{3}$ radians above the horizontal. Ignore the effects of air resistance.

6. Find the length of the curve with parametric equations $x = \cos t$ and $y = \sin t$ from $t = 0$ to $t = 2\pi$. What type of curve is this?

7. Find the length of the curve with parametric equations $x = 4\cos t$ and $y = 3\sin t$ from $t = 0$ to $t = 2\pi$. What type of curve is this?

8. Find a formula in terms of a and b for the circumference of an ellipse with equation $\frac{x^2}{a^2} + \frac{y^2}{b^2} = 1$.

Lab 15 *Length of Path*

In this lab we investigate further the length of the path for the motion of a projectile.

1. Assume a projectile is fired across a flat plain at an initial angle θ with the horizontal and an initial velocity of 100 meters per second. Ignore the effects of air resistance. Calculate the length of the path for various values of θ. What value of θ gives the longest path?

2. Introduce air resistance into the model in part 1 for the path of the projectile. Assume that the force due to air resistance is proportional to the velocity squared, with a proportionality constant equal to 0.001. Also, assume that the mass of the projectile is 1, so that force is equal to acceleration.

   a. Write the parametric equations for the horizontal and vertical components of acceleration.

   b. Use the linear Euler's method to find approximate values for the horizontal and vertical components of the velocity.

   c. Use the linear Euler's method to find approximate values for the position coordinates of the projectile.

   d. For an initial firing angle of $\theta = \frac{\pi}{4}$ radians, what is the horizontal displacement of the projectile from the firing point to the landing point?

   e. For an initial firing angle of $\theta = \frac{\pi}{4}$ radians, what is the length of the path of the projectile from the firing point to the landing point?

   f. What value of θ gives the greatest horizontal displacement for the flight of the projectile?

   g. What value of θ gives the longest path for the flight of the projectile?

Write a report summarizing your results and including your calculations. Discuss your results, commenting on the reasonableness of your results and whether your results are what you expected.

END OF CHAPTER EXERCISES

Review and Extensions

1. Evaluate the following integrals

   a. $\int xe^{x^2}\,dx$

   b. $\int_1^2 xe^x\,dx$

   c. $\int_3^\infty \dfrac{dx}{(x-1)^2}$

   d. $\int x\sin 2x\,dx$

   e. $\int_{-\infty}^\infty \dfrac{x}{(x^2+3)^2}\,dx$

   f. $\int \sin^{-1}x\,dx$

2. Use the fact that the integral $\int_1^\infty x^{-3/2}\,dx$ converges to explain why the series $\sum_{n=1}^\infty n^{-3/2}$ converges. Enhance your explanation with an illustration.

3. The expected value of an exponential distribution is 10. Find the median value of this distribution.

4. Set up an integral to find the volume of the solid formed when the region bound by $f(x)=\sin x$ and $g(x)=\cos x$ on the interval $\left[0,\frac{\pi}{4}\right]$ is revolved around the x-axis.

5. a. The weight of a particular cereal (sold in boxes) is distributed normally with a mean of 12 ounces and a standard deviation of 0.6 oz. What is the probability of getting a box that weighs at least 13 oz.?

   b. The machine that fills cereal boxes can be set to any mean but has a standard deviation of 0.6 oz. If the manufacturer wants at *most* 5% of the boxes to weigh less than 12 oz., at what value should the mean of the machine be set? Give the smallest possible value.

6. How is our study of geometric probability at the beginning of this chapter related to the work we did with probability distributions?

7. a. Find the average value of $f(x)=x^2$ on the interval $[0,3]$.

   b. Find c so that $f(c)$ equals the average value of f on $[0,3]$.

   c. Sketch $f(x)=x^2$ on $[0,3]$. Now sketch a rectangle whose base is the interval $[0,3]$ and whose area is equal to the area under $f(x)=x^2$ from 0 to 3.

Taylor Series and Fourier Series

7.1 Taylor Series and Intervals of Convergence

Throughout this text, we have used calculus to help us approximate functions about points at which the functions are differentiable. Recall that one linear approximation of the function f about the point $(a, f(a))$ is a tangent line with equation

$$l(x) = f'(a) \cdot (x - a) + f(a).$$

In Section 4.5, we improved this linear approximation by using quadratic functions to approximate other functions. The parabola with equation

$$q(x) = \frac{f''(a)}{2} \cdot (x - a)^2 + f'(a) \cdot (x - a) + f(a)$$

is one quadratic approximation for f about the point $(a, f(a))$.

In Section 4.8, we used higher-degree polynomials known as Taylor polynomials to approximate functions. For example, we found that $p(x) = 1 - x + x^2 - x^3 + x^4$ approximates $f(x) = \frac{1}{x+1}$ about $(0,1)$. Taylor polynomials are determined by setting the approximating polynomial equal to the function value and the derivatives of the function at a particular point. The nth-degree Taylor polynomial for f about $x = a$ has equation

$$p_n(x) = f(a) + f'(a) \cdot (x - a) + \frac{f''(a)}{2!} \cdot (x - a)^2 + \ldots + \frac{f^{(n)}(a)}{n!} \cdot (x - a)^n,$$

where the subscript of p indicates the degree of the polynomial. The function f must be n times differentiable at $x = a$ for this polynomial to exist.

In Section 4.8, we showed that the eleventh-degree Taylor polynomial for $\sin x$ about $x = 0$ is

$$p_{11}(x) = x - \frac{1}{3!}x^3 + \frac{1}{5!}x^5 - \frac{1}{7!}x^7 + \frac{1}{9!}x^9 - \frac{1}{11!}x^{11}.$$

When we graphed the Taylor polynomials of varying degrees about $x = 0$ for $\sin x$, we noticed two consequences of adding more terms to the Taylor polynomial. The accuracy of the approximation increased for x values around 0, and the interval of x values over which the Taylor polynomial appeared to approximate $\sin x$ well increased in width. As we increase n, the nth-degree Taylor polynomial appears to approximate $\sin x$ well over more and more of its entire domain.

The tenth-degree Taylor polynomial for $f(x) = \frac{1}{x+1}$ about $x = 0$ is

$$p_{10}(x) = 1 - x + x^2 - x^3 + x^4 - x^5 + x^6 - x^7 + x^8 - x^9 + x^{10}.$$

When we graphed the Taylor polynomials of varying degrees for $\frac{1}{x+1}$, the situation was different from the Taylor polynomial for $\sin x$. The accuracy of the approximation increased for x values around 0; however, the length of the interval of x values over which the Taylor polynomial appeared to approximate $\frac{1}{x+1}$ did not increase without bound. The nth-degree Taylor polynomial for $\frac{1}{x+1}$ about $x = 0$ works well for x values close to 0, but it does not appear to approximate $\frac{1}{x+1}$ for all values in its domain. These observations are illustrated in the graphs of the Taylor polynomials of fourth degree and tenth degree shown in Figure 7.1.

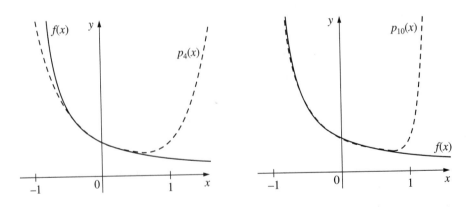

Figure 7.1 Graphs of $f(x) = \frac{1}{x+1}$ and Taylor polynomials of degrees 4 and 10

If we can continue taking derivatives of f so that the degree of the Taylor polynomial increases without bound, then we can write an "infinite" polynomial which is called a **Taylor series** and which is written as

Taylor series

$$p(x) = f(a) + f'(a) \cdot (x - a) + \frac{f''(a)}{2!} \cdot (x - a)^2 + \cdots + \frac{f^{(n)}(a)}{n!} \cdot (x - a)^n + \cdots.$$

For the Taylor series to exist, the function f must be infinitely differentiable at $x = a$, meaning $f^{(n)}(a)$ must exist for all n no matter how large. A Taylor series about $x = 0$ is also known as a ***Maclaurin series***. Taylor series and Maclaurin series are special cases of a ***power series***, which is an infinite polynomial of the form

Maclaurin series
power series

$$a_0 + a_1 x + a_2 x^2 + a_3 x^3 + \cdots,$$

where a_0, a_1, a_2, and so on, are the coefficients of the power series.

The Maclaurin series for $f(x) = \frac{1}{x+1}$ is

$$p(x) = 1 - x + x^2 - x^3 + x^4 - x^5 + \cdots + (-1)^n x^n + \cdots. \qquad (1)$$

The notation "..." has the same implication as with infinite geometric series and means that the polynomial continues for an unlimited number of terms. If a pattern exists in the terms of the series, we often include an expression for the general term, which is $(-1)^n x^n$ in this example.

We have seen that adding terms to the Taylor polynomial for $\frac{1}{x+1}$ about $x = 0$ improves the approximation for x values near 0. By adding an infinite number of terms, which gives a Maclaurin series, is the infinite polynomial in (1) actually equal to the function $\frac{1}{x+1}$ over some interval?

From our previous work and the graphs in Figure 7.1, we suspect that the series $p(x)$ in (1) equals $\frac{1}{x+1}$ at some, but not all, x values. If $x = 0$, the series in (1) and $\frac{1}{x+1}$ both equal 1, so $p(x) = \frac{1}{x+1}$ for $x = 0$. For what other x values is $p(x) = \frac{1}{x+1}$ true? Notice that for any value of x, $p(x)$ is an infinite geometric series with common ratio $-x$. We know that an infinite geometric series converges if the common ratio r has absolute value less than 1. Thus, the Maclaurin series $p(x)$ converges to $\frac{1}{x+1}$ for all x values in the interval $-1 < x < 1$. The series $p(x)$ diverges if $x \le -1$ or $x \ge 1$, so only x values in the interval $-1 < x < 1$ make $p(x) = \frac{1}{x+1}$ true.

The set of x values for which a power series converges is called the ***interval of convergence*** of the series. The interval of convergence of the Maclaurin series for $\frac{1}{x+1}$ is $-1 < x < 1$. This interval of convergence is the domain over which the Maclaurin series is equal to $\frac{1}{x+1}$, so we can write

interval of convergence

$$\frac{1}{x+1} = 1 - x + x^2 - x^3 + x^4 - x^5 + \cdots + (-1)^n x^n + \cdots$$

if $-1 < x < 1$.

If we write an equation stating that a function f and the Taylor series about $x = a$ are equal, such as

$$f(x) = f(a) + f'(a) \cdot (x - a) + \frac{f''(a)}{2!} \cdot (x - a)^2 + \cdots + \frac{f^{(n)}(a)}{n!} \cdot (x - a)^n + \cdots, \qquad (2)$$

then we are saying that two conditions are satisfied.

1. For any x value in the interval of convergence, the infinite sum on the right side of (2) converges.

2.   For any x value in the interval of convergence, the infinite sum on the right side of (2) converges to a value that is the same as the value on the left side of (2).

For the functions we will study, if the Taylor series for f converges, then it converges to f.

We wish to determine the intervals of convergence for other power series. For example, recalling our work with Taylor polynomials about $x = 0$ for $\sin x$, the Maclaurin series for $\sin x$ appears to converge to $\sin x$ for all values of x. To investigate this and other questions related to intervals of convergence, we need to investigate some theoretical results about the conditions under which a series of numbers converges.

The Divergence Test

When we introduced infinite series in Chapter 6, we said that an infinite series converges if the sequence of the sum of the terms approaches a finite number. Let us assume that the series $\sum_{k=1}^{\infty} a_k$ is convergent. If S_n is the sum of the first n terms, so that $S_n = a_1 + a_2 + \cdots + a_n$, then $\lim_{n \to \infty} S_n = S$, where S is some finite number that is called the sum of the series. Notice that if $\lim_{n \to \infty} S_n = S$, then the difference between S_n and S goes to zero as n increases without bound. This implies that a_n approaches zero as n increases without bound. Therefore, if an infinite series of numbers $\sum_{k=1}^{\infty} a_k$ converges, then the sequence of terms of the series must approach zero, that is, $\lim_{n \to \infty} a_n = 0$. Consequently, if the nth term of an infinite series does not approach zero as n increases without bound, then the series must diverge. This result, which often is

divergence test

used to show that a series diverges, is called the ***divergence test***.

The converse of the divergence test, which is that if a series diverges then the nth term does not approach zero, is not true. We will see that some series diverge even though the nth term approaches zero.

The Ratio Test

An important characteristic of a series of numbers is the behavior of the ratio of consecutive terms. For an infinite geometric series

$$a + ar + ar^2 + ar^3 + ar^4 + \cdots,$$

the ratio of consecutive terms is r. We know that an infinite geometric series converges for r values between -1 and 1, and diverges for other r values.

One series that is not geometric is

$$\frac{1}{3} + \frac{4}{9} + \frac{9}{27} + \frac{16}{81} + \frac{25}{243} + \cdots + \frac{n^2}{3^n} + \cdots. \tag{3}$$

The nth term of this series approaches 0 as n increases without bound, so the series may converge. Even though this series is not geometric, the ratio of consecutive terms of the series gives us important information about how the series behaves. The ratio of the $(n+1)$st term to the nth term is

$$\frac{(n+1)^2/3^{n+1}}{n^2/3^n},$$

which can be simplified as

$$\frac{(n+1)^2}{n^2} \cdot \frac{3^n}{3^{n+1}} = \frac{1}{3}\left(\frac{n+1}{n}\right)^2.$$

The ratio of the $(n+1)$st term to the nth term is not constant, but the limit of the ratio as n increases without bound is $\frac{1}{3}$, that is, $\lim_{n\to\infty} \frac{1}{3}\left(\frac{n+1}{n}\right)^2 = \frac{1}{3}$. This means that when n is sufficiently large, the terms in (3) behave like a geometric series whose ratio is $\frac{1}{3}$. Since the series in (3) eventually behaves like a geometric series whose ratio is between -1 and 1, the series in (3) converges.

Another non-geometric series is

$$\frac{2}{1} + \frac{2^2}{2} + \frac{2^3}{3} + \frac{2^4}{4} + \cdots + \frac{2^n}{n} + \cdots. \tag{4}$$

As n increases without bound, the nth term of the series in (4) increases without bound. Thus we can conclude from the divergence test that the series diverges. We can arrive at the same conclusion if we examine the limit of the ratio of the $(n+1)$st term to the nth term, which is

$$\lim_{n\to\infty} \frac{2^{n+1}/(n+1)}{2^n/n} = \lim_{n\to\infty} \frac{2^{n+1}}{2^n} \cdot \frac{n}{n+1} = \lim_{n\to\infty} 2 \cdot \frac{n}{n+1}.$$

As $n \to \infty$, the ratio of consecutive terms approaches 2. This means that the series in (4) eventually behaves like a geometric series whose ratio is 2. We know a geometric series whose ratio is 2 diverges; this tells us that (4) diverges.

The harmonic series, which is

$$1 + \frac{1}{2} + \frac{1}{3} + \frac{1}{4} + \cdots \frac{1}{n} + \cdots,$$

is another non-geometric series, but with terms that approach 0. By considering the integral $\int_1^\infty \frac{dx}{x}$ in Example 2 of Section 6.4, we determined that this series diverges. The ratio of consecutive terms is

$$\frac{1/(n+1)}{1/n} = \frac{n}{n+1}$$

which approaches 1 as $n \to \infty$. We also determined in Section 6.5 that the p-series $\sum \frac{1}{n^p}$ converges for $p > 1$ and diverges for $p \le 1$. For example, the p-series with $p = 3$ is

$$1 + \frac{1}{8} + \frac{1}{27} + \cdots + \frac{1}{n^3} + \cdots,$$

and it converges. The ratio of consecutive terms of the p-series with $p = 3$ is

$$\frac{1/(n+1)^3}{1/n^3} = \frac{n^3}{(n+1)^3},$$

which approaches 1 as $n \to \infty$. The harmonic series and the p-series with $p = 3$ both have a limiting value of 1 for the ratio of consecutive terms, yet one diverges and the other converges.

The examples we have looked at illustrate the following test for convergence of an infinite series, called the ***ratio test***:

ratio test

Let $\sum_{n=0}^{\infty} a_n$ be a series with positive terms.

1. The series converges if $\displaystyle \lim_{n \to \infty} \frac{a_{n+1}}{a_n} < 1$.

2. The series diverges if $\displaystyle \lim_{n \to \infty} \frac{a_{n+1}}{a_n} > 1$.

3. The test is inconclusive if $\displaystyle \lim_{n \to \infty} \frac{a_{n+1}}{a_n} = 1$.

Alternating Series

Notice that the series

$$1 - \frac{1}{2} + \frac{1}{3} - \frac{1}{4} + \cdots + (-1)^{n+1} \frac{1}{n} + \cdots \qquad (5)$$

is similar to the harmonic series, except that the terms alternate in sign. A series in which the terms alternate between positive and negative is called an ***alternating series***, and the series in (5) often is called the ***alternating harmonic series***. We would like to know if the series in (5) converges or diverges. The series is not geometric, and the ratio test does not apply since the terms in the series are not all positive. Instead, we can investigate the sum of the infinite series by thinking about how the sum of the first n terms behaves as n increases without bound. If we use S_n to represent the sum of the first n terms, then

alternating series
alternating harmonic series

$$S_n = 1 - \frac{1}{2} + \frac{1}{3} - \frac{1}{4} + \cdots + (-1)^{n+1} \frac{1}{n}.$$

The series converges if S_n approaches a finite number as n increases without bound.

We can write the first four values of S_n as

$$S_1 = 1,$$

$$S_2 = 1 - \frac{1}{2} = S_1 - \frac{1}{2},$$

$$S_3 = 1 - \frac{1}{2} + \frac{1}{3} = S_2 + \frac{1}{3},$$

$$S_4 = 1 - \frac{1}{2} + \frac{1}{3} - \frac{1}{4} = S_3 - \frac{1}{4}.$$

In general, $S_n = S_{n-1} \pm \frac{1}{n}$, with the choice of + or − depending on whether n is odd or even. The number line in Figure 7.2 shows the behavior of the first four values of S_n.

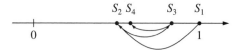

Figure 7.2 Number line showing values of S_n

The pattern in Figure 7.2 indicates that the difference between S_n and S_{n-1} diminishes as n increases. Furthermore, we can see that the position of S_{99} on the number line will be to the right of S_{98}, and S_{100} will be to the left of S_{99} but to the right of S_{98}. Therefore, S_{100} is between S_{98} and S_{99}, and, in general, the limiting value S is trapped between S_{n-1} and S_n. The difference between S_n and S_{n-1} is $\pm \frac{1}{n}$, and this difference goes to zero as $n \to \infty$. Since S_{n-1} and S_n get closer together as n increases, S is trapped in an interval with a size that approaches zero as $n \to \infty$. These observations lead us to conclude that the alternating harmonic series approaches some limiting value, which means that the alternating harmonic series converges.

We might generalize from our analysis of the alternating harmonic series that any alternating series of the form

$$a_1 - a_2 + a_3 - a_4 + \cdots$$

converges provided the sequence of positive numbers $a_1, a_2, a_3, \ldots$ is decreasing and $\lim_{n \to \infty} a_n = 0$. In fact, this generalization can be proven to be true. This result is known as the **alternating series test**.   *alternating series test*

Example 1

Determine whether the series

$$1 - \frac{1}{3} + \frac{1}{5} - \frac{1}{7} + \frac{1}{9} + \cdots + (-1)^n \frac{1}{2n+1} + \cdots$$

converges or diverges.

Solution   We can use the alternating series test and determine whether the series satisfies two conditions. First, do the absolute values of the terms in the series decrease? Second, do the terms in the series approach zero? Since $1 > \frac{1}{3} > \frac{1}{5} > \frac{1}{7}...$, the terms in the series are decreasing. Since $\lim_{n \to \infty} \frac{1}{2n+1} = 0$, the terms approach zero. We therefore can conclude that the series converges. ■

We know that if the terms in an alternating series go to zero, then the series converges. However, if the terms decrease to zero in a series of positive terms, then the series may or may not converge. For example, previously in this section we noted that the series $\sum_{n=1}^{\infty} \frac{1}{n^3}$ converges and the harmonic series $\sum_{n=1}^{\infty} \frac{1}{n}$ diverges, yet both are series of positive terms with the terms decreasing to zero. Even though the harmonic series diverges, the alternating harmonic series converges. In summary, if the terms of a series go to zero, convergence is guaranteed only if the series is alternating.

With the harmonic series as an example, we see that sometimes an alternating series converges even though a divergent series is formed by the series of the absolute values of the terms of the alternating series. In contrast, if a series of positive terms $a_1 + a_2 + a_3 + a_4 + \cdots$ converges, then the related alternating series $a_1 - a_2 + a_3 - a_4 + \cdots$ converges. This result generalizes to any series $c_1 + c_2 + c_3 + c_4 + \cdots$ consisting of a combination of positive and negative c_i's. If the series with absolute values $|c_1| + |c_2| + |c_3| + |c_4| + \cdots$ converges, then the series without absolute values also must converge. However, if the series with absolute values diverges, then the series without absolute values may or may not converge.

If the series $|c_1| + |c_2| + |c_3| + |c_4| + \cdots$ converges, then the series ***absolutely convergent*** $c_1 + c_2 + c_3 + c_4 + \cdots$ is called ***absolutely convergent***. If the series $|c_1| + |c_2| + |c_3| + |c_4| + \cdots$ diverges and the series $c_1 + c_2 + c_3 + c_4 + \cdots$ converges, then ***conditionally convergent*** the series $c_1 + c_2 + c_3 + c_4 + \cdots$ is called ***conditionally convergent***. Based on our previous discussion, we know that the alternating harmonic series is conditionally convergent.

Example 2

Determine whether the series

$$1 - 3 + \frac{9}{2} - \frac{27}{6} + \frac{81}{24} - \frac{243}{120} + \cdots + (-1)^n \frac{3^n}{n!} + \cdots$$

converges absolutely, converges conditionally, or diverges.

Solution   We will first check for absolute convergence. To determine whether the series converges absolutely, we can use the ratio test with the absolute values of the terms of the series. The ratio of the absolute values of consecutive terms is

$$\frac{3^{n+1}/(n+1)!}{3^n/n!} = \frac{3^{n+1}}{(n+1)!} \cdot \frac{n!}{3^n},$$

which is $\frac{3}{n+1}$. As n increases without bound, the limit of this ratio is zero. According to the ratio test, this means that the series

$$1+3+\frac{9}{2}+\frac{27}{6}+\frac{81}{24}+\frac{243}{120}+\cdots+\frac{3^n}{n!}+\cdots$$

converges. Since the series converges when all terms are positive, we know that the original series with alternating terms also converges. We therefore conclude that the alternating series is not only convergent, but absolutely convergent. ■

Example 3

Determine whether the series

$$1-\frac{4}{5}+\frac{6}{10}-\frac{8}{17}+\cdots+(-1)^{n+1}\frac{2n}{n^2+1}+\cdots$$

converges absolutely, converges conditionally, or diverges.

Solution   If we take the absolute value of the terms we have the series $\sum_{n=1}^{\infty}\frac{2n}{n^2+1}$. Notice that the numerator is the derivative of the denominator, so we can find an antiderivative for the expression. The integral $\int_1^{\infty}\frac{2x}{x^2+1}\,dx$ diverges, so it follows by the integral test that the series of positive terms diverges. Since the original series is alternating and the terms decrease with a limit of zero, we know that the alternating series converges. Thus the original series converges conditionally. ■

Convergence of Power Series

We are now prepared to return to the question posed at the beginning of this section: what is the interval of convergence for a particular power series? In other words, given a power series for a function, for what x values does the power series converge to the function?

We begin with the Maclaurin series for $f(x)=e^x$, which is

$$p(x)=1+x+\frac{x^2}{2!}+\frac{x^3}{3!}+\frac{x^4}{4!}+\cdots+\frac{x^n}{n!}+\cdots. \tag{6}$$

What is the interval of convergence for this series? We know that $p(0)=1$ and $f(0)=1$. This tells us that $x=0$ is in the interval of convergence. What about $x=1$? Does

$$p(1)=1+1+\frac{1^2}{2!}+\frac{1^3}{3!}+\frac{1^4}{4!}+\cdots+\frac{1^n}{n!}+\cdots \tag{7}$$

equal $f(1)$? First, does the series for $p(1)$ converge? The ratio of consecutive terms for the series in (7) is

$$\frac{1/(n+1)!}{1/n!}=\frac{n!}{(n+1)!},$$

which equals $\frac{1}{n+1}$. Since all of the terms are positive and $\lim_{n\to\infty} \frac{1}{n+1} = 0$, which is less than 1, by the ratio test the series converges. We stated previously that for the series we consider in this text, if a power series for f converges at x, then it converges to $f(x)$. Therefore, the Maclaurin series in (6) with $x = 1$ converges to $f(1)$, so $x = 1$ is in the interval of convergence.

Now we will examine the Maclaurin series in (6) with $x = -1$. We need to determine if

$$p(-1) = 1 + (-1) + \frac{(-1)^2}{2!} + \frac{(-1)^3}{3!} + \frac{(-1)^4}{4!} + \cdots + \frac{(-1)^n}{n!} + \cdots \tag{8}$$

converges. The series in (8) is an alternating series that is absolutely convergent. We know this because the series of absolute values of the terms is the series in (7), and we have shown that the series in (7) converges. Therefore, the Maclaurin series in (6) with $x = -1$ converges to $f(-1)$, so $x = -1$ is in the interval of convergence.

So far we know that $x = 0$, $x = 1$, and $x = -1$ are in the interval of convergence of the Maclaurin series in (6). Rather than repeatedly substituting a value for x and applying a convergence test to the resulting series of numbers, we can apply the ratio test directly to the Maclaurin series without specifying an x value. We do not know whether or not all the terms in (6) are positive, so we will apply the ratio test to the absolute values of consecutive terms to determine the interval of absolute convergence. For the Maclaurin series in (6), the limit of the ratio of the absolute values of consecutive terms is

$$\lim_{n\to\infty} \frac{\left|x^{n+1}/(n+1)!\right|}{\left|x^n/n!\right|} = \lim_{n\to\infty} \left|\frac{x^{n+1}}{x^n}\right| \cdot \left|\frac{n!}{(n+1)!}\right|$$

$$= \lim_{n\to\infty} \left|\frac{x}{n+1}\right|.$$

No matter what value we choose for x, the limit of $\left|\frac{x}{n+1}\right|$ is 0 as n increases without bound. Therefore, the ratio test states that the Maclaurin series in (6) converges absolutely for all x values. This means that the Maclaurin series for $f(x) = e^x$ equals $f(x)$ at all x values.

To say that a Taylor series or Maclaurin series converges for all x values has important implications for Taylor polynomials. A Taylor polynomial is exact only at the point about which it is defined; however, if the Taylor series converges for all x, then the difference between the value of the Taylor polynomial and the value of the function can be made as small as desired for any value of x by increasing the number of terms in the Taylor polynomial.

Example 4

Find the interval of convergence for the power series

$$p(x) = x - 2x^2 + 4x^3 - 8x^4 + \cdots + (-1)^{n-1}2^{n-1}x^n + \cdots, \tag{9}$$

which is the Taylor series for $f(x) = \frac{x}{1+2x}$ about $x = 0$.

Solution We can determine the interval of absolute convergence by using the ratio test. To apply this test, we need to find the x values for which the ratio of the absolute values of consecutive terms has a limit that is less than 1 as n increases without bound. This limit is

$$\lim_{n \to \infty} \left| \frac{2^n \cdot x^{n+1}}{2^{n-1} \cdot x^n} \right| = |2x|,$$

so we need to determine the x values for which $|2x| < 1$. We can write the inequality $|2x| < 1$ without the absolute value symbol as

$$-1 < 2x < 1,$$

then divide by 2 so that

$$-\frac{1}{2} < x < \frac{1}{2}.$$

This shows that the series in (9) converges absolutely, and therefore converges, for x values such that $-\frac{1}{2} < x < \frac{1}{2}$. The ratio test tells us that the series of *absolute values* of the terms in (9) diverges for x values such that $x > \frac{1}{2}$ or $x < -\frac{1}{2}$. However, the ratio test does not give us information about the series in (9) itself for x values such that $x > \frac{1}{2}$ or $x < -\frac{1}{2}$. If $x > \frac{1}{2}$ or $x < -\frac{1}{2}$, then the limit of the ratio of consecutive terms, which is $\pm 2x$, is greater than 1 or less than -1. This implies that if $x > \frac{1}{2}$ or $x < -\frac{1}{2}$, then the series behaves like a geometric series with ratio greater than 1 or less than -1 and therefore diverges. We can now say that the series in (9) converges if $-\frac{1}{2} < x < \frac{1}{2}$ and diverges if $x < -\frac{1}{2}$ or $x > \frac{1}{2}$.

Since the ratio test is inconclusive if the limit of the ratio of consecutive terms is equal to 1, we need to investigate separately whether the series converges if $x = -\frac{1}{2}$ or $x = \frac{1}{2}$. Substituting $x = -\frac{1}{2}$ in (9) yields

$$p\left(-\frac{1}{2}\right) = -\frac{1}{2} - \frac{1}{2} - \frac{1}{2} - \cdots,$$

which diverges. Substituting $x = \frac{1}{2}$ in (9) yields

$$p\left(\frac{1}{2}\right) = \frac{1}{2} - \frac{1}{2} + \frac{1}{2} - \frac{1}{2} + \cdots,$$

which also diverges since the sum of the first n terms alternates between $\frac{1}{2}$ and 0 as n increases without bound. The series in (9) diverges at both $x = -\frac{1}{2}$ and $x = \frac{1}{2}$; therefore, the interval of convergence for the series in (9) is $-\frac{1}{2} < x < \frac{1}{2}$. ∎

We have seen in this section that elementary functions that are infinitely differentiable can be written as infinite polynomials, or power series. We have also looked at examples of ways to determine the interval of convergence for a power series, so that we can determine the domain over which a power series equals the original function. So far, we have written a power series that equals a given function by constructing a Taylor series about some value $x = a$. In the next section, we will examine alternative methods for writing a power series that can be applied to functions for which we would have great difficulty constructing a Taylor series directly from the nth-order derivatives.

A power series can be used to replace the function to which it converges. For example, e^x is equal to the power series $1 + x + \frac{x^2}{2!} + \frac{x^3}{3!} + \cdots + \frac{x^n}{n!} + \cdots$ for all x, and so this series can be used to compute values of e^x. In addition, the first few terms of a power series for f can be substituted for f in an equation, which will result in an approximation for the equation, but may simplify the equation or allow it to be solved by algebraic manipulations. Finally, power series are a consistent, theoretical way of defining all of the elementary functions. For example, the sine and cosine functions can be defined as the coordinates of a point on the unit circle. For most x values, however, evaluating $\sin x$ and $\cos x$ by this definition involves geometric measurements. In contrast, the power series for $\sin x$ and $\cos x$ allow us to evaluate these functions to whatever accuracy we desire by using arithmetic. Power series therefore provide a theoretical basis for our calculus work with elementary functions.

Exercise Set 7.1

1. Show that

$$\sin x = x - \frac{x^3}{3!} + \frac{x^5}{5!} - \frac{x^7}{7!} + \cdots + (-1)^n \frac{x^{2n+1}}{(2n+1)!} + \cdots$$

and

$$\cos x = 1 - \frac{x^2}{2!} + \frac{x^4}{4!} - \frac{x^6}{6!} + \cdots + (-1)^n \frac{x^{2n}}{(2n)!} + \cdots$$

for all real numbers.

2. Write the Maclaurin series for $\ln(1+x)$ and determine the interval of convergence of this series.

3. Find the interval of convergence for each power series.

a. $1 - x^2 + \frac{x^4}{2!} - \frac{x^6}{3!} + \frac{x^8}{4!} - \cdots + (-1)^n \frac{x^{2n}}{n!} + \cdots$

b. $x - \frac{x^2}{2} + \frac{x^3}{3} - \frac{x^4}{4} + \cdots$

c. $x^2 - \dfrac{x^6}{3!} + \dfrac{x^{10}}{5!} - \dfrac{x^{14}}{7!} + \cdots$ d. $t - \dfrac{t^3}{3} + \dfrac{t^5}{5} - \dfrac{t^7}{7} + \cdots$

e. $\displaystyle\sum_{n=0}^{\infty} (-1)^n t^{2n}$ f. $\displaystyle\sum_{n=0}^{\infty} \dfrac{x^2}{n+1}$

g. $\displaystyle\sum_{n=0}^{\infty} \dfrac{\pi^n (x-1)^{2n}}{(2n+1)}$ h. $\displaystyle\sum_{n=1}^{\infty} \dfrac{5^n}{n^2} x^n$

i. $\displaystyle\sum_{n=1}^{\infty} (-1)^{n+1} \dfrac{(x+1)^n}{n}$

4. To what functions do the power series in 3a and 3e converge?

7.2 Operations with Power Series

Since a power series represents a function, a power series can be manipulated just as we manipulate a function. Operations on one power series are helpful in producing another power series, often without the effort required to match derivatives as we do to construct a Taylor series.

One way we can generate a new power series is to make a substitution in a known series. For example, using the Maclaurin series for e^x, which is

$$e^x = 1 + x + \frac{x^2}{2!} + \frac{x^3}{3!} + \frac{x^4}{4!} + \cdots,$$

we can substitute $-x^2$ for x, which yields

$$e^{-x^2} = 1 + \left(-x^2\right) + \frac{\left(-x^2\right)^2}{2!} + \frac{\left(-x^2\right)^3}{3!} + \frac{\left(-x^2\right)^4}{4!} + \cdots \tag{10}$$
$$= 1 - x^2 + \frac{x^4}{2!} - \frac{x^6}{3!} + \frac{x^8}{4!} - \frac{x^{10}}{5!} + \cdots.$$

The right side of equation (10) is a power series for e^{-x^2}.

In Exercise Set 7.1 we determined a Maclaurin series for $g(x) = \ln(1+x)$ by matching function values and derivatives, and this Maclaurin series is

$$p(x) = x - \frac{x^2}{2} + \frac{x^3}{3} - \frac{x^4}{4} + \frac{x^5}{5} - \cdots. \tag{11}$$

Compare the series in (11) with the Maclaurin series for $g'(x) = \frac{1}{1+x}$, which is

$$q(x) = 1 - x + x^2 - x^3 + x^4 - x^5 + \cdots. \tag{12}$$

Notice that term-by-term differentiation of the series in (11) produces the series in (12). This suggests that we can generate a new power series if we differentiate a known power series term-by-term. For example, if we differentiate the Maclaurin series for $\sin x$, we have

$$\frac{d}{dx}(\sin x) = \frac{d}{dx}\left(x - \frac{x^3}{3!} + \frac{x^5}{5!} - \frac{x^7}{7!} + \cdots\right)$$

$$= 1 - \frac{3x^2}{3!} + \frac{5x^4}{5!} - \frac{7x^6}{7!} + \cdots$$

$$= 1 - \frac{x^2}{2!} + \frac{x^4}{4!} - \frac{x^6}{6!} + \cdots,$$

which is the Maclaurin series for $\cos x$. This result shows that we can construct the Maclaurin series for $\cos x$ by differentiating the Maclaurin series for $\sin x$.

Just as we can use term-by-term differentiation to generate a new power series, we can likewise use term-by-term integration. The Maclaurin series for $f(x) = \frac{1}{1+x}$ is given by

$$\frac{1}{1+x} = 1 - x + x^2 - x^3 + x^4 - x^5 + \cdots + (-1)^n x^n + \cdots. \tag{13}$$

If we substitute x^2 for x in equation (13), then we have

$$\frac{1}{1+x^2} = 1 - x^2 + x^4 - x^6 + x^8 - x^{10} + \cdots + (-1)^n x^{2n} + \cdots. \tag{14}$$

Since $\int \frac{1}{1+x^2}\,dx = \tan^{-1} x + C$, we can find a power series for $\tan^{-1} x$ by integrating term-by-term the right side of equation (14), which yields

$$\int \frac{1}{1+x^2}\,dx = \int \left(1 - x^2 + x^4 - x^6 + x^8 - x^{10} + \cdots + (-1)^n x^{2n} + \cdots\right)dx. \tag{15}$$

Evaluating the integrals in equation (15) gives

$$\tan^{-1} x = C + x - \frac{x^3}{3} + \frac{x^5}{5} - \frac{x^7}{7} + \frac{x^9}{9} - \frac{x^{11}}{11} + \cdots + (-1)^n \frac{x^{2n+1}}{2n+1} + \cdots. \tag{16}$$

The constant of integration in equation (16) can be determined by substituting $x = 0$ on both sides of the equation, which yields $C = 0$. This means that a power series for $\tan^{-1} x$ is given by

$$\tan^{-1} x = x - \frac{x^3}{3} + \frac{x^5}{5} - \frac{x^7}{7} + \frac{x^9}{9} - \frac{x^{11}}{11} + \cdots + (-1)^n \frac{x^{2n+1}}{2n+1} + \cdots.$$

When we create a new series from a known series using term-by-term differentiation or integration, we need to investigate how the interval of convergence of the new series is related to that of the original series. The following exercise set includes questions related to how the interval of convergence is affected by differentiation and integration.

Exercise Set 7.2

1. Use power series and term-by-term differentiation to confirm that $\frac{d}{dx}\sin(x^2) = \cos(x^2) \cdot 2x$.

2. Write a power series for each function by manipulating a known series. Determine the interval of convergence of the new series. How does this interval of convergence compare with the interval of convergence for the original series?

   a. $f(x) = \sin(-2x)$ b. $f(x) = xe^x$

   c. $f(x) = \dfrac{1}{1+2x^2}$ d. $f(x) = \dfrac{x}{1+4x}$

3. Write a power series to represent $g(x) = \int e^{-x^2}\, dx$.

4. Use a known power series and term-by-term integration to write a power series for each integral.

   a. $\displaystyle\int \sin x^2 dx$ b. $\displaystyle\int \cos \sqrt{x}\, dx$

   c. $\displaystyle\int \frac{1}{1+x^4}\, dx$

5. Use the series you wrote in problem 4 to find approximate values for the following definite integrals.

   a. $\displaystyle\int_0^1 \sin x^2 dx$ b. $\displaystyle\int_0^1 \cos \sqrt{x}\, dx$

   c. $\displaystyle\int_0^{1/2} \frac{1}{1+x^4}\, dx$

6. Investigate the interval of convergence of the series for $f(x)$ and of the series for $\frac{d}{dx} f(x)$. How does differentiation appear to influence the interval of convergence?

   a. $f(x) = \ln(1+x)$ b. $f(x) = \ln(1+x^2)$

   c. $f(x) = \dfrac{1}{1-x^2}$

7. Find the interval of convergence of the power series for $\frac{1}{1+x^2}$ and the interval of convergence of the power series for $\tan^{-1} x$. How does integration appear to influence the interval of convergence of a power series? Also refer to problem 6 in which pairs of series are related by differentiation, which means the pairs are also related by integration.

8. Use the Maclaurin series for e^x, $\sin x$, and $\cos x$ to verify the identity

   $$e^{i\theta} = \cos\theta + i\sin\theta,$$

   where θ is a real number and $i = \sqrt{-1}$. Use this identity to show that $e^{i\pi} + 1 = 0$, which is an equation derived by Euler that relates five of the most important numbers in mathematics.

7.3 Series Solutions to Differential Equations

One important application of power series is determining solutions to differential equations. The process of using a power series to solve a differential equation requires two assumptions. First, we assume that a function f that satisfies the given differential equation has a power series representation p. This requires the function f to be infinitely differentiable. A second assumption is that the power series p converges on the interval in which we are interested. Before we introduce the technique of using a power series to solve a differential equation, consider the following problem.

Suppose we form a pendulum by hanging a ball from the end of a string of length L with the other end attached to the top of a door frame. The pendulum is at rest when the ball hangs so that the string forms a right angle with the top of the door frame. When the ball is moved from the rest position with the string kept taut and then released, its motion depends upon the forces that are acting on it. We will ignore the effects of the forces due to friction and air resistance. The force resulting from gravity acts on the ball in the downward direction with magnitude mg, where m is the mass of the ball and g is the acceleration due to gravity. The component of this force tangent to the path of the ball, which is perpendicular to the string, causes the motion of the ball when it is released. Figure 7.3 shows the force of gravity and its components parallel to and perpendicular to the string. The angle θ is defined as the angle of the swing of the pendulum from rest position, and the positive direction is counterclockwise. From the right triangle definition of the sine function, we see that the magnitude of the force acting to change the position of the ball is $-mg \sin \theta$. The negative sign indicates that the direction of the force is opposite to the direction of increasing θ.

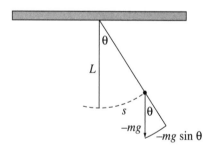

Figure 7.3 Force diagram for the pendulum

Newton's second law tells us that $F = ma$, so in this situation

$$ma = -mg \sin \theta. \tag{17}$$

The acceleration a is the second derivative of the position s of the ball. Since $a = \frac{d^2 s}{dt^2}$, we substitute for a in equation (17) to obtain

$$m \cdot \frac{d^2 s}{dt^2} = -mg \cdot \sin \theta. \tag{18}$$

The path of the ball is an arc of a circle with radius L, so we can specify the position of the ball by the length of the arc from the ball to the rest position. Equation (18) involves the three variables s, θ, and t. We can eliminate s from this equation by using the relationship $s = L\theta$ to calculate the second derivative of s with respect to t, which is

$$\frac{d^2 s}{dt^2} = L \cdot \frac{d^2 \theta}{dt^2}.$$

By substitution for $\frac{d^2 s}{dt^2}$, we can rewrite equation (18) as

$$mL \cdot \frac{d^2 \theta}{dt^2} = -mg \cdot \sin \theta,$$

or

$$\frac{d^2 \theta}{dt^2} = -\frac{g}{L} \sin \theta. \tag{19}$$

Equation (19) is the differential equation for the motion of the pendulum.

If the angle θ is small, then we can replace $\sin \theta$ in the right side of equation (19) with the linear approximation for $\sin \theta$ about $\theta = 0$, which is $\sin \theta \approx \theta$. The reasons for using the linear approximation for $\sin \theta$ will become clear as we solve this problem. Note that the relative error for this approximation is less than 10% when $-0.749 < \theta < 0.749$, or in degrees, $-42.9° < \theta < 42.9°$. Substituting θ for $\sin \theta$ in equation (19) yields

$$\frac{d^2 \theta}{dt^2} = -\frac{g}{L} \cdot \theta.$$

Before solving this differential equation, we will first look at two simpler examples that illustrate the technique of power series solutions of differential equations.

Example 1

We know that the differential equation $\frac{dy}{dx} = y$ has the general solution $y = Ce^x$. Show how a power series can be used to find this solution.

Solution We represent the function that is a solution to $\frac{dy}{dx} = y$ with the power series

$$y = a_0 + a_1 x + a_2 x^2 + a_3 x^3 + \cdots. \tag{20}$$

We can differentiate both sides of (20) to obtain

$$\frac{dy}{dx} = a_1 + 2a_2x + 3a_3x^2 + 4a_4x^3 + \cdots. \tag{21}$$

By substituting (20) and (21) into the differential equation $\frac{dy}{dx} = y$, we have

$$a_1 + 2a_2x + 3a_3x^2 + 4a_4x^3 + \cdots = a_0 + a_1x + a_2x^2 + a_3x^3 + \cdots. \tag{22}$$

Equation (22) is true for all values of x for which the series in (20) and (21) both converge. You might recall that for two polynomials to be equal, the coefficients of like terms must be equal. Setting the coefficients of like terms on opposite sides of equation (22) equal to each other gives us the system of equations

$$a_1 = a_0$$
$$2a_2 = a_1$$
$$3a_3 = a_2$$
$$\vdots$$
$$na_n = a_{n-1}$$
$$\vdots$$

Solving for each coefficient in terms of a_0, we have

$$a_1 = a_0$$
$$a_2 = \frac{1}{2}a_1 = \frac{1}{2}a_0$$
$$a_3 = \frac{1}{3}a_2 = \frac{1}{6}a_0$$
$$\vdots$$
$$a_n = \frac{1}{n}a_{n-1} = \frac{1}{n!}a_0$$
$$\vdots$$

Substituting for the coefficients in (20) we have

$$y = a_0 + a_0x + \frac{1}{2!}a_0x^2 + \frac{1}{3!}a_0x^3 + \frac{1}{4!}a_0x^4 + \cdots + \frac{1}{n!}a_0x^n + \cdots,$$

which can be written as

$$y = a_0\left(1 + x + \frac{x^2}{2!} + \frac{x^3}{3!} + \frac{x^4}{4!} + \cdots + \frac{x^n}{n!} + \cdots\right). \tag{23}$$

We now have a power series that represents the function that is the solution to the differential equation $\frac{dy}{dx} = y$. We recognize the power series in (23) as the Maclaurin series for e^x. Thus, we have used power series to show that $y = a_0e^x$ is the general solution to the differential equation $\frac{dy}{dx} = y$. ∎

Example 2

Determine the particular solution with $y(0) = 1$ for the differential equation

$$\frac{dy}{dx} = x + y.$$

Solution As in Example 1, we begin by writing the solution as a power series for y in terms of x

$$y = a_0 + a_1 x + a_2 x^2 + a_3 x^3 + \cdots,$$

and then the derivative of y with respect to x is

$$\frac{dy}{dx} = a_1 + 2a_2 x + 3a_3 x^2 + 4a_4 x^3 + \cdots.$$

Substituting into $\frac{dy}{dx} = x + y$ we obtain the equation

$$a_1 + 2a_2 x + 3a_3 x^2 + 4a_4 x^3 + \cdots = x + a_0 + a_1 x + a_2 x^2 + a_3 x^3 + \cdots.$$

Setting coefficients of like terms equal to each other, we have

$$a_1 = a_0$$
$$2a_2 = a_1 + 1$$
$$3a_3 = a_2$$
$$\vdots$$
$$na_n = a_{n-1}$$
$$\vdots$$

Solving for each coefficient in terms of a_0 yields

$$a_1 = a_0$$
$$a_2 = \frac{1}{2}(a_1 + 1) = \frac{1}{2}(a_0 + 1)$$
$$a_3 = \frac{1}{3}a_2 = \frac{1}{6}(a_0 + 1)$$
$$\vdots$$
$$a_n = \frac{1}{n}a_{n-1} = \frac{1}{n!}(a_0 + 1)$$
$$\vdots$$

Substituting for the coefficients in the power series for y gives the solution

$$y = a_0 + a_0 x + \frac{1}{2!}(a_0 + 1)x^2 + \frac{1}{3!}(a_0 + 1)x^3 + \cdots + \frac{1}{n!}(a_0 + 1)x^n + \cdots. \quad (24)$$

Except for the first two terms, the right side of (24) looks somewhat like the power series for $(a_0 + 1)e^x$, which is

$$(a_0 + 1)e^x = (a_0 + 1) + (a_0 + 1)x + \frac{1}{2!}(a_0 + 1)x^2 + \frac{1}{3!}(a_0 + 1)x^3 + \cdots. \quad (25)$$

We can expand the first two terms in the right side of (25) to see how the right side of (25) includes the right side of (24), which gives

$$(a_0 + 1)e^x = 1 + x + a_0 + a_0 x + \frac{1}{2!}(a_0 + 1)x^2 + \frac{1}{3!}(a_0 + 1)x^3 + \cdots. \quad (26)$$

The right side of (26) is the sum of $1 + x$ and the series for y as written in equation (24), so by substitution equation (26) becomes

$$(a_0 + 1)e^x = 1 + x + y.$$

Solving for y gives

$$y = (a_0 + 1)e^x - x - 1. \quad (27)$$

The function with the equation given in (27) is the general solution to the differential equation $\frac{dy}{dx} = x + y$. With the condition $y(0) = 1$, we find that $a_0 = 1$. The particular solution with $y(0) = 1$ is $y = 2e^x - x - 1$. ∎

We now return to the differential equation that we wish to solve to find an equation that models the motion of the pendulum, which is

$$\frac{d^2\theta}{dt^2} = -\frac{g}{L} \cdot \theta. \quad (28)$$

To solve differential equation (28), we assume that θ can be written as the power series

$$\theta = a_0 + a_1 t + a_2 t^2 + a_3 t^3 + a_4 t^4 + a_5 t^5 + \cdots. \quad (29)$$

If we differentiate both sides of equation (29) with respect to t, we get

$$\frac{d\theta}{dt} = a_1 + 2a_2 t + 3a_3 t^2 + 4a_4 t^3 + 5a_5 t^4 + 6a_6 t^5 + \cdots.$$

Differentiating again we get

$$\frac{d^2\theta}{dt^2} = 2a_2 + 6a_3 t + 12a_4 t^2 + 20a_5 t^3 + 30a_6 t^4 + 42a_7 t^5 + \cdots. \quad (30)$$

Substituting the series for θ in (29) and the series for $\frac{d^2\theta}{dt^2}$ in (30) into differential equation (28) yields

$$2a_2 + 6a_3 t + 12a_4 t^2 + 20a_5 t^3 + \cdots = -\frac{g}{L}\left(a_0 + a_1 t + a_2 t^2 + a_3 t^3 + \cdots\right). \quad (31)$$

(If we had not used the linear approximation for $\sin\theta$, we would not have been able to find an expression for the differential equation with which we can work so easily.)

Equation (31) is true for all values of t for which the series in (29) and (30) both converge, so the coefficients of like powers of t must be equal. Setting coefficients of like terms on opposite sides of equation (31) equal to each other, we have the system of equations

$$2a_2 = -\frac{g}{L}a_0$$

$$6a_3 = -\frac{g}{L}a_1$$

$$12a_4 = -\frac{g}{L}a_2$$

$$20a_5 = -\frac{g}{L}a_3$$

$$\vdots$$

The even-numbered coefficients all depend on a_0, and the odd-numbered coefficients depend upon a_1. Solving for the coefficients in terms of a_0 and a_1 yields

$$a_2 = -\frac{g}{2L} \cdot a_0$$

$$a_3 = -\frac{g}{6L} \cdot a_1$$

$$a_4 = -\frac{g}{12L} \cdot a_2 = \frac{-g}{12L}\left(\frac{-g}{2L}a_0\right) = \frac{1}{4!}\left(\frac{g}{L}\right)^2 a_0$$

$$a_5 = -\frac{g}{20L} \cdot a_3 = \frac{-g}{20L}\left(\frac{-g}{6L}a_1\right) = \frac{1}{5!}\left(\frac{g}{L}\right)^2 a_1$$

$$\vdots$$

Substituting for the coefficients in the power series in (29), θ can now be written as

$$\theta = a_0 + a_1 t - \frac{g}{2L}a_0 t^2 - \frac{g}{6L}a_1 t^3 + \frac{1}{4!}\left(\frac{g}{L}\right)^2 a_0 t^4 +$$

$$\frac{1}{5!}\left(\frac{g}{L}\right)^2 a_1 t^5 - \frac{1}{6!}\left(\frac{g}{L}\right)^3 a_0 t^6 - \frac{1}{7!}\left(\frac{g}{L}\right)^3 a_1 t^7 + \cdots.$$

(32)

If we group the terms with factor a_0 separately from the terms with factor a_1, then we can rewrite equation (32) as

$$\theta = a_0 \left(1 - \frac{g}{2!\,L} t^2 + \frac{1}{4!} \left(\frac{g}{L} \right)^2 t^4 - \frac{1}{6!} \left(\frac{g}{L} \right)^3 t^6 + \cdots \right) +$$
$$a_1 \left(t - \frac{g}{3!\,L} t^3 + \frac{1}{5!} \left(\frac{g}{L} \right)^2 t^5 - \frac{1}{7!} \left(\frac{g}{L} \right)^3 t^7 + \cdots \right), \tag{33}$$

which is the power series solution for the differential equation $\frac{d^2\theta}{dt^2} = -\frac{g}{L} \cdot \theta$.

The solution in (33) can also be expressed in terms of elementary functions by determining what functions are represented by the two series on the right side of (33). The series that is multiplied by a_0 is similar to the Maclaurin series for $\cos t$, whereas the series that is multiplied by a_1 is similar to the Maclaurin series for $\sin t$.

Notice that in the series in (33) that is multiplied by a_0, the power of $\frac{g}{L}$ in each term is half the power of t. We can rewrite this series as

$$a_0 \left[1 - \frac{1}{2!} \left(\sqrt{\frac{g}{L}}\, t \right)^2 + \frac{1}{4!} \left(\sqrt{\frac{g}{L}}\, t \right)^4 - \frac{1}{6!} \left(\sqrt{\frac{g}{L}}\, t \right)^6 + \cdots \right],$$

which is a power series for $a_0 \cdot \cos \left(\sqrt{\frac{g}{L}}\, t \right)$.

Notice that in each term of the series in (33) that is multiplied by a_1, the power of $\frac{g}{L}$ is 0.5 less than half the power of t. If we multiply each term by $\sqrt{\frac{g}{L}}$, and also multiply the entire series by $\sqrt{\frac{L}{g}}$ so that its value remains unchanged, then we can rewrite the series as

$$a_1 \sqrt{\frac{L}{g}} \left[\sqrt{\frac{g}{L}}\, t - \frac{1}{3!} \sqrt{\frac{g}{L}} \left(\frac{g}{L} \right) t^3 + \frac{1}{5!} \sqrt{\frac{g}{L}} \left(\frac{g}{L} \right)^2 t^5 - \frac{1}{7!} \sqrt{\frac{g}{L}} \left(\frac{g}{L} \right)^3 t^7 + \cdots \right],$$

or

$$a_1 \sqrt{\frac{L}{g}} \left[\left(\frac{g}{L} \right)^{1/2} t - \frac{1}{3!} \left(\frac{g}{L} \right)^{3/2} t^3 + \frac{1}{5!} \left(\frac{g}{L} \right)^{5/2} t^5 - \frac{1}{7!} \left(\frac{g}{L} \right)^{7/2} t^7 + \cdots \right]. \tag{34}$$

In the series in (34), the power of $\frac{g}{L}$ is half the power of t. We can rewrite this series as

$$a_1 \sqrt{\frac{L}{g}} \left[\sqrt{\frac{g}{L}}\, t - \frac{1}{3!} \left(\sqrt{\frac{g}{L}}\, t \right)^3 + \frac{1}{5!} \left(\sqrt{\frac{g}{L}}\, t \right)^5 - \frac{1}{7!} \left(\sqrt{\frac{g}{L}}\, t \right)^7 + \cdots \right],$$

which is the Maclaurin series for $a_1 \sqrt{\frac{L}{g}} \sin \left(\sqrt{\frac{g}{L}}\, t \right)$.

The series solution in (33) can now be written as

$$\theta = a_0 \cos\left(\sqrt{\frac{g}{L}}\, t\right) + a_1 \sqrt{\frac{L}{g}} \sin\left(\sqrt{\frac{g}{L}}\, t\right). \tag{35}$$

The interpretation of this solution is investigated in the exercises that follow this section.

In the pendulum problem and in examples 1 and 2, we were able to rewrite the power series solution to the differential equations in terms of elementary functions. The power series solution technique does not always lead to a power series that we can recognize as an elementary function, as the following example illustrates.

Example 3

Find the particular solution to $y'' = -xy$ with $y(0) = 1$ and $y'(0) = 0$.

Solution   Similar to the pendulum problem, we begin with

$$y = a_0 + a_1 x + a_2 x^2 + a_3 x^3 + \cdots,$$

$$y' = a_1 + 2a_2 x + 3a_3 x^2 + 4a_4 x^3 + \cdots,$$

and

$$y'' = 2a_2 + 6a_3 x + 12a_4 x^2 + 20a_5 x^3 + 30a_6 x^4 + \cdots.$$

Substitution into $y'' = -xy$ gives

$$2a_2 + 6a_3 x + 12a_4 x^2 + 20a_5 x^3 + 30a_6 x^4 + \cdots = -a_0 x - a_1 x^2 - a_2 x^3 - a_3 x^4 - \cdots.$$

Setting coefficients of like terms equal to each other, we have

$$2a_2 = 0$$
$$6a_3 = -a_0$$
$$12a_4 = -a_1$$
$$20a_5 = -a_2$$
$$30a_6 = -a_3$$
$$42a_7 = -a_4$$
$$56a_8 = -a_5$$
$$\vdots$$
$$(n)(n-1)a_n = -a_{n-3}$$
$$\vdots$$

Rewriting each of the equations in terms of a_0, a_1, or a_2, we have

$$a_2 = 0$$

$$a_3 = -\frac{a_0}{6}$$

$$a_4 = -\frac{a_1}{4 \cdot 3}$$

$$a_5 = -\frac{a_2}{5 \cdot 4} = 0$$

$$a_6 = -\frac{a_3}{6 \cdot 5} = \frac{a_0}{(6 \cdot 5)(3 \cdot 2)}$$

$$a_7 = -\frac{a_4}{7 \cdot 6} = \frac{a_1}{(7 \cdot 6)(4 \cdot 3)}$$

$$a_8 = -\frac{a_5}{8 \cdot 7} = 0$$

$$a_9 = -\frac{a_6}{9 \cdot 8} = -\frac{a_0}{(9 \cdot 8)(6 \cdot 5)(3 \cdot 2)}$$

$$a_{10} = -\frac{a_7}{10 \cdot 9} = -\frac{a_1}{(10 \cdot 9)(7 \cdot 6)(4 \cdot 3)}$$

$$a_{11} = -\frac{a_8}{11 \cdot 10} = 0$$

$$\vdots$$

Substituting the coefficients in the power series for y gives

$$y = a_0 + a_1 x - a_0 \frac{x^3}{3 \cdot 2} - a_1 \frac{x^4}{4 \cdot 3} + a_0 \frac{x^6}{(6 \cdot 5)(3 \cdot 2)} +$$

$$a_1 \frac{x^7}{(7 \cdot 6)(4 \cdot 3)} - a_0 \frac{x^9}{(9 \cdot 8)(6 \cdot 5)(3 \cdot 2)} - a_1 \frac{x^{10}}{(10 \cdot 9)(7 \cdot 6)(4 \cdot 3)} + \cdots, \tag{36}$$

which is the general solution to the differential equation $y'' = -xy$. The solution in (36) does not include any power series that we recognize. In the particular solution with $y(0) = 1$ and $y'(0) = 0$, we know that $a_0 = 1$ and $a_1 = 0$, which yields

$$y = 1 - \frac{x^3}{3 \cdot 2} + \frac{x^6}{(6 \cdot 5)(3 \cdot 2)} - \frac{x^9}{(9 \cdot 8)(6 \cdot 5)(3 \cdot 2)} +$$

$$\cdots + (-1)^n \frac{x^{3n}}{[(3n) \cdot (3n-1)][(3n-3) \cdot (3n-4)] \cdots (6 \cdot 5)(3 \cdot 2)} + \cdots. \tag{37}$$

In Exercise Set 7.3, you will use the ratio test to show that this series converges for all x. Therefore, we have a series solution to our differential equation, even though we are not able to write the solution in terms of elementary functions. ∎

Exercise Set 7.3

1. In the solution to the pendulum problem given by (35), what do a_0 and a_1 represent? What is the physical situation associated with the conditions $a_0 = 0$ and $a_1 > 0$? What is the physical situation associated with the conditions $a_0 > 0$ and $a_1 = 0$?

2. Solve the following differential equations analytically using power series.

    a. $y' = x + y + 1$, $y(0) = 2$

    b. $y'' = 4y$, $y(0) = 1$ and $y'(0) = 2$

    c. $y' - x^2 = x + 2y$, $y(0) = 0$

    d. $xy'' = y$

3. Use power series to show that the particular solution to the differential equation
    $$xy'' + 2y' + xy = 0$$
    with $y(0) = 1$ and $y'(0) = 0$ is $y = \begin{cases} \dfrac{\sin x}{x} & \text{if } x \neq 0, \\ 1 & \text{if } x = 0. \end{cases}$

4. Use the ratio test to show that the solution (37) converges for all x.

5. Power series can also be used to solve a system of differential equations. Use power series to solve the system
    $$\frac{dA}{ds} = B,$$
    $$\frac{dB}{ds} = -A.$$

7.4 Fourier Series

For much of this text we have been approximating functions with other simpler functions. These approximations are used in mathematical proofs and to generate values for a function. Early in the text we used the linear approximation. This was later followed by the quadratic approximation, which led to Taylor polynomials. Each of these approximations has the property that it is equal to the value of the function, and its derivatives are equal to the function's derivatives, at only one point. The accuracy of each approximation deteriorates away from that one point.

   In the previous section we extended Taylor polynomials to Taylor series. In contrast to finite approximations, the Taylor series for a particular function actually is equal to the function over the entire interval of convergence in the sense that the series converges to the function for each x value in the interval of convergence.

We now introduce Fourier series, a function approximation that is inherently different from a Taylor series. We will begin with series of period 2π. Except for the constant term, the terms that are added in a Fourier series are sine and cosine functions of all possible periods of the form $\frac{2\pi}{k}$, where k is a positive integer. A *Fourier series* of period 2π has the form

Fourier series

$$F(x) = a_0 + a_1 \cos(x) + b_1 \sin(x) + \cdots + a_k \cos(kx) + b_k \sin(kx) + \cdots,$$

which can be written using summation notation as

$$F(x) = a_0 + \sum_{k=1}^{\infty} \left[a_k \cos(kx) + b_k \sin(kx) \right]. \tag{38}$$

Each term of the summation in (38) has period $\frac{2\pi}{k}$, where k is a positive integer, so each term has exactly k cycles over an interval of length 2π. Therefore, the entire sum in (38) has period 2π.

Fourier series were first used in the early 1800s by Joseph Fourier (1768–1830) to describe complicated periodic phenomena. Since a Fourier series is periodic, it is applied to the study of heat flows, oscillations, vibrations, sound, and other wave forms. Today, processes associated with Fourier series also are used in speech recognition, music analysis and synthesis, ultra-sound acoustic imaging, and CAT scans.

Our goal at this stage is to approximate a given function f with the Fourier series F in (38) by choosing appropriate values for a_k and b_k. Our initial work with this approximation will be on the interval $-\pi \leq x \leq \pi$. We will learn later how to modify our work if we want a Fourier series approximation on some other interval.

With Taylor series, our goal was to equate derivatives at a point. Our goal here is to approximate f over an interval, which suggests using definite integrals instead. Thus, we require that

$$\int_{-\pi}^{\pi} f(x)\, dx = \int_{-\pi}^{\pi} F(x)\, dx. \tag{39}$$

Substituting for $F(x)$ allows us to rewrite equation (39) as

$$\int_{-\pi}^{\pi} f(x)dx = \int_{-\pi}^{\pi} [a_0 + a_1 \cos(x) + b_1 \sin(x) + a_2 \cos(2x) + $$
$$b_2 \sin(2x) + \cdots + a_n \cos(nx) + b_n \sin(nx) + \cdots]dx. \tag{40}$$

We can evaluate the integral on the right side of (40) by noting that for any integer k, the definite integrals of both $\sin(kx)$ and $\cos(kx)$ over the interval $[-\pi, \pi]$ are zero. We know this because the number of cycles that $\sin(kx)$ and $\cos(kx)$ have over the interval $[-\pi, \pi]$ is an integer, so the area above the x-axis equals the area below the *x*-axis over this interval. The graphs in Figure 7.4 illustrate this characteristic for $k = 1$ and $k = 2$.

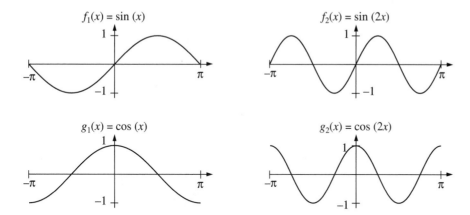

Figure 7.4 Illustrations of $\int_{-\pi}^{\pi} \sin(kx)\, dx = 0$ and $\int_{-\pi}^{\pi} \cos(kx)\, dx = 0$ for $k = 1$ and $k = 2$

Therefore, the right side of (40) is equal to $\int_{-\pi}^{\pi} a_0 dx$, and we have

$$\int_{-\pi}^{\pi} f(x)\, dx = \int_{-\pi}^{\pi} a_0\, dx = a_0\, x \Big|_{-\pi}^{\pi} = a_0[\pi - (-\pi)] = 2\pi\, a_0. \qquad (41)$$

Solving (41) for a_0 yields

$$a_0 = \frac{1}{2\pi} \int_{-\pi}^{\pi} f(x)dx.$$

If f is a function whose antiderivative is known, we can find the value of a_0 by using the Fundamental Theorem to evaluate $\frac{1}{2\pi} \int_{-\pi}^{\pi} f(x)dx$. If f is a function whose antiderivative is not known, we can find the value of a_0 by using a numerical technique to evaluate the integral $\frac{1}{2\pi} \int_{-\pi}^{\pi} f(x)dx$. In either case, we can find the value of a_0, the first term of the Fourier series.

We would like to continue using integrals to determine the values of the coefficients a_k and b_k, $k = 1, 2, 3, \ldots$. As we have just seen, integrating $\sin(kx)$ and $\cos(kx)$ over the interval $[-\pi, \pi]$ produces integrals with value zero. To solve for a_1, we will create an expression involving $F(x)$ so that when we integrate over the interval $[-\pi, \pi]$, the term containing a_1 is non-zero and the terms containing a_k, $k \neq 1$, and b_k, $k = 1, 2, 3, \ldots$, are all equal to zero.

One way to accomplish this is to have $a_1 \cos^2 x$, instead of $a_1 \cos x$, in the integrand. We are sure that the definite integral $\int_{-\pi}^{\pi} a_1 \cos^2(x) dx$ is not zero, because the function $\cos^2 x$ has only positive values. If we multiply the integrands on both sides of equation (40) by $\cos x$, we have

$$\int_{-\pi}^{\pi} f(x) \cos(x) \, dx =$$

$$\int_{-\pi}^{\pi} [a_0 \cos x + a_1 \cos^2 x + b_1 \cos x \sin x + a_2 \cos x \cos(2x) + \qquad (42)$$

$$b_2 \cos x \sin(2x) + \cdots + a_n \cos x \cos(nx) + b_n \cos x \sin(nx) + \cdots] dx.$$

We know that the definite integral of $a_0 \cos x$ over $\left[-\pi, \pi\right]$ is zero. In the exercise set at the end of this section, you are asked to show that the integrals of almost all of the other terms in the right side of (42) are also zero on $\left[-\pi, \pi\right]$. The only non-zero integral comes from the $a_1 \cos^2 x$ term. Therefore, equation (42) becomes

$$\int_{-\pi}^{\pi} f(x) \cos(x) \, dx = \int_{-\pi}^{\pi} a_1 \cos^2(x) \, dx.$$

We can solve for a_1 by evaluating $\int_{-\pi}^{\pi} a_1 \cos^2(x) dx$ using the trigonometric identity $\cos^2 x = \frac{1}{2} \cos(2x) + \frac{1}{2}$. By substitution,

$$\int_{-\pi}^{\pi} a_1 \cos^2(x) \, dx = a_1 \int_{-\pi}^{\pi} \left[\frac{1}{2} \cos(2x) + \frac{1}{2} \right] dx,$$

which can be evaluated as follows:

$$a_1 \int_{-\pi}^{\pi} \left[\frac{1}{2} \cos(2x) + \frac{1}{2} \right] dx = \left[a_1 \cdot \frac{1}{4} \sin(2x) + \frac{1}{2} x \right]_{-\pi}^{\pi}$$

$$= a_1 \left[\left(0 + \frac{\pi}{2} \right) - \left(0 - \frac{\pi}{2} \right) \right]$$

$$= a_1 \pi.$$

We can conclude that $\int_{-\pi}^{\pi} f(x) \cos(x) \, dx = a_1 \pi$, and solving for a_1 we find that

$$a_1 = \frac{1}{\pi} \int_{-\pi}^{\pi} f(x) \cos(x) \, dx.$$

To find the value of a_2, we need to modify the process used to find the value of a_1. Now we multiply both integrands in equation (40) by $\cos(2x)$ to generate the equation

$$\int_{-\pi}^{\pi} f(x)\cos(2x)\,dx =$$

$$\int_{-\pi}^{\pi} \left(a_0 \cos(2x) + \sum_{k=1}^{\infty} \left[a_k \cos(kx)\cos(2x) + b_k \sin(kx)\cos(2x) \right] \right) dx.$$

From our earlier work we know that $\int_{-\pi}^{\pi} a_0 \cos(2x)\,dx = 0$. In the exercise set at the end of this section, you are asked to show that $\int_{-\pi}^{\pi} b_k \sin(kx)\cos(2x)\,dx = 0$ for all k and that $\int_{-\pi}^{\pi} a_k \cos(kx)\cos(2x)\,dx = 0$ for all k, $k \neq 2$. Thus, we know that

$$\int_{-\pi}^{\pi} f(x)\cos(2x)\,dx = \int_{-\pi}^{\pi} a_2 \cos^2(2x)\,dx.$$

We can use the identity $\cos^2(2x) = \frac{1}{2}\cos(4x) + \frac{1}{2}$ to evaluate this integral as follows:

$$\int_{-\pi}^{\pi} a_2 \cos^2(2x)\,dx = a_2 \int_{-\pi}^{\pi} \left[\frac{1}{2}\cos(4x) + \frac{1}{2} \right] dx$$

$$= a_2 \left[\frac{1}{8}\sin(4x) + \frac{1}{2}x \right]_{-\pi}^{\pi}$$

$$= a_2 \pi.$$

We can conclude that $\int_{-\pi}^{\pi} f(x)\cos(2x)\,dx = a_2\pi$, and solving for a_2 we find that

$$a_2 = \frac{1}{\pi} \int_{-\pi}^{\pi} f(x)\cos(2x)\,dx.$$

We can generalize the technique for determining a_1 and a_2 to find an expression for a_k for all k. Repeating the process of multiplying the integrands in equation (40) by $\cos(kx)$ and integrating over the interval $[-\pi, \pi]$ generates an equation of the form

$$\int_{-\pi}^{\pi} f(x)\cos(kx)\,dx = \int_{-\pi}^{\pi} a_k \cos^2(kx)\,dx.$$

For every integer k, $\int_{-\pi}^{\pi} a_k \cos^2(kx)\,dx = a_k\pi$, so that

$$a_k = \frac{1}{\pi} \int_{-\pi}^{\pi} f(x)\cos(kx)\,dx, \quad k = 1, 2, 3, \ldots.$$

The values of the coefficients b_k are found using a similar strategy. We multiply the integrands on both sides of equation (40) by $\sin(kx)$ and integrate over the interval $[-\pi, \pi]$, which yields the equation

$$\int_{-\pi}^{\pi} f(x)\sin(kx)\,dx = \int_{-\pi}^{\pi} a_k \sin^2(kx)\,dx.$$

The result of the strategy for finding the coefficients b_k is

$$b_k = \frac{1}{\pi} \int_{-\pi}^{\pi} f(x)\sin(kx)\, dx, \ k = 1, 2, 3, \ldots.$$

We now know that the Fourier series $F(x)$ for a given function $y = f(x)$ over the interval $[-\pi, \pi]$ is

$$F(x) = a_0 + \sum_{k=1}^{\infty} \left[a_k \cos(kx) + b_k \sin(kx) \right],$$

where

$$a_0 = \frac{1}{2\pi} \int_{-\pi}^{\pi} f(x)\, dx,$$

$$a_k = \frac{1}{\pi} \int_{-\pi}^{\pi} f(x)\cos(kx)\, dx, \quad k = 1, 2, 3\ldots,$$

$$b_k = \frac{1}{\pi} \int_{-\pi}^{\pi} f(x)\sin(kx)\, dx, \quad k = 1, 2, 3\ldots. \tag{43}$$

The equations in (43) define the coefficients a_k and b_k in terms of definite integrals, which in practice generally are computed numerically, rather than analytically.

Applications of Fourier series usually do not use all terms in the infinite series. The ***Fourier series of order n***, symbolized by F_n, is the finite sum of the first $2n + 1$ terms of the infinite Fourier series and therefore includes the constant term a_0 and the terms $a_k \cos(kx)$ and $b_k \sin(kx)$ for $k = 1, 2, 3, \ldots, n$. Note that some of the $2n + 1$ terms in a Fourier series of order n may be zero.

Fourier series of order n

To illustrate how to use the equations in (43) to find the coefficients in a Fourier series, we first will write the Fourier series for a non-periodic function.

Example 1

Find the Fourier series for $f(x) = 1 - x^2$ on the interval $[-\pi, \pi]$.

Solution   To find the Fourier series for $f(x) = 1 - x^2$ on the interval $[-\pi, \pi]$, we must compute values of a_k and b_k using the formulas in (43) with $f(x) = 1 - x^2$.

We first write a Fourier series of order 1. Using numerical integration and technology, we find that the value of a_0 is $\frac{1}{2\pi} \int_{-\pi}^{\pi} (1 - x^2)\, dx \approx -2.29$, the value of a_1 is $\frac{1}{\pi} \int_{-\pi}^{\pi} (1 - x^2)\cos(x)\, dx = 4.00$, and the value of b_1 is $\frac{1}{\pi} \int_{-\pi}^{\pi} (1 - x^2)\sin x\, dx = 0$. The Fourier series of order 1 is therefore $F_1(x) = -2.29 + 4.00 \cos x$. The graphs of f and F_1 are shown in Figure 7.5.

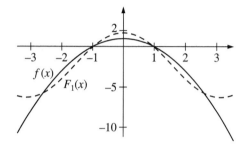

Figure 7.5 Graphs of $f(x) = 1 - x^2$ and $F_1(x) = -2.29 + 4.00 \cos x$

Calculating terms in the Fourier series for $k = 2$, we find that $a_2 = -1.00$ and $b_2 = 0$, so the Fourier series of order 2 is $F_2(x) = -2.29 + 4.00 \cos x - 1.00 \cos(2x)$. The graphs of f and F_2 are shown in Figure 7.6. We see that the second order Fourier series is a better approximation for f than the first order Fourier series.

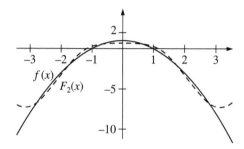

Figure 7.6 Graphs of $f(x) = 1 - x^2$ and $F_2(x) = -2.29 + 4.00 \cos x - 1.00 \cos(2x)$

If we continue to calculate values of a_k and b_k through $k = 5$, then we have the Fourier series of order 5, illustrated in Figure 7.7:

$$F_5(x) = -2.29 + 4.00 \cos x - 1.00 \cos(2x) + 0.44 \cos(3x) - 0.25 \cos(4x) + 0.16 \cos(5x). \quad \blacksquare$$

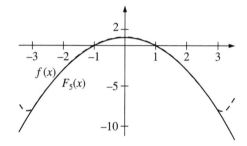

Figure 7.7 Graphs of $f(x)$ and $F_5(x)$ on $[-\pi, \pi]$

Notice in Example 1 that F_1, F_2, and F_5 contain only a constant and cosine terms, and each sine term has coefficient 0. This is due to the fact that f is an even function. In general, a Fourier series for an even function contains only a constant and cosine terms. A Fourier series for an odd function contains only sine terms.

The graphs shown in Example 1 illustrate that the accuracy of a Fourier series approximation does not deteriorate significantly across the interval of interest. Rather, a Fourier series approximation will oscillate between being too large at some x values and too small at other x values. That is, the series alternately overestimates and underestimates the function it is designed to approximate. As the number of terms in the Fourier series increases, these oscillations generally decrease and the Fourier series deviates less and less from the function it is designed to approximate.

Because they are based upon sines and cosines, Fourier series are always periodic. Whether or not the function we are trying to approximate is periodic, the Fourier series is periodic. Thus, if we approximate any function (periodic or non-periodic) over the interval $a \leq x \leq b$, the Fourier series will include repetition of the portion of the series in the interval from a to b, creating a periodic function of period $b - a$. Figure 7.8 shows the Fourier series of order 5 generated in Example 1 graphed over $[-10, 10]$. Note that the Fourier series is periodic with period 2π.

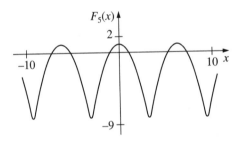

Figure 7.8 Graph of $F_5(x)$ on $[-10, 10]$

Example 2

The function g whose graph is shown in Figure 7.9 is periodic. The graph of g consists of the graph of $f(x) = x^2 - 2$ on $-\pi \leq x \leq \pi$ repeated with period 2π. Write the Fourier series of order 3 for g.

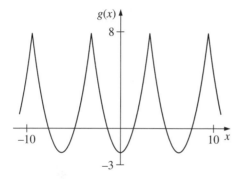

Figure 7.9 Periodic function g for Example 2

Solution   To write a Fourier series for the function g shown in Figure 7.9, we need to approximate the function on the interval $[-\pi, \pi]$. That is, we need to write the Fourier series for $f(x) = x^2 - 2$ on $[-\pi, \pi]$. Using the formulas $a_0 = \frac{1}{2\pi} \int_{-\pi}^{\pi} f(x)\, dx$, $a_k = \frac{1}{\pi} \int_{-\pi}^{\pi} f(x)\cos(kx)\, dx$, and $b_k = \frac{1}{\pi} \int_{-\pi}^{\pi} f(x)\sin(kx)\, dx$, we find that $a_0 \approx 1.29$, $a_1 = -4.00$, $b_1 = 0$, $a_2 = 1.00$, $b_2 = 0$, $a_3 \approx -0.44$, and $b_3 = 0$. Therefore, the Fourier series of order 3 for $f(x) = x^2 - 2$ on $[-\pi, \pi]$ is

$$F_3(x) = 1.29 - 4.00 \cos x + 1.00 \cos(2x) - 0.44 \cos(3x).$$

If we graph this Fourier series on a larger interval of x values, say $[-20, 20]$, as shown in Figure 7.10, we see that $F_3(x)$ is periodic with period 2π. Thus, $F_3(x)$ is the Fourier series of order 3 for the function g graphed in Figure 7.9. ∎

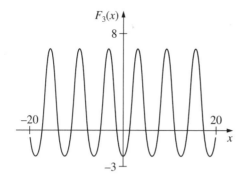

Figure 7.10 Graph of $F_3(x)$ on $[-20, 20]$

Exercise Set 7.4

1. Write the Fourier series of order 3 for each of the following functions over the interval $[-\pi, \pi]$. Use a numerical technique to evaluate the integrals that determine the coefficients in the Fourier series.

    a. $f(x) = \begin{cases} x+1 & \text{if } -\pi \le x \le 0, \\ -x+1 & \text{if } 0 < x \le \pi \end{cases}$

    b. $f(x) = \begin{cases} 0 & \text{if } -\pi \le x < 0, \\ 1 & \text{if } 0 \le x \le \pi \end{cases}$

    c. $f(x) = \begin{cases} x+\pi & \text{if } -\pi \le x < 0, \\ x & \text{if } 0 \le x \le \pi \end{cases}$

2. Graph each of the Fourier series from problem 1 on the interval $[-10, 10]$. What is the period of each Fourier series?

3. a. Write the Fourier series of order 4, 7, and 10 for $f(x) = (x-\pi)^2(x+\pi)^2$ on $[-\pi, \pi]$.

    b. Graph $E_n(x) = f(x) - F_n(x)$, which represents the error in the Fourier series of order n, on $[-\pi, \pi]$ for $n = 4$, 7, and 10.

    c. Describe how $E_n(x)$ in part b changes across the interval $[-\pi, \pi]$.

    d. Describe how $E_n(x)$ in part b changes as n increases.

4. Write the Fourier series of order 5 for $f(x) = \sin^3(x)$ on the interval $[-\pi, \pi]$.

5. a. Use a graphical argument to show that $\int_{-a}^{a} f(x)dx = 0$ for any odd function f and any number a.

    b. Show that the product of an odd function and an even function is an odd function. Recall that if f is an odd function, then $f(-x) = -f(x)$, and if f is an even function, then $f(-x) = f(x)$.

    c. Use part b to show that $\sin(mx) \cdot \cos(nx)$ is an odd function when m and n are integers.

    d. Use the results of parts a–c to show that $\int_{-\pi}^{\pi} \sin(mx)\cos(nx)dx = 0$ when m and n are integers.

6. Use the table of integrals in the appendix to show that $\int_{-\pi}^{\pi} \sin(nx)\sin(mx)dx = 0$ and $\int_{-\pi}^{\pi} \cos(nx)\cos(mx)dx = 0$ when m and n are integers and $m \ne n$.

7.5 Changing the Interval of Approximation

In the previous section we used the interval $-\pi \le x \le \pi$ to compute the Fourier coefficients. As a consequence, we have created Fourier series whose period is 2π. In many situations we want to write a Fourier series whose period is something other than 2π. For instance, in Chapter 8 one investigation requires the use of Fourier series to model sound waves whose period is not 2π.

In this section we will learn how to write Fourier series with any period. If we are modeling a periodic phenomenon, we will need to choose an interval $c \le x \le d$ that represents one cycle of the phenomenon we are modeling.

To create a Fourier series for $f(x)$ on the interval $c \le x \le d$ rather than on $-\pi \le x \le \pi$, we need to create a series with period $d-c$ rather than period 2π. Notice that the period of both $\sin(\beta x)$ and $\cos(\beta x)$ are $d-c$ provided $\beta = \frac{2\pi}{d-c}$. This implies that we can produce the appropriate horizontal stretch or compression in the Fourier series we developed in the previous section if we change all the sine and cosine functions to $\sin(k\beta x)$ and $\cos(k\beta x)$, so that

$$F(x) = a_0 + \sum_{k=1}^{\infty} \left[a_k \cos(k\beta x) + b_k \sin(k\beta x) \right].$$

We also need to modify the equations that define a_0, a_k, and b_k. Recall that for a Fourier series on the interval $[-\pi, \pi]$, $a_0 = \frac{1}{2\pi} \int_{-\pi}^{\pi} f(x)\, dx$. The term $\frac{1}{2\pi}$ is based on the fact that the length of the interval $[-\pi, \pi]$ is 2π. If we want to work with the interval $[c,d]$, then $a_0 = \frac{1}{d-c} \int_c^d f(x)\, dx$. The integrals that define a_k and b_k must be modified in a similar way. The factor $\frac{1}{\pi}$ in the equations for a_k and b_k is equal to $\frac{2}{2\pi}$, so in the Fourier series for $f(x)$ on the interval $c \le x \le d$, we have $a_k = \frac{2}{d-c} \int_c^d f(x)\cos(k\beta x)\, dx$ and $b_k = \frac{2}{d-c} \int_c^d f(x)\sin(k\beta x)\, dx$. Therefore, the Fourier series for $f(x)$ on the interval $[c,d]$ is given by

$$F(x) = a_0 + \sum_{k=1}^{\infty} \left[a_k \cos(k\beta x) + b_k \sin(k\beta x) \right],$$

where

$$\beta = \frac{2\pi}{d-c},$$

$$a_0 = \frac{1}{(d-c)} \int_c^d f(x)\, dx,$$

$$a_k = \frac{2}{d-c} \int_c^d f(x)\cos(k\beta x)\, dx \quad \text{for } k = 1,2,3,\ldots, \text{ and}$$

$$b_k = \frac{2}{d-c} \int_c^d f(x)\sin(k\beta x)\, dx \quad \text{for } k = 1,2,3,\ldots.$$

The graph of $f(x) = 1 - x^2$ on the interval $[0, 1]$ is shown in Figure 7.11. On the left, the Fourier series for $f(x)$ of order 3, F_3, is graphed with f. On the right, the Fourier series for $f(x)$ of order 12, F_{12}, is graphed with f. The coefficients for these series were determined by evaluating the following integrals:

$$a_0 = \frac{1}{1} \int_0^1 \left(1 - x^2\right) dx \approx 0.667,$$

$$a_1 = \frac{2}{1} \int_0^1 \left(1 - x^2\right) \cos(2\pi x)\, dx \approx -0.101, \quad b_1 = \frac{2}{1} \int_0^1 \left(1 - x^2\right) \sin(2\pi x)\, dx \approx 0.318,$$

$$a_2 = \frac{2}{1} \int_0^1 \left(1 - x^2\right) \cos(4\pi x)\, dx \approx -0.025, \quad b_2 = \frac{2}{1} \int_0^1 \left(1 - x^2\right) \sin(4\pi x)\, dx \approx 0.159,$$

$$a_3 = \frac{2}{1} \int_0^1 \left(1 - x^2\right) \cos(6\pi x)\, dx \approx -0.011, \quad b_3 = \frac{2}{1} \int_0^1 \left(1 - x^2\right) \sin(6\pi x)\, dx \approx 0.106,$$

$$\vdots \qquad\qquad\qquad\qquad \vdots$$

$$a_{12} = \frac{2}{1} \int_0^1 \left(1 - x^2\right) \cos(24\pi x)\, dx \approx -0.001, \quad b_{12} = \frac{2}{1} \int_0^1 \left(1 - x^2\right) \sin(24\pi x)\, dx \approx 0.027.$$

The Fourier series of order 3 is

$$F_3(x) = 0.667 - 0.101\cos(2\pi x) + 0.318\sin(2\pi x) - 0.025\cos(4\pi x) +$$
$$0.159\sin(4\pi x) - 0.011\cos(6\pi x) + 0.106\sin(6\pi x).$$

Since none of the coefficients a_k or b_k are equal to zero, the Fourier series of order 3 has 7 terms, and the Fourier series of order 12 has 25 terms.

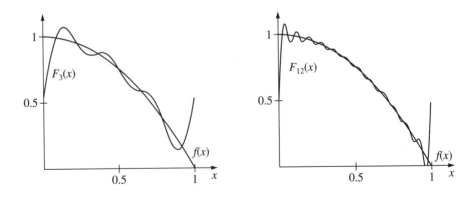

Figure 7.11 Graphs of $f(x) = 1 - x^2$ with $F_3(x)$ and $F_{12}(x)$ on $[0, 1]$

Exercise Set 7.5

1. Write the Fourier series of order 3 for each of the following functions over the interval $[0, 1]$. Use a numerical technique to evaluate the integrals that determine the coefficients for each Fourier series to the hundredths place.

   a. $f(x) = x$

   b. $f(x) = \begin{cases} x + 1 & \text{if } 0 \le x \le 0.5, \\ -x + 2 & \text{if } 0.5 < x \le 1 \end{cases}$

   c. $f(x) = \begin{cases} 0 & \text{if } 0 \le x < 0.75, \\ 1 & \text{if } 0.75 \le x \le 1 \end{cases}$

   d. $f(x) = \begin{cases} -x & \text{if } 0 \le x < 0.5, \\ x - 1 & \text{if } 0.5 \le x \le 1 \end{cases}$

2. Graph each of the Fourier series created in problem 1 on the interval $-4 \le x \le 4$. What is the period of each Fourier series you wrote?

3. Determine the Fourier series for

$$f(x) = \begin{cases} 1 & \text{if } -\frac{\pi}{2} \le x \le \frac{\pi}{2}, \\ 0 & \text{if } \frac{\pi}{2} < x \le \frac{3\pi}{2} \end{cases}$$

   on the interval $\left[-\frac{\pi}{2}, \frac{3\pi}{2} \right]$.

4. In Exercise Set 7.4 you may have found that the order 3 Fourier series for $f(x) = \sin^3 x$ on $[-\pi, \pi]$ is $F_3(x) = 0.75 \sin x - 0.25 \sin(3x)$.

   a. Graph $F_3(x)$ on the interval $[-10, 10]$. What is the period of this Fourier series?

   b. Write the order 3 Fourier series for $f(x) = \sin^3 x$ on the interval $\left[-\frac{\pi}{2}, \frac{\pi}{2} \right]$. Graph this series on the interval $[-10, 10]$. What is the period of the series? Does the series behave like $f(x)$ on the interval $[-10, 10]$? Why or why not?

   c. Write the order 3 Fourier series for $f(x) = \sin^3 x$ on the interval $[0, 4\pi]$. Graph this series on the interval $[-10, 10]$. What is the period of the series? Does the series behave like $f(x)$ on the interval $[-10, 10]$? Why or why not?

7.6 Using a Discrete Fourier Series to Model Data

In the previous sections we have used a Fourier series to approximate a function over an interval. Fourier series also can be used to model data on an interval. A Fourier series that is based on discrete data, rather than on a piecewise continuous function, is called a *discrete Fourier series*.

discrete Fourier series

Example

Consider the set of data that consists of ten equally-spaced x and y values of the function $f(x) = 1 - x^2$ on the interval $[-\pi, \pi]$. If we divide the interval $[-\pi, \pi]$ into nine subintervals, each of length $\frac{2\pi}{9}$, we generate the ten x values that are paired with y values such that $y_i = f(x_i)$ for $i = 1, 2, \ldots 10$. These values are shown rounded to hundredths and graphed in Figure 7.12. Write a discrete Fourier series of order 4 to model these data. Note that the data in the table below have been rounded to 2 decimal places. The Fourier coefficients have been calculated using more decimal places.

x	−3.14	−2.44	−1.75	−1.05	−0.35	0.35	1.05	1.75	2.44	3.14
y	−8.87	−4.97	−2.05	−0.10	0.88	0.88	−0.10	−2.05	−4.97	−8.87

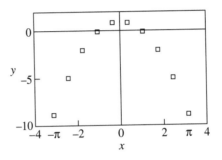

Figure 7.12 Ten data points from $f(x) = 1 - x^2$

Solution   We need to find a way to create this Fourier series based only on the data from our sample, without knowing the function f. Recall that when we know the equation of the function f, the Fourier coefficients are determined as follows:

$$a_0 = \frac{1}{2\pi} \int_{-\pi}^{\pi} f(x)\,dx,$$

$$a_k = \frac{1}{\pi} \int_{-\pi}^{\pi} f(x)\cos(kx)\,dx, \text{ and}$$

$$b_k = \frac{1}{\pi} \int_{-\pi}^{\pi} f(x)\sin(kx)\,dx.$$

The Fourier series based on these coefficients is

$$F(x) = a_0 + \sum_{k=1}^{\infty} \left[a_k \cos(kx) + b_k \sin(kx) \right].$$

Since we are assuming that we do not know f, we will use these ten data values to generate the required values of a_k and b_k. We accomplish this by rewriting definite integrals as summations. For instance, the definite integral used to define a_0, $\frac{1}{2\pi}\int_{-\pi}^{\pi} f(x)\,dx$, can be approximated by the Riemann sum $\frac{1}{2\pi}\sum_{i=1}^{9} y_i\,\Delta x$. The ten data points $x_1, x_2, \ldots, x_{10}$ divide the interval $[-\pi, \pi]$ into nine equal subintervals. Since the interval length is 2π we know that $\Delta x = \frac{2\pi}{9}$. We can compute a_0 with the sum $a_0 = \frac{1}{2\pi}\sum_{i=1}^{9} y_i \frac{2\pi}{9}$. Rounded to hundredths, this sum is -2.37. Therefore, we conclude that $a_0 = -2.37$ is the constant term in the discrete Fourier series.

The value of a_1 in a Fourier series is defined by $\frac{1}{\pi}\int_{-\pi}^{\pi} f(x)\cos(x)\,dx$, which can be approximated with the Riemann sum $\frac{1}{\pi}\sum_{i=1}^{9} y_i \cos(x_i)\,\Delta x$. Again, $\Delta x = \frac{2\pi}{9}$, so the value of a_1 in the discrete Fourier series is

$$a_1 = \frac{1}{\pi}\sum_{i=1}^{9} y_i \cos(x_i)\cdot\left(\frac{2\pi}{9}\right).$$

Evaluating this sum and rounding to hundredths, we find that $a_1 = 4.17$. Continuing in a similar fashion with higher order terms, we find that $a_2 = -1.18$, $a_3 = 0.65$, and $a_4 = -0.50$.

The values of the coefficients b_k in a discrete Fourier series are determined in a similar way by replacing $b_k = \frac{1}{\pi}\int_{-\pi}^{\pi} f(x)\sin(kx)\,dx$ with $b_k = \frac{1}{\pi}\sum_{i=1}^{N-1} y_i \sin(kx_i)\,\Delta x$, where N is the number of data points. In this example, the computed values of b_1, b_2, b_3, and b_4 are all less than 10^{-10}. These values are relatively small compared to the a_i's, so we can leave the corresponding terms out of the series without losing significant accuracy. Therefore, the discrete Fourier series of order 4 based on the ten data points is

$$F_4(x) = -2.37 + 4.17\cos x - 1.18\cos(2x) + 0.65\cos(3x) - 0.50\cos(4x).$$

The discrete Fourier series is graphed with the data in Figure 7.13. ■

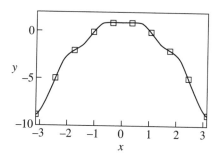

Figure 7.13 Graphs of ten data points and $F_4(x)$

In general, if N is the number of data points we have on the interval $[-\pi, \pi]$, then the spacing between data points is $\Delta x = \frac{2\pi}{N-1}$. When we calculate the value of any a_k or b_k in a discrete Fourier series, we use the ordered pairs (x_i, y_i) in the data set we are modeling rather than the values of a function as in previous sections. The coefficient a_0 in the discrete Fourier series is defined by

$$a_0 = \frac{1}{2\pi} \sum_{i=1}^{N-1} y_i \frac{2\pi}{N-1} = \frac{1}{N-1} \sum_{i=1}^{N-1} y_i.$$

In general, the coefficients a_k in the discrete Fourier series are defined by

$$a_k = \frac{1}{\pi} \sum_{i=1}^{N-1} y_i \cos(kx_i) \frac{2\pi}{N-1} = \frac{2}{N-1} \sum_{i=1}^{N-1} y_i \cos(kx_i).$$

Similarly, the coefficients b_k in the discrete Fourier series are defined by

$$b_k = \frac{1}{\pi} \sum_{i=1}^{N-1} y_i \sin(kx_i) \frac{2\pi}{N-1} = \frac{2}{N-1} \sum_{i=1}^{N-1} y_i \sin(kx_i).$$

To summarize, we know that the discrete Fourier series for a data set containing N points of the form (x_i, y_i) on $[-\pi, \pi]$ is defined by

$$F(x) = a_0 + \sum_{k=1}^{\infty} \left[a_k \cos(kx) + b_k \sin(kx) \right],$$

where

$$a_0 = \frac{1}{N-1} \sum_{i=1}^{N-1} y_i,$$

$$a_k = \frac{2}{N-1} \sum_{i=1}^{N-1} y_i \cos(kx_i) \quad \text{for } k = 1, 2, 3, \ldots, \text{ and}$$

$$b_k = \frac{2}{N-1} \sum_{i=1}^{N-1} y_i \sin(kx_i) \quad \text{for } k = 1, 2, 3, \ldots.$$

If the interval of interest is $[c, d]$ rather than $[-\pi, \pi]$, we make the same type of transformations as in Section 7.2. The discrete Fourier series for a data set containing N points (x_i, y_i) on $[c, d]$ is defined by

$$F(x) = a_0 + \sum_{k=1}^{\infty} \left[a_k \cos(k\beta x) + b_k \sin(k\beta x) \right],$$

where $\beta = \dfrac{2\pi}{d-c}$ and

$$a_0 = \frac{1}{N-1} \sum_{i=1}^{N-1} y_i,$$

$$a_k = \frac{2}{N-1} \sum_{i=1}^{N-1} y_i \cos(k\beta x_i) \quad \text{for } k = 1, 2, 3, \ldots, \text{ and}$$

$$b_k = \frac{2}{N-1} \sum_{i=1}^{N-1} y_i \sin(k\beta x_i) \quad \text{for } k = 1, 2, 3, \ldots.$$

Exercise Set 7.6

1. Our work in the example in this section showed that the discrete Fourier series of order 4 based on 10 data points from $f(x) = 1 - x^2$ on $[-\pi, \pi]$ is

    $$F_4(x) = -2.37 + 4.17 \cos x - 1.18 \cos(2x) + 0.65 \cos(3x) - 0.50 \cos(4x).$$

    We know from our previous work that the Fourier series of order 4 for $f(x) = 1 - x^2$ on $[-\pi, \pi]$ is

    $$F_4(x) = -2.29 + 4.00 \cos x - 1.00 \cos(2x) + 0.44 \cos(3x) - 0.25 \cos(4x) .$$

    a. Calculate the relative error in the coefficients of successive terms. That is, by what percent does each coefficient in the discrete Fourier series differ from its counterpart in the original Fourier series?

    b. Create a set of twenty equally-spaced data points from the function $f(x) = 1 - x^2$ on the interval $[-\pi, \pi]$. Determine the coefficients in the discrete Fourier series of order 4 for this data set. Repeat this procedure for a data set of forty equally-spaced points. How do these coefficients compare to the coefficients in the series created from the function itself?

2. The temperatures taken every hour in Durham, North Carolina, on a day in April are shown in the following table. Write a discrete Fourier series to approximate this data set over a 24 hour time interval. What order Fourier series is needed for a good fit? Is a series of higher order necessarily better?

Time P.M.	Degrees F	Time A.M.	Degrees F
1	65	1	55
2	66	2	55
3	65	3	55
4	65	4	56
5	64	5	56
6	63	6	57
7	61	7	59
8	59	8	61
9	59	9	63
10	57	10	64
11	56	11	66
Midnight	56	Noon	66

3. The average monthly precipitation in Osaka, Japan is given in the following table. Write a discrete Fourier series that fits the given data.

Month	Precipitation (mm)	Month	Precipitation (mm)
January	45.8	July	156.9
February	60.4	August	94.8
March	102.0	September	171.5
April	133.8	October	107.5
May	139.4	November	65.1
June	206.4	December	34.4

Reference
Asahi Shimbun Japan Almanac, 1994. Tokyo: Asahi Shimbun Publishing Company, 1993.

7.7 Using Fourier Series to Filter Data

An important application of discrete Fourier series is "filtering" a set of data, such as a radio signal. Suppose we want to transmit the data set of 256 points shown in Figure 7.14 to a distant location.

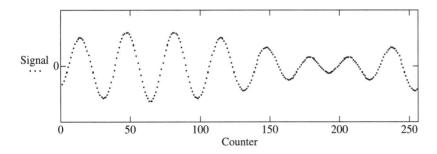

Figure 7.14 Data set to be transmitted

After the 256 data points are transmitted, the distant receiver can find a discrete Fourier series to model the signal. Based on the techniques of the previous section, the discrete Fourier series is given by the equation

$$F(x) = a_0 + \sum_{k=1}^{\infty} \left[a_k \cos(k\beta x) + b_k \sin(k\beta x) \right],$$

where $\beta = \frac{2\pi}{256}$, $N = 256$, and

$$a_0 = \frac{1}{N-1} \sum_{i=1}^{N-1} y_i,$$

$$a_k = \frac{2}{N-1} \sum_{i=1}^{N-1} y_i \cos(k\beta x_i) \quad \text{for } k = 1, 2, 3, \ldots, \text{ and}$$

$$b_k = \frac{2}{N-1} \sum_{i=1}^{N-1} y_i \sin(k\beta x_i) \quad \text{for } k = 1, 2, 3, \ldots.$$

The discrete Fourier series of order 15 is shown in Figure 7.15. The series does an excellent job of modeling the original signal.

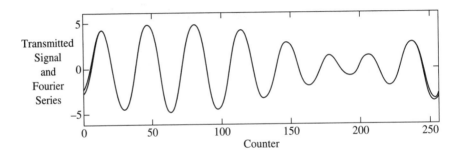

Figure 7.15 The discrete Fourier series of order 15 plotted with the original data

Unfortunately, interference often occurs when data is transmitted over a distance, which is an effect called *noise*. Thus, a distant receiver of the data set in Figure 7.14 would not actually receive the signal that we transmitted, but might instead receive the signal shown in Figure 7.16. The data set shown in Figure 7.16 was created by adding a random error to each point of the original data set in Figure 7.14. The errors added are normally distributed with a mean of 0 and a standard deviation of 1.3. These errors act as noise, for they interfere with the transmission of the signal we send.

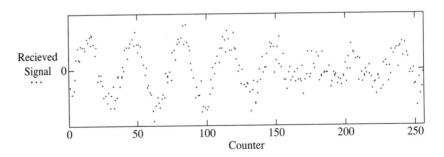

Figure 7.16 The signal with random noise

How much of the original signal can be recovered from the "noisy" data set shown in Figure 7.16? Figure 7.17 shows a discrete Fourier series of order 8 fit to the noisy data in Figure 7.16. The discrete Fourier series seems to ignore the noise in the data, which is desirable, since we prefer to transmit a signal without transmitting noise.

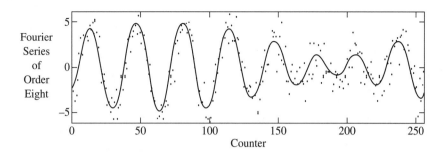

Figure 7.17 The discrete Fourier series of order 8 with the noisy signal

Figure 7.18 compares the discrete Fourier series of order 8 generated from the noisy data and the original signal before it was contaminated by noise. The correspondence is not exact, but most of the character of the original data set has been preserved by the discrete Fourier series generated from the noisy data.

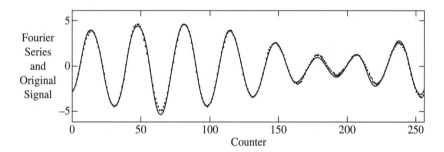

Figure 7.18 Discrete Fourier series generated from noisy data and the original signal

What happens if we use a discrete Fourier series of order higher than 8? Figure 7.19 shows the discrete Fourier series of order 25 with the noisy signal. Notice that the higher order series is sensitive to more of the noise in the signal. In this example, the frequency of the noise is higher than the frequency of the signal. Since frequency is the reciprocal of period, the period of the noise is shorter than the period of the signal. The period of the higher order terms $a_k \cos(k\beta x)$ and $b_k \sin(k\beta x)$ in the discrete Fourier series decreases as the order of the series increases; therefore, a higher order series will more closely model the noise in the signal.

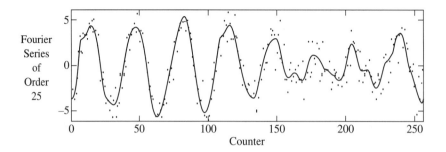

Figure 7.19 Discrete Fourier series of order 25 with the noisy signal

Figure 7.20 shows the discrete Fourier series of order 50 with the noisy signal. This series is even more sensitive to the noise than the series of order 25, and thus does a worse job of modeling the original data set. Since the noise in this data set has a high frequency and therefore a short period, a lower order discrete Fourier series gives a better representation of the actual signal than a higher order discrete Fourier series.

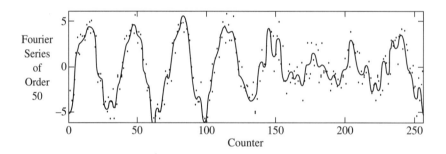

Figure 7.20 Discrete Fourier series of order 50 with the noisy signal

A discrete Fourier series can be used to filter high frequency random errors from a data set by leaving out terms above a certain frequency, which are the higher order terms. Similar methods can be used to filter out low frequency noise, or some specific frequency, as would be the case if a signal were contaminated by 60 Hertz power-line interference. Another filtering technique involves leaving out all the low amplitude terms from the discrete Fourier series. Fourier filters are also used as one of the steps in data smoothing techniques. One standard technique to smooth data is to first remove any linear trend in the data by subtracting from each data point the value of the line passing through the first and last data points, then using a Fourier filter on the re-expressed data, and finally adding the linear trend back to the filtered data.

The process of filtering and cleaning up data is an important application of Fourier series. In practice, more sophisticated methods than the one illustrated here are often used, but the basic principle behind them is the same as in this section.

Exercise Set 7.7

1. In this exercise you will need to have a partner. Each of you will independently
    generate a set of data, exchange the data with your partner, then independently try
    to find a Fourier series that fits the data well and filters out the noise.

    a. Use a function of the form $g(x) = A\sin(Kx) + (2 \cdot r(1) - 1)$, where $A > 0$, K is
        an integer between 1 and 10, and $r(1)$ represents a random number between
        0 and 1. (Since $r(1)$ represents a random number between 0 and 1, $2 \cdot r(1) - 1$
        represents a random number between -1 and 1.) Pick your own values for A
        and K. Generate at least 30 ordered pairs of this function on the interval
        $[0, 2\pi]$. Be sure to write down your values for A and K.

    b. Exchange this data with your partner. Now find the discrete Fourier series of
        least order that models the noisy data generated by your partner. As you
        increase the order of the discrete Fourier series, what order is needed to give
        a reasonable representation without noise of the underlying function
        $g(x) = A\sin(Kx)$? What value did you determine for A and K? Compare
        these to the values your partner used to generate the data.

2. How is the fundamental period of a signal related to the number of data points
    needed to find the Fourier series? Would there ever be a situation where you
    might gather too few data points? Could you ever gather too many points?

CHAPTER 8

Investigations

This chapter is a compilation of challenging and thought-provoking problems for student investigation. These problems involve many different concepts of calculus and encompass a variety of real-world situations. You may not know at the beginning of a problem exactly where or how the concepts of calculus will be used to solve the problem. One goal of this chapter is to learn to recognize when various principles of calculus can be used to your advantage. The problems often are stated loosely and may require some research to understand the problem setting. Such background research is an important part of mathematical modeling and applying mathematics in a particular context.

Many of the problems lend themselves to several different solution paths. None of the problems have a specific desired result. In other words, there is no single right answer for any of the problems. Just as with the lab activities in previous chapters, more thoughtful and reflective work leads to more creative and insightful use of mathematics. This in turn often leads to better solutions.

You are encouraged to struggle with these problems and not to give up when faced with difficulties. Talk them over with other students. Do not be afraid to test your ideas. Often, you cannot tell immediately which directions will yield useful results and which will not. Be creative. Be inventive. One good approach is to start with a simple version of the problem and work on it. See what insight you can obtain by playing with a specific example of the problem and varying the parameters. Above all, have some fun in these investigations using the techniques of calculus developed so far.

For several of the problem settings, specific questions are suggested to help you with your solution. They are designed to give you a start. If they do not help you and you decide to follow a different path, then you may ignore these questions. If you do answer the suggested questions, understand that your investigation is not limited to these questions. You should consider questions of your own as well. In all cases, in your report you must explain with mathematics why your solution works and what your mathematical model has revealed about the problem situation.

8.1 The Battle of Trafalgar

The French Revolution, which began in 1789, created circumstances that brought Napoleon Bonaparte to power in France. His armies overran much of Europe. The security of Britain at that time depended on the ability of the British navy to control the seas.

In 1805 the British navy fought against the combined French and Spanish fleets in the waters off Cape Trafalgar. At this period in history, naval warfare followed certain rules that seem rather formal to us today. The ships in each fleet lined up in a row facing the enemy, then the ships in these two parallel rows sailed past each other and fired their cannons. These maneuvers were repeated until one of the fleets was destroyed. According to these rules of engagement, the rate at which one navy lost ships depended on two factors: the number of ships in the opposing navy and the effectiveness of these ships in destroying their opponents. In planning for a battle, a commander had to make the assumption that the effectiveness of each navy would remain constant for the duration of any particular battle. This implied that the rate at which one navy lost ships was proportional to the number of ships that the other force had engaged in the battle.

The commander of the British fleet was Admiral Nelson. In the now-famous Battle of Trafalgar, he exhibited cunning military strategy. When Nelson fought the French-Spanish fleet at Trafalgar, he believed that the fighting effectiveness of the two opposing navies was equal. As a consequence, the navy that had the most ships would inevitably win. He expected to have only 27 ships at his disposal, while he expected his enemy to have a fleet of 33 ships. Nelson was destined to lose unless he tried a new strategy. Nelson planned to meet the French-Spanish fleet in the open waters and sail through the middle of their line, cutting their fleet into two segments and engaging them in two separate smaller battles. He hoped to have a superior force in one battle, which he would win, and an inferior force in the other battle, which he would lose. If the battle went according to plan, Nelson expected to have more ships remaining in his victorious fleet than were left in the victorious French-Spanish fleet. In the final battle between the two forces that survived the two preliminary battles, the British were indeed victorious, although Nelson himself was fatally wounded.

Investigate outcomes of the British fleet battling the French-Spanish fleet using the rules of engagement described above. Assume that the effectiveness constants in this conflict are 0.02 for both the British and the French-Spanish. In other words, each British ship can destroy 0.02 French-Spanish ships per hour, and each ship in the French-Spanish fleet can destroy 0.02 British ships per hour. Develop a model that predicts the results of a battle between two fleets of sizes A and B, where $A > B$, and with the effectiveness constants equal. How many ships will remain in the larger fleet when the last ship in the smaller one is destroyed? By using your model, show how Nelson could divide his fleet so that Nelson's 27 ships might defeat the 33 French-Spanish ships. If you find several strategies, determine whether one is more desirable and explain why it is more desirable.

Investigate other questions such as how long the battles last, how many ships remain at the end, and how different effectiveness constants would influence the outcome. If the effectiveness constants for each fleet are not equal, what will be the outcomes of the battles? Do the outcomes of the battles seem to be more dependent on the effectiveness constants or on the initial number of ships in each fleet?

Questions to Consider

As you investigate the above problem, you should consider, but not limit yourself to, the following questions:

1. Write and solve a differential equation for $\frac{dA}{dB}$. Compare the information you get concerning the relationship of A and B from the analytical solution to the information you have gotten numerically.

2. Write and solve differential equations for $\frac{d}{dt}(A - B)$ and $\frac{d}{dt}(A + B)$.

3. Use your results from question 2 to find closed form expressions for A in terms of t and B in terms of t. Compare information about how A and B change over time that you get from these expressions to the information you have gotten numerically.

Reference

Nash, David H. 1985. Differential equations and the Battle of Trafalgar. *The College Mathematics Journal* 6: 98–102.

8.2 Disease or Drug Testing

Many of the blood tests used to detect disease and the urine tests used to detect the presence of a drug are rather expensive. The plan outlined below has historically been used to reduce the total cost of blood testing while detecting all of the people in a population who have a particular disease. The same plan can be used to test urine samples for a particular drug.

A blood sample is taken from each person in the population. The population is divided into groups of a fixed size. Blood from each individual's sample is mixed with blood from the samples of the other members of the same group. This group sample is tested for the disease. If the test for the group sample is negative, then we assume that no one in the group has the disease. That is, we assume that the test always produces accurate results. If, on the other hand, the group sample tests positive, then we assume that at least one person in the group has the disease. If this is the case, the blood sample for each individual in the group must be tested separately. Alternatively, the individuals in this group can be divided into smaller groups, and a mixed sample for each of the smaller groups can be tested. If a smaller group sample tests positive, then the sample for each individual in the smaller group must be tested separately.

Use the above principles to devise a testing program for a large community that minimizes cost. Assume that you know from previous data that the probability that an individual in the given population has a particular disease is p.

References
Edwards, Dilwyn and Mike Hanson. *Guide to Mathematical Modeling*. Boca Raton: CRC Press, 1989. pp. 191–208.
Meyer, Paul L. *Introducing Probability and Statistical Applications*. Reading, MA: Addison-Wesley Publishing Company, 1970. pp. 131–134.

8.3 Investigating National Income

One model for the national income Y_t in any accounting period t uses three primary components: consumer expenditure C_t, induced private investment I_t, and government expenditure G_t. Consumer expenditure is the total of the funds used by individuals for the purchase of consumer goods. Induced private investment is the total of the funds used by companies to increase production. For example, induced private investment includes costs associated with buying machinery or improving efficiency. Government expenditure is the total of the funds spent by the government in the national economy. The model describes the national income as the sum of these three components. We will make the following three assumptions concerning these variables.

1. Consumer expenditure in any period is proportional to the national income of the preceding period. The constant of proportionality α is called the *marginal propensity to consume*. We assume that when someone earns additional income, then that individual will spend part, but not all, of the additional income. For example, if a person who receives a bonus of $300 spends $250 and saves the rest, then this person's marginal propensity to consume is $\alpha = \frac{250}{300} \approx 0.83$. With this interpretation, we assume that $0 \leq \alpha \leq 1$. The marginal propensity to consume has an important effect on the national income. This effect is known as the "acceleration principle" in consumption because an increase in consumption in one area of the economy is felt throughout the economy. Suppose you hire a firm for $8000 to add a deck onto your house. This $8000 is part of the national income. The carpenters, hardware manufacturers, and lumber producers now have $8000 that they can spend or invest. However, if the marginal propensity to consume α for these workers is 0.75, then they will spend $(0.75)(\$8000) = \$6000$ of it; this is income to the businesses which receive it, so this adds $6000 to the national income. The recipients of this $6000 will then spend 0.75 of the $6000, which adds another $4500 to the national income. This process continues, creating an infinite series of dollar amounts added to the national income in the category of consumer expenditure, all due to your $8000 deck.

2. Induced private investment in any period is proportional to the difference between consumer expenditure in that period and in the preceding period, with positive constant of proportionality β. We interpret negative induced private investment as a withdrawal of funds committed to investment purposes. This might occur, for example, if depreciated machinery is not replaced as planned. Without the commitment of new funds to upgrade machinery, the value of the investment in machinery will decline. β is known as the "multiplier".

3. To keep the model manageable, we assume that government expenditure is the same in all periods. For convenience, we will choose our units so that this expenditure equals 1.

Initial Investigation

Develop an iterative model of national income based on the preceding information. These assumptions do not involve rates of change, and therefore the model does not require the use of differential equations. Assume that $Y_1 = 2$, $Y_2 = 3$, $\alpha = 0.6$, and $\beta = 1.2$. Investigate the curve defined by (t, Y_t). What are the characteristics of this curve? What is the behavior of national income under these assumptions?

Vary α and β. In what way is the behavior of national income independent of α and β? What effects can the values of α and β produce in the national income? In what way does the long-term behavior of the national income depend upon Y_1 and

Y_2? How does the national income change if government expenditure is set at a different value of G?

Secondary Investigation

Studies have shown that the actual value of α depends upon personal income, rising steadily from a value of 0.53 for families in the highest income bracket to 0.98 for families in the lowest income bracket. From our work with the Gini index in Section 5.4, we know that income is not evenly distributed in the United States. Table 8.1 gives the percent distribution of aggregate income for U.S. households in 1992 (U.S. Bureau of the Census, Current population reports, Series p 60–184. *Money Income of Households, Families, and Persons in the United States: 1992*, U.S. Government Printing Office, Washington, D.C., 1993).

Fifth of population	Percent of income
Lowest fifth	3.8
Second fifth	9.4
Third fifth	15.8
Fourth fifth	24.2
Highest fifth	46.9

Table 8.1 Percent distribution of aggregate income in 1992

The value of α for the lowest fifth of the population is 0.98 and for the highest fifth is 0.53. Use the information in Table 8.1 to estimate the value of α for the second, third, and fourth fifths, and then determine a value of α for the country as a whole. Refer to the data for income distribution in Lab 10, The Gini Index, to compare the values of α from different years.

Extension

What historical aspects of the American economy are explained by this model? How well does the model reflect our economic history? Can you determine realistic, historical values of α and β?

References

Chaing, Alpha C. *Fundamental Methods of Mathematical Economics*. New York: McGraw-Hill Book Company, 1967. pp. 540–546.

Goldberg, Samuel. *Introduction to Difference Equations*. New York: Dover Publications, 1986. pp. 5–8.

Samuelson, Paul A. *Economics*, 10th ed. New York: McGraw-Hill Book Company, 1976. pp. 223–229.

8.4 Cardiac Output as Measured by Thermal Dilution

When a patient has serious heart or circulation problems, one of the things a cardiologist needs to determine is the rate at which the heart is circulating blood through the body. This rate is called the *cardiac output*. The most widely used method to determine cardiac output is the thermal dilution method. In this procedure a cold saline solution is injected into a vein near the heart. A detecting catheter, called a thermistor, is floated in the pulmonary artery, and the temperature of the blood is measured at this point. Before reaching the thermistor, the saline solution begins mixing with the blood and passes through the heart.

Calculation of cardiac output by this method is based on the conservation of energy principle, which states that the energy gained by the cold saline will equal the energy lost by the blood. This principle results in the equation

$$V_{saline} \cdot (T_{mix} - T_{saline}) \cdot c_{saline} \cdot D_{saline} = V_{blood} \cdot (T_{blood} - T_{mix}) \cdot c_{blood} \cdot D_{blood}, \quad (1)$$

in which the left side gives the energy in calories gained by the saline solution and the right side gives the energy lost by the blood. Because the volume of blood is so much greater than the volume of saline, the change in temperature represented by $(T_{mix} - T_{saline})$ on the left side of equation (1) can be approximated by the quantity $(T_{blood} - T_{saline})$. Equation (1) applies to an interval of time in which each quantity in the equation is constant and the saline solution has mixed with the blood. In equation (1), V_{saline}, T_{saline}, c_{saline}, and D_{saline} are respectively the volume, initial temperature, specific heat, and density of the saline. The quantities V_{blood}, T_{blood}, c_{blood}, and D_{blood} are respectively the volume, initial temperature, specific heat, and density of the blood. The quantity T_{mix} is the temperature of the blood/saline mixture. The values of the constants c_{blood}, D_{blood}, c_{saline}, and D_{saline} are 0.87 calories$/(g \cdot^{\circ} C)$, 1.045 g/ml, 0.965 calories$/(g \cdot^{\circ} C)$, and 1.018 g/ml, respectively.

If we mix a small volume of saline with a large volume of blood then equation (1) can be used to determine the volume of the blood that flows through the heart in an interval of time. A cardiologist, however, is not interested in a volume of blood, but rather in the rate at which blood is flowing through the heart. The quantity V_{blood} in equation (1) can be replaced with the product of rate and time. The rate is the cardiac output, and the time is the interval of time over which measurements are taken.

The injection of the cold saline takes several seconds. When passing the thermistor, the blood-saline mixture is not thoroughly mixed or confined to one section in the flow of the blood. Instead, the saline mixes with some blood and enters the heart, where some of the saline remains for a while. As the heart pumps, all of the blood-saline mixture eventually passes through the heart and past the thermistor in the artery. The data in Table 8.2 is taken from a patient who has received a 10 ml injection of saline solution at 1°C. (Time is in seconds and temperature is in °C.) According to the data collected, the temperature of the mixture drops to about 36.5°C before it begins to increase to normal body temperature. Based on the data provided, estimate this patient's cardiac output in liters per minute.

Time	0.0	1.5	2.8	3.5	4.1	5.5	6.5	7.9	8.8	10.3
Temp	37.00	36.75	36.57	36.51	36.48	36.55	36.62	36.68	36.72	36.75

Time	11.9	12.6	13.5	14.2	15.5	17.1	18.9	22.3	25.5	30.0
Temp	36.76	36.80	36.82	36.85	36.86	36.89	36.90	36.92	36.94	36.95

Table 8.2 Data for patient

Normally, not all the data shown in Table 8.2 are used in calculations because of error from the effects of recirculation and heat loss over time. In addition, the initial readings are not used because of incomplete mixing. The data used in the calculation of cardiac output are typically in the interval where the temperature is between 80% and 30% of the maximum deviation from normal. Determine a strategy for calculating cardiac output that uses only the data in Table 8.2 in this interval. Compare this new result with your previous results.

References
Bowdle, T. Andrew; Peter R. Freund; and G. Alec Rooke. *Biophysical Measurement Series: Cardiac Output.* Redmond, WA: SpaceLabs, Inc., 1991.
Branthwaite, M. A. and R. D. Bradley. 1968. Measurement of cardiac output by thermal dilution in man. *Journal of Applied Physiology* 24: 434–438.
Levett, James M. and Robert L. Replogle. 1979. Thermodilution cardiac output: a critical analysis and review of the literature. *Journal of Surgical Research* 27: 392–404.
Sanmarco, Miguel E.; Charles M. Philips; *et al.* 1971. Measuring cardiac output by thermal dilution. *The American Journal of Cardiology* 28: 54-58.

8.5 Predicting the Spread of an Infectious Disease

What happens when a small group of people with an infectious disease is introduced into a larger population? Does everyone in the larger population get the disease? Will the disease die out before it spreads to a large number of people? Is there some critical mass of contagious people that must exist for the disease to become an epidemic? When will the number of infected people reach its maximum number, and what fraction of the total population will have the disease at that time? In this investigation, you will build a model for the spread of a disease, and you will use your model to explore these questions.

Suppose a disease has a very short incubation period, so that immediately after contracting the disease, the infected person can pass it on. Individuals who contract the disease either die or become permanently immune and cannot catch it again. We will ignore births, deaths due to factors unrelated to the disease, immigration, and emigration. Therefore, the size N of the population remains constant throughout the duration of the disease. This allows us to divide the population into three groups: the infected group, the susceptible group, and the recovered group. The infected group consists of those individuals who presently have the disease and can transmit it to others. The susceptible group consists of those individuals who are not infected, but are capable of becoming infected. The recovered group consists of those who have had the disease and thus are unable either to transmit or to contract the disease. For potentially fatal diseases, those who die from the disease are included in this group.

To investigate how a particular disease spreads, you will need to determine differential equations for $\frac{dS}{dt}$, $\frac{dI}{dt}$, and $\frac{dR}{dt}$, where $S(t)$, $I(t)$, and $R(t)$ represent the number of susceptible, infected, and recovered individuals, respectively. Use the following assumptions to write the differential equations.

1. The rate of change of the susceptible group is proportional, with constant of proportionality k, to the product of the number of people in the susceptible group and the number of people in the infected group.

2. Individuals are removed from the infected group at a rate proportional to the size of the infected group, with constant of proportionality r.

Questions to Consider

1. Investigate how $S(t)$, $I(t)$, and $R(t)$ change over time. How do the values of k, r, S_0, and I_0 affect the spread of the disease? Which factor seems to be more important to the spread of the disease, the initial number of people who are susceptible or the initial number of people infected? Which has a greater impact on the cessation of the disease, the initial number of people who are susceptible or the initial number of people infected?

2. Write a differential equation for $\frac{dI}{dS}$ and solve this equation for $I(S)$. Use information from $\frac{dI}{dS}$ and $\frac{d^2I}{dS^2}$ to describe the graph of $I(S)$. Where is the maximum value of I located? What implication does this have for the spread of the disease? What information do the graph of $I(S)$ and the derivative $\frac{dS}{dt}$ give you about the graph of $I(t)$? Is this consistent with the graphs generated in question 1?

3. For a fatal disease, the typical result of infection is death. In this case, the function R counts the number of deaths due to the disease. With a fatal disease, scientists want to use a daily death toll to estimate the values of k and r. This would allow them to make predictions about the behavior of the disease in the long run. The

data in Table 8.3 give the number of deaths per day for a plague similar to the one that struck Bombay in 1905. Use this data to estimate the values of k and r for this disease. Use the estimated values of k and r to determine the behavior of S and I.

Day	1	2	3	4	5	6	7	8	9	10
Deaths	20	20	30	40	60	50	110	160	280	390

Day	11	12	13	14	15	16	17	18	19	20
Deaths	450	750	770	700	690	860	920	810	590	400

Day	21	22	23	24	25	26	27	28	29	30
Deaths	370	220	100	80	60	40	60	50	50	30

Table 8.3 Number of deaths per day from the plague

Extensions

4. Use the fact that $S(t) + I(t) + R(t) = N$ to write a differential equation for $\frac{dR}{dt}$. Your differential equation should be in the form $\frac{dR}{dt} = f(R(t))$. That is, $\frac{dR}{dt}$ should be expressed as a function of $R(t)$. You will need to find $\frac{dS}{dR}$ and solve this differential equation for $S(R)$. You will also need to express $I(t)$ in terms of $\frac{dR}{dt}$. Compare the graph of $\frac{dR}{dt} = f(R(t))$ to the graph of $I(t)$. What information do these graphs give you?

5. Use the data in Table 8.3 and your differential equation for $\frac{dR}{dt}$ to estimate the values of k and r for this plague. How do these estimates compare to the ones you made previously?

6. How does the model for the spread of a disease change if a recovered individual can become infected again? How does this modification influence the spread of the disease?

References

Braun, Martin. *Differential Equations and their Applications.* New York: Springer-Verlag, 1983. pp. 456–462.

Maki, Daniel and Maynard Thompson. *Mathematical Models and Applications.* Englewood Cliffs, NJ: Prentice-Hall, Inc., 1973. pp. 359–370.

Olinick, Michael. *An Introduction to Mathematical Models in the Social and Life Sciences.* Reading, MA: Addison-Wesley Publishing Company, 1978. pp. 347–365.

8.6 Population Models

Throughout this text we have studied population models in which we assumed that the growth rate is proportional to the existing population. In 1798, Thomas Malthus made the same assumption. He theorized that while the world population was growing exponentially, food production would increase linearly, causing eventual food shortages and famine. The Malthus model, also known as the unconstrained growth model, is represented by the differential equation

$$\frac{dP}{dt} = kP, \tag{2}$$

where k is a positive constant of proportionality. This constant k is equal to the birth rate minus the death rate, and is called the *fertility rate*. Malthus assumed that the fertility rate is constant no matter how large the population.

About 40 years after Malthus made his predictions, P.F. Verhulst proposed a modification of the Malthus model that imposed an upper limit on the size of the population. Verhulst attributed this limiting size to environmental factors that limit the food supply and other resources needed to sustain a population. He assumed that the rate at which the world population is changing is jointly proportional to the existing population and to the fraction of the maximum sustainable population that the existing population has left to grow. This assumption leads to the Verhulst model, also known as the constrained or logistic growth model, represented by the differential equation

$$\frac{dP}{dt} = kP\left(1 - \frac{P}{M}\right), \tag{3}$$

where k is a positive constant of proportionality and M is the maximum sustainable population. As with the Malthus model, k is interpreted as the fertility rate.

In the early 1960s, Heinz von Foerster, an electrical engineer, proposed a modification of the Malthus model for world population. Von Foerster proposed that the fertility rate, rather than being constant, is an increasing function of the size of the population. He justified this assumption by citing improvements in technology that have increased the birth rate, increased life expectancy, and provided more efficient ways to produce food and shelter. The fertility rate he proposed is kP^r, where k and r are positive constants. Substituting the fertility rate kP^r for the constant fertility rate k in the Malthus model in (2) yields

$$\frac{dP}{dt} = kP^r \cdot P$$
$$= kP^{r+1},$$

which is the differential equation for the von Foerster model.

A decade after von Foerster proposed his model, A.L. Austin and J.W. Brewer, both mechanical engineers, proposed a combination of the Verhulst model and the von Foerster model. They assumed that there must be an upper limit to the fertility rate as well as an upper limit to the size of the population. They defined the fertility rate as

$$A\left(1 - e^{-\frac{k P^r}{A}}\right) \tag{4}$$

where k and r are positive constants. The Austin-Brewer fertility rate approaches an upper bound A as the population P increases. Substituting the expression in (4) for the constant fertility rate k in the Verhulst model in (3) yields

$$\frac{dP}{dt} = A\left(1 - e^{-\frac{k P^r}{A}}\right) P\left(1 - \frac{P}{M}\right),$$

which is the differential equation for the Austin-Brewer model.

Analyze the long term trends in population implied by each model. The Malthus, Verhulst, and von Foerster models involve differential equations that can be solved analytically. We cannot solve the Austin-Brewer model analytically, so population values must be approximated with a numerical method. For the Malthus model you might start with a world population of 100 million in year 0 and experiment with various growth rates. In the Verhulst model, estimates for the maximum sustainable population range from 10 billion to well over 50 billion. Austin and Brewer used the parameter values $A = 0.10$, $k = 5.0 \times 10^{-12}$, $r = 1.0$, and $M = 50$ billion.

Use the world population figures provided in Table 8.4 to determine which models, if any, accurately describe the world population. You may want to reconsider each model using only figures for world population since 1750, which are more reliable than earlier estimates.

Date (A.D.)	Population (millions)
0	100
1000	200
1650	545
1750	728
1800	906
1850	1171
1900	1608
1920	1834
1930	2070
1940	2295
1950	2517
1960	3005

Table 8.4 World population data

Find more recent data for the world population to test your models. In addition, you may want to apply these models to the populations of single nations or continents such as the United States, Japan, Europe, or Africa.

References

Austin, Arthur L. and John W. Brewer. 1970. World population growth and related technical problems. *IEEE Spectrum* 7 (Dec): 43–54; 8 (Mar): 10.

McEvedy, Colin and Richard Jones. 1960. *Atlas of World Population History*. Great Britain: Penguin Books, 1979.

von Foerster, H.; P. M. Mora; and L. W. Amiot. 1960. Doomsday: Friday, 13 November, A.D. 2026. *Science* 132: 1291–295.

8.7 Permeability of Red Blood Cells

For medical researchers to diagnose and treat certain blood diseases, they need to understand as much as possible about the structure of blood cells. One property of blood cells that gives researchers information about the health of an individual is the permeability of the cell's surface to potassium ions (K^+).

One method used to determine permeability is to inject a radioactive isotope of potassium, K^{42+}, into the bloodstream and then determine the rates at which potassium ions move in and out of red blood cells. Initially, the potassium isotope is entirely in the blood plasma and then begins to move through the walls of the red blood cells. Over time these ions pass in and out of the cells, usually at different rates. We will assume for now that we have a closed system; that is, at any time some of the potassium ions are found in the blood plasma and some are found in the red blood cells. We also assume that the rate at which potassium flows into one component of the blood is proportional to the amount in the other component. To test for permeability, samples of blood plasma are taken over a relatively short time span (approximately one hour), and the radioactivity of the plasma is used to determine the amount of the isotope in the plasma.

The data in Table 8.5 represent the amount of the radioactive potassium in the plasma for a particular patient over time. Time t is measured in seconds and K^{42+} is measured in milligrams. Determine the rates at which potassium passes in and out of the red blood cells of this patient.

t	0	250	500	750	1000	1250	1500	1750	2000	2250	2500
K^{42+}	5.00	2.96	2.01	1.49	1.14	1.01	0.97	0.92	0.87	0.85	0.85

Table 8.5 Amount of potassium in blood plasma over time

Questions to Consider

1. Write the differential equation describing the movement of potassium ions to and from the plasma as a function of the amount of potassium in the plasma. Use analytic and numerical methods to solve this differential equation. Use the data to determine the constants of proportionality governing the movement between the plasma and the red blood cells.

2. Use difference quotients to determine a model for the amount of potassium in the plasma as a function of time. Which technique gives better results, modeling directly from the data or using difference quotients?

3. Compare the values for the constants of proportionality obtained when a model is fit to the differential equation using difference quotients and when a model is fit to the original data.

4. The potassium used to determine red blood cell permeability is radioactive; its half life is 12.5 hours. Incorporate this information into your differential equation and use Euler's methods to solve the differential equation. Does the fact that the potassium decays influence the values you previously determined for the constants of proportionality?

Reference
Horelick, Brindell and Sinan Koont. *Tracer Methods in Permeability*. UMAP Module 74. Lowell, MA: COMAP, 1980.

8.8 Sound, Music, and Fourier Series

From the time of the early Greek mathematicians, the study of music has been linked to the study of mathematics. The mathematical investigation of music begun by Pythagoras culminated in the investigations of the French mathematician Joseph Fourier (1768–1830). Fourier discovered that any sound, however complex, could be written as a series of simple sine functions whose frequencies are all multiples of a fundamental frequency.

Any sound can be distinguished by three characteristics, *loudness*, *pitch*, and *quality*. The amplitude of the sine waves is determined by the loudness and is a measure of the energy of the wave. The louder the sound, the larger the amplitude of the sine waves. The period of the fundamental sine wave is determined by the pitch. The shorter the period (or the higher the frequency), the higher the pitch of the sound. When two different instruments play a note with the same pitch and loudness, they produce two noticeably different sounds. Although the pitch and loudness are the same, the two sounds differ in quality. Different characteristics of the sounds produce the qualities of sound that we describe as soft, piercing, braying, hollow, rich, dull, bright, or crisp. This investigation will use Fourier series to investigate these characteristics of sound.

The 38 data points listed in Table 8.6 and graphed in Figure 8.7 were generated by sampling the sound of a trumpet playing a C. The sound wave was captured by an oscilloscope, and the amplitude was then sampled approximately every 0.104 milliseconds for one complete cycle. (When possible, it is preferable to capture a large data set and enter it into a spreadsheet or data software electronically, thereby automating the whole process.) Use the techniques from Chapter 7 to find a Fourier series that approximates this sound wave.

Time	Amplitude	Time	Amplitude	Time	Amplitude
0.764	−0.146	2.116	0.202	3.468	−0.051
0.868	−1.289	2.220	−0.037	3.572	−0.051
0.972	−1.967	2.324	−0.122	3.676	−0.084
1.076	−1.378	2.428	−0.052	3.780	−0.130
1.180	0.478	2.532	0.051	3.884	−0.158
1.284	1.942	2.636	0.048	3.988	−0.146
1.388	1.772	2.740	−0.046	4.092	−0.117
1.492	1.425	2.844	−0.117	4.196	−0.093
1.596	0.187	2.948	−0.117	4.300	0.087
1.700	−0.475	3.052	−0.058	4.404	0.389
1.804	−0.247	3.156	−0.011	4.508	0.328
1.908	0.197	3.260	−0.024	4.612	−0.286
2.012	0.363	3.364	−0.043		

Table 8.6 Data for a trumpet playing C

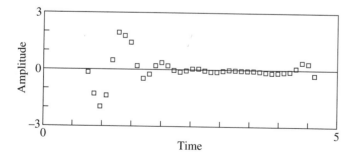

Figure 8.7 Graph of trumpet data

If you have access to the necessary electronic equipment, capture sound data from a trumpet, a flute, and an organ making three different sounds, for example, each playing A, C, and F. Find a discrete Fourier series, based on one complete cycle, for each of these sounds. (If you do not have access to the necessary equipment, data sets are provided in the appendix.) What characteristics do the graphs of the series for the same note share? What characteristics do the graphs of the series for the same instrument share?

Certain characteristics of a sound are determined by the magnitude of a quantity known as a *harmonic*. The magnitude of the kth harmonic H_k is defined as

$$H_k = \sqrt{(a_k)^2 + (b_k)^2} \, ,$$

where a_k is the coefficient of $\cos(k\beta x)$ and b_k is the coefficient of $\sin(k\beta x)$ in the Fourier series. Therefore, the magnitude of a harmonic is a measure of the total contribution to a Fourier series of the sine and cosine terms with a particular period. The larger the magnitude of a harmonic, the more important that harmonic is in producing particular characteristics of the sound. Each instrument has certain harmonics that stand out among the others. Harmonics associated with higher order terms of a Fourier series are essential if the sound is to be bright, but if the magnitudes of higher harmonics higher than the sixth or seventh order are very large compared to the lower order harmonics, the tone is perceived as piercing and rough.

Determine the magnitude of the first ten harmonics for the trumpet, flute, and organ sounds you have already investigated. Characterize the harmonics of each instrument.

Determine which of these instruments is represented by the graph and data set in Figure 8.8. Time is in milliseconds and rounded to 3 decimal places.

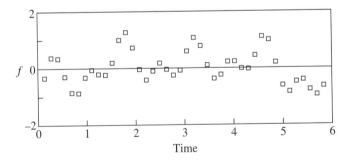

Time	Amplitude	Time	Amplitude	Time	Amplitude
0.139	−0.333	2.083	−0.040	4.028	0.231
0.278	0.377	2.222	−0.412	4.167	−0.011
0.417	0.344	2.361	−0.100	4.305	−0.033
0.556	−0.292	2.500	0.181	4.444	0.442
0.695	−0.854	2.639	−0.042	4.583	1.108
0.833	−0.870	2.778	−0.248	4.722	0.994
0.972	−0.319	2.917	−0.075	4.861	0.196
1.111	−0.056	3.055	0.596	5.000	−0.613
1.250	−0.207	3.194	1.072	5.139	−0.837
1.389	−0.231	3.333	0.787	5.278	−0.481
1.528	0.194	3.472	0.106	5.417	−0.415
1.667	1.001	3.611	−0.370	5.555	−0.760
1.806	1.273	3.750	−0.235	5.694	−0.941
1.944	0.725	3.889	0.222	5.833	−0.637

Table 8.8 Data from unknown instrument

One contemporary application of Fourier series is voice recognition. Based on what you have seen in the harmonics, explain how it is possible for a computer to distinguish one speaker from another and one command from another.

Extensions

1. Using a single instrument, move up the scale one half tone at a time, from C to C sharp to D to D sharp, and so on. What difference do you notice in the graph of the series as you move up a half tone? Could you predict the shape of an E from the shapes of the notes preceding it? (Data sets are provided in the appendix.)

2. The Fourier series you have written so far have been in terms of both sines and cosines. Sound waves, however, are generally thought of as being composed of sine waves only. We would like to write the Fourier series

$$F(x) = a_0 + \sum_{k=1}^{\infty} \left[a_k \cos(k\beta x) + b_k \sin(k\beta x) \right]$$

in the form

$$F(x) = A_0 + \sum_{k=1}^{\infty} A_k \cdot \sin(k\beta x + \theta_k).$$

Rewrite the Fourier series as a function of sine waves only, using the substitutions $a = A\sin\theta$ and $b = A\cos\theta$ in the expression $a\cos(nx) + b\sin(nx)$. Compare the amplitudes A_k of the successive components of these sine waves to the magnitude of the harmonics defined previously as $H_k = \sqrt{(a_k)^2 + (b_k)^2}$.

3.   When sounds are produced by a musical instrument, the perceived sound is greatly influenced by how the sound is initiated, how it is sustained, and how it decays. These characteristics are given in what is known as the *envelope* for the sound. For an instrument like a piano, the sound begins quickly as the hammer hits the piano wire. This makes the *attack* phase of the envelope very steep, followed by a gradual *release* or *decay*. Other instruments have different envelope patterns such as the *attack-sustain-release* curve and the *attack-decay-sustain-release* curve. The sounds we have investigated so far, such as the trumpet data given in Table 8.6, are in the sustain region of an envelope. Figure 8.9 illustrates these typical envelope curves. Use Fourier series to create an envelope of each of these types. Make the amplitude of each wave 1 and the period 2π.

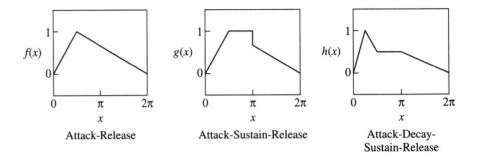

Figure 8.9 Typical envelope patterns

References

Berg, Richard E. and David G. Stork. *The Physics of Sound.* Englewood Cliffs, NJ: Prentice-Hall, Inc., 1995. pp. 91–116.

Moravcsik, Michael. *Musical Sound.* New York: Paragon House Publishers, 1987.

Rigden, John S. *Physics and the Sound of Music.* John Wiley & Sons, 1977.

Rossing, Thomas D. *The Science of Sound.* Reading, MA: Addison-Wesley Publishing Company, 1982.

8.9 Advanced Predator-Prey Models

In Lab 9, we investigated the predator-prey model and some of its variations. In this investigation, we will consider other modifications of the basic model.

Predator-Prey Model with Seasonal Fishing

Suppose one species of fish preys on another with rate constants equal to those in the fox-rabbit predator-prey model in Lab 9. The system of differential equations for this predator-prey model is

$$\frac{dS}{dt} = 0.05S - 0.001SR,$$

$$\frac{dR}{dt} = -0.03R + 0.0002SR,$$

where R and S are the populations of the predator fish and the prey fish, respectively, as functions of time. Modify this model by introducing fishing into the situation. Assume that fishing is cyclic and depletes fish populations at a rate proportional to the fish population with proportionality multiplier equal to a sinusoidal function f. The function f has equation $f(t) = f_0 + a \cdot \sin(p(t - t_0))$, where f_0 is an average value about which the fishing level oscillates, a is the amplitude of the fishing cycle with $a \le f_0$ (so $f(t) \ge 0$ for all t), p is a parameter that controls the length of the fishing cycle, and t_0 is a phase shift that determines the timing of the fishing cycle.

Compare the dynamics of the predator-prey model with fishing and without fishing under three different scenarios.

1. Only the predators are fished. What effect does the timing of the fishing season have on the dynamics of the system? When is the "best" time to fish, within the season?

2. Only the prey are fished. Investigate the timing of the fishing season and its effect on the dynamics of the system.

3. Both species are fished. Investigate the timing of the fishing season for each species.

Competition

If two species X and Y are in competition for a common food source, but neither population preys on the other (squirrels and chipmunks, for example), a modification of the predator-prey model can be used to investigate the populations over time. In this situation, the assumptions for the model are:

1. Each species grows exponentially if the other species is absent.

2. The competition between the species is proportional to the product of the populations of the two species.

Write differential equations to model this system and investigate the dynamics of the system.

The Russian biologist G.F. Gause studied the interaction of competitive species in the early 1900s. Based on laboratory experiments with competitive species, Gause claimed that two species cannot indefinitely coexist in the same locality if they have identical ecological requirements. Does your model support or reject Gause's claim?

Four-Species Interaction

Devise a system of differential equations to model a four-species interaction. Include a predator/prey interaction and an interaction in which different species are in competition for the same food source. Explain the meaning of each coefficient in the equations. Investigate the dynamics of the system.

Reference
Danby, J.M.A. *Computing Applications to Differential Equations: Modeling in the Physical and Social Sciences*. Reston, VA: Reston Publishing Company, Inc., 1985. pp. 64–75.

Complex Numbers

Nearly all of the applications of calculus that we have studied in this text are related to real numbers, but in the section on Newton's Method we saw that when the iteration defined by this method was applied to complex numbers, interesting things happened. If you are not familiar with complex numbers, this brief introduction should give you enough information to understand how the images in Section 3.8, Investigation 6 are formed.

A complex number is a number of the form $a + bi$, where $i = \sqrt{-1}$ and a and b are real numbers. The first place you may have encountered such numbers was in algebra, when you used the quadratic formula to find solutions to a quadratic equation. The first term, a, is called the ***real part*** and the second term, bi, is called the ***imaginary part*** of the complex number.

real part

imaginary part

Example 1

Find the solutions to $x^2 - 4x + 13 = 0$.

Solution   Using the quadratic formula, we get $x = \frac{4 \pm \sqrt{16 - 52}}{2} = \frac{4 \pm \sqrt{-36}}{2} = \frac{4 \pm 6\sqrt{-1}}{2} = 2 \pm 3\sqrt{-1}$. Mathematicians in the 17th Century decided to define $i = \sqrt{-1}$. This allows us to write the solutions to the quadratic equation as $2 \pm 3i$. ∎

The systems of numbers of the form $a + bi$, where a and b are real numbers and $i = \sqrt{-1}$, turns out to have most of the same properties as the real numbers. The only significant loss is that of ordering. (For example, which is greater, $1 + 2i$ or $2 + i$?) The operations of addition and multiplication are defined logically as

$$(a + bi) + (c + di) = (a + c) + (b + d)i$$

and

$$(a + bi)(c + di) = ac + adi + bci + bdi^2 = ac + (ad + bc)i - bd = (ac - bd) + (ad + bc)i$$

Notice that both addition and multiplication are performed as if the numbers $(a + bi)$ and $(c + di)$ were binomials. For addition you add like terms and for multiplication you use the distributive property. Notice also that the term bdi^2 simplifies to $-bd$ since $i^2 = -1$. It is not difficult to show that the complex numbers with the operations defined above are closed under addition and multiplication (that is, if you add or multiply two complex numbers the result is a complex number) and have identities for both operations. Also, it can be shown that additive inverses exist for complex numbers and multiplicative inverses exist for all non-zero numbers, that both operations are commutative and associative, and that multiplication still distributes over addition.

Example 2

Add the complex numbers $2 + 3i$ and $3 - 7i$. Find the product of these numbers.

Solution   Addition is simply $(2 + 3i) + (3 - 7i) = (2 + 3) + (3 - 7)i = 5 - 4i$. The product of these two numbers is $(2 + 3i)(3 - 7i) = (2 \cdot 3) + 2(-7)i + 3(3)i + (-7)(3)i^2 = 6 - 14i + 9i - 21i^2 = 27 - 5i.$ ∎

Subtraction of complex numbers is completely analogous to subtraction of real numbers, as is division. However, with division simplification of answers may be preferred.

Example 3

Determine the quotient of $3 + 4i$ and $4 - 9i$.

Solution   We could write the quotient as $\frac{3+4i}{4-9i}$ and this is indeed correct, but we can write the quotient in a simpler form. In order to do this, we use the same technique you may have used in algebra to remove radicals from denominators of fractions. If we had $3 + \sqrt{5}$ in the denominator of a fraction, we could multiply both the numerator and the denominator of the fraction by the conjugate $3 - \sqrt{5}$. Since the i in the denominator of our complex quotient is actually a radical, we will do the same thing to simplify the quotient.

$$\frac{3 + 4i}{4 - 9i} = \frac{(3 + 4i)(4 + 9i)}{(4 - 9i)(4 + 9i)} = \frac{-24 + 43i}{97} = -\frac{24}{97} + \frac{43}{97}i$$ ∎

We can graph all of the real numbers on a number line, but complex numbers have two parts, the real part and the imaginary part, so we cannot graph them on a line. A set of perpendicular lines, or axes, are used to define a plane, and complex numbers are graphed in this plane. The horizontal axis is used to locate the real part of the number and the vertical axis is used to locate the imaginary part. Along the horizontal axis the units are simply real numbers, but along the vertical or imaginary axis, the units are in terms of i.

Example 4

Graph the complex numbers $3 + 2i$ and $-2 - i$.

Solution The real part of $3 + 2i$ is 3 and the imaginary part is $2i$. The coordinates for $3 + 2i$ are (3, 2). Similarly, the coordinates for $-2-i$ are (−2, −1).

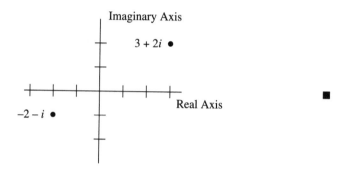

A Brief Table of Integrals

1. $\int u\,dv = uv - \int v\,du$

2. $\int u^n\,du = \frac{1}{n+1}u^{n+1} + C$, if $n \neq 1$

3. $\int \frac{du}{u} = \ln|u| + C$

4. $\int e^u\,du = e^u + C$

5. $\int a^u\,du = \frac{a^u}{\ln a} + C$

6. $\int \sin u\,du = -\cos u + C$

7. $\int \cos u\,du = \sin u + C$

8. $\int \sec^2 u\,du = \tan u + C$

9. $\int \csc^2 u\,du = -\cot u + C$

10. $\int \sec u \tan u = \sec u + C$

11. $\int \csc u \cot u\,du = -\csc u + C$

12. $\int \tan u\,du = \ln|\sec u| + C$

13. $\int \cot u\,du = \ln|\sin u| + C$

14. $\int \sec u\,du = \ln|\sec u + \tan u| + C$

15. $\int \csc u\,du = \ln|\csc u - \cot u| + C$

16. $\int \frac{du}{\sqrt{a^2 - u^2}} = \sin^{-1}\frac{u}{a} + C$

17. $\int \frac{du}{a^2 + u^2} = \frac{1}{a}\tan^{-1}\frac{u}{a} + C$

18. $\int \frac{du}{a^2 - u^2} = \frac{1}{2a}\ln\left|\frac{u+a}{u-a}\right| + C$

19. $\int \frac{du}{u\sqrt{u^2 - a^2}} = \frac{1}{a}\sec^{-1}\left|\frac{u}{a}\right| + C$

20. $\int \sin(au)\sin(bu)du = \dfrac{\sin(a-b)u}{2(a-b)} - \dfrac{\sin(a+b)u}{2(a+b)} + C$, if $a \ne \pm b$

21. $\int \cos(au)\cos(bu)du = \dfrac{\sin(a-b)u}{2(a-b)} + \dfrac{\sin(a+b)u}{2(a+b)} + C$, if $a \ne \pm b$

22. $\int \sin(au)\cos(bu)du = -\dfrac{\cos(a-b)u}{2(a-b)} - \dfrac{\cos(a+b)u}{2(a+b)} + C$, if $a \ne \pm b$

23. $\int \sin^n u\, du = -\dfrac{1}{n}\sin^{n-1} u\cos u + \dfrac{n-1}{n}\int \sin^{n-2} u\, du$, if $n \ge 2$

24. $\int \cos^n u\, du = \dfrac{1}{n}\cos^{n-1} u\sin u + \dfrac{n-1}{n}\int \cos^{n-2} u\, du$, if $n \ge 2$

25. $\int \sqrt{u^2 \pm a^2}\, du = \dfrac{u}{2}\sqrt{u^2 \pm a^2} \pm \dfrac{a^2}{2}\ln\left|u + \sqrt{u^2 \pm a^2}\right| + C$

26. $\int \dfrac{du}{\sqrt{u^2 \pm a^2}} = \ln\left|u + \sqrt{u^2 \pm a^2}\right| + C$

27. $\int \dfrac{\sqrt{u^2 + a^2}}{u}\, du = \sqrt{u^2 + a^2} - a\ln\left|\dfrac{a + \sqrt{u^2 + a^2}}{u}\right| + C$

28. $\int u^2\sqrt{u^2 \pm a^2}\, du = \dfrac{u}{8}(2u^2 \pm a^2)\sqrt{u^2 \pm a^2} - \dfrac{a^4}{8}\ln\left|u + \sqrt{u^2 \pm a^2}\right| + C$

29. $\int \dfrac{\sqrt{u^2 - a^2}}{u}\, du = \sqrt{u^2 - a^2} - a\sec^{-1}\left|\dfrac{u}{a}\right| + C$

30. $\int \dfrac{u^2\, du}{\sqrt{u^2 \pm a^2}} = \dfrac{u}{2}\sqrt{u^2 \pm a^2} \mp \dfrac{a^2}{2}\ln\left|u + \sqrt{u^2 \pm a^2}\right| + C$

31. $\int \dfrac{du}{u^2\sqrt{u^2 \pm a^2}} = \mp \dfrac{\sqrt{u^2 \pm a^2}}{a^2 u} + C$

32. $\int \dfrac{\sqrt{u^2 \pm a^2}}{u^2}\, du = -\dfrac{\sqrt{u^2 \pm a^2}}{u} + \ln\left|u + \sqrt{u^2 \pm a^2}\right| + C$

33. $\int \sqrt{a^2 - u^2}\, du = \dfrac{u}{2}\sqrt{a^2 - u^2} + \dfrac{a^2}{2}\sin^{-1}\dfrac{u}{a} + C$

34. $\int \dfrac{u^2\, du}{\sqrt{a^2 - u^2}} = -\dfrac{u}{2}\sqrt{a^2 - u^2} + \dfrac{a^2}{2}\sin^{-1}\dfrac{u}{a} + C$

35. $\int \dfrac{\sqrt{a^2 - u^2}}{u}\, du = \sqrt{a^2 - u^2} - a\ln\left|\dfrac{a + \sqrt{a^2 - u^2}}{u}\right| + C$

36. $\int u^2 \sqrt{a^2 - u^2}\, du = \frac{u}{8}(2u^2 - a^2)\sqrt{a^2 - u^2} + \frac{a^4}{8}\sin^{-1}\frac{u}{a} + C$

37. $\int \frac{du}{u^2 \sqrt{a^2 - u^2}} = -\frac{\sqrt{a^2 - u^2}}{a^2 u} + C$

38. $\int \frac{\sqrt{a^2 - u^2}}{u^2}\, du = -\frac{\sqrt{a^2 - u^2}}{u} - \sin^{-1}\frac{u}{a} + C$

39. $\int \frac{du}{u\sqrt{a^2 - u^2}} = -\frac{1}{a}\ln\left|\frac{a + \sqrt{a^2 - u^2}}{u}\right| + C$

40. $\int \frac{du}{(a^2 - u^2)^{\frac{3}{2}}} = \frac{u}{a^2 \sqrt{a^2 - u^2}} + C$

41. $\int (a^2 - u^2)^{\frac{3}{2}}\, du = \frac{u}{8}(5a^2 - 2u^2)\sqrt{a^2 - u^2} + \frac{3a^4}{8}\sin^{-1}\frac{u}{a} + C$

Table of Standard Normal Probabilities

$$P(-\infty < x \le z) = \int_{-\infty}^{z} \frac{1}{\sqrt{2\pi}} e^{-\frac{x^2}{2}} dx$$

z	0.00	0.01	0.02	0.03	0.04	0.05	0.06	0.07	0.08	0.09
−3.4	0.0003	0.0003	0.0003	0.0003	0.0003	0.0003	0.0003	0.0003	0.0003	0.0002
−3.3	0.0005	0.0005	0.0005	0.0004	0.0004	0.0004	0.0004	0.0004	0.0004	0.0003
−3.2	0.0007	0.0007	0.0006	0.0006	0.0006	0.0006	0.0006	0.0005	0.0005	0.0005
−3.1	0.0010	0.0009	0.0009	0.0009	0.0008	0.0008	0.0008	0.0008	0.0007	0.0007
−3.0	0.0014	0.0013	0.0013	0.0012	0.0012	0.0011	0.0011	0.0011	0.0010	0.0010
−2.9	0.0019	0.0018	0.0018	0.0017	0.0016	0.0016	0.0015	0.0015	0.0014	0.0014
−2.8	0.0026	0.0025	0.0024	0.0023	0.0023	0.0022	0.0021	0.0021	0.0020	0.0019
−2.7	0.0035	0.0034	0.0033	0.0032	0.0031	0.0030	0.0029	0.0028	0.0027	0.0026
−2.6	0.0047	0.0045	0.0044	0.0043	0.0041	0.0040	0.0039	0.0038	0.0037	0.0036
−2.5	0.0062	0.0060	0.0059	0.0057	0.0055	0.0054	0.0052	0.0051	0.0049	0.0048
−2.4	0.0082	0.0080	0.0078	0.0075	0.0073	0.0071	0.0069	0.0068	0.0066	0.0064
−2.3	0.0107	0.0104	0.0102	0.0099	0.0096	0.0094	0.0091	0.0089	0.0087	0.0084
−2.2	0.0139	0.0136	0.0132	0.0129	0.0126	0.0122	0.0119	0.0116	0.0113	0.0110
−2.1	0.0177	0.0174	0.0170	0.0166	0.0162	0.0158	0.0154	0.0150	0.0146	0.0142
−2.0	0.0228	0.0222	0.0217	0.0212	0.0207	0.0202	0.0197	0.0192	0.0188	0.0183
−1.9	0.0287	0.0281	0.0274	0.0268	0.0262	0.0256	0.0250	0.0244	0.0239	0.0233
−1.8	0.0359	0.0352	0.0344	0.0336	0.0329	0.0321	0.0314	0.0307	0.0301	0.0293
−1.7	0.0446	0.0436	0.0427	0.0418	0.0409	0.0401	0.0392	0.0384	0.0375	0.0367
−1.6	0.0548	0.0537	0.0526	0.0516	0.0505	0.0495	0.0485	0.0475	0.0465	0.0455
−1.5	0.0668	0.0655	0.0643	0.0630	0.0618	0.0606	0.0594	0.0582	0.0571	0.0559
−1.4	0.0808	0.0793	0.0778	0.0764	0.0749	0.0735	0.0722	0.0708	0.0694	0.0681
−1.3	0.0968	0.0951	0.0934	0.0918	0.0901	0.0885	0.0869	0.0853	0.0838	0.0823
−1.2	0.1151	0.1131	0.1112	0.1093	0.1075	0.1056	0.1038	0.1020	0.1003	0.0985
−1.1	0.1357	0.1335	0.1314	0.1292	0.1271	0.1251	0.1230	0.1210	0.1190	0.1170

Table of Standard Normal Probabilities, continued

z	0.00	0.01	0.02	0.03	0.04	0.05	0.06	0.07	0.08	0.09
−1.0	0.1587	0.1562	0.1539	0.1515	0.1492	0.1469	0.1446	0.1423	0.1401	0.1379
−0.9	0.1841	0.1814	0.1788	0.1762	0.1736	0.1711	0.1685	0.1660	0.1635	0.1611
−0.8	0.2119	0.2090	0.2061	0.2033	0.2005	0.1977	0.1949	0.1922	0.1894	0.1867
−0.7	0.2420	0.2389	0.2358	0.2327	0.2296	0.2266	0.2236	0.2206	0.2177	0.2148
−0.6	0.2743	0.2709	0.2676	0.2643	0.2611	0.2578	0.2546	0.2514	0.2483	0.2451
−0.5	0.3085	0.3050	0.3015	0.2981	0.2946	0.2912	0.2877	0.2843	0.2810	0.2776
−0.4	0.3446	0.3409	0.3372	0.3336	0.3300	0.3264	0.3228	0.3192	0.3156	0.3121
−0.3	0.3821	0.3783	0.3745	0.3707	0.3669	0.3632	0.3594	0.3557	0.3520	0.3483
−0.2	0.4207	0.4168	0.4129	0.4090	0.4052	0.4013	0.3974	0.3936	0.3897	0.3859
−0.1	0.4602	0.4562	0.4522	0.4483	0.4443	0.4404	0.4364	0.4325	0.4286	0.4247
0.0	0.5000	0.5040	0.5080	0.5120	0.5160	0.5199	0.5239	0.5279	0.5319	0.5359
0.1	0.5398	0.5438	0.5478	0.5517	0.5557	0.5596	0.5636	0.5675	0.5714	0.5753
0.2	0.5793	0.5832	0.5871	0.5910	0.5948	0.5987	0.6026	0.6064	0.6103	0.6141
0.3	0.6179	0.6217	0.6255	0.6293	0.6331	0.6368	0.6406	0.6443	0.6480	0.6517
0.4	0.6554	0.6591	0.6628	0.6664	0.6700	0.6736	0.6772	0.6808	0.6844	0.6879
0.5	0.6915	0.6950	0.6985	0.7019	0.7054	0.7088	0.7123	0.7157	0.7190	0.7224
0.6	0.7257	0.7291	0.7324	0.7357	0.7389	0.7422	0.7454	0.7486	0.7517	0.7549
0.7	0.7580	0.7611	0.7642	0.7673	0.7704	0.7734	0.7764	0.7794	0.7823	0.7852
0.8	0.7881	0.7910	0.7939	0.7967	0.7995	0.8023	0.8051	0.8078	0.8106	0.8133
0.9	0.8159	0.8186	0.8212	0.8238	0.8264	0.8289	0.8315	0.8340	0.8365	0.8389
1.0	0.8413	0.8438	0.8461	0.8485	0.8508	0.8531	0.8554	0.8577	0.8599	0.8621
1.1	0.8643	0.8665	0.8686	0.8708	0.8729	0.8749	0.8770	0.8790	0.8810	0.8830
1.2	0.8849	0.8869	0.8888	0.8907	0.8925	0.8944	0.8962	0.8980	0.8997	0.9015
1.3	0.9032	0.9049	0.9066	0.9082	0.9099	0.9115	0.9131	0.9147	0.9162	0.9177
1.4	0.9192	0.9207	0.9222	0.9236	0.9251	0.9265	0.9279	0.9292	0.9306	0.9319
1.5	0.9332	0.9345	0.9357	0.9370	0.9382	0.9394	0.9406	0.9418	0.9429	0.9441
1.6	0.9452	0.9463	0.9474	0.9484	0.9495	0.9505	0.9515	0.9525	0.9535	0.9545
1.7	0.9554	0.9564	0.9573	0.9582	0.9591	0.9599	0.9608	0.9616	0.9625	0.9633
1.8	0.9641	0.9649	0.9656	0.9664	0.9671	0.9678	0.9686	0.9693	0.9699	0.9706
1.9	0.9713	0.9719	0.9726	0.9732	0.9738	0.9744	0.9750	0.9756	0.9761	0.9767
2.0	0.9772	0.9778	0.9783	0.9788	0.9793	0.9798	0.9803	0.9808	0.9812	0.9817
2.1	0.9821	0.9826	0.9830	0.9834	0.9838	0.9842	0.9846	0.9850	0.9854	0.9857
2.2	0.9861	0.9864	0.9868	0.9871	0.9875	0.9878	0.9881	0.9884	0.9887	0.9890
2.3	0.9893	0.9896	0.9898	0.9901	0.9904	0.9906	0.9909	0.9911	0.9913	0.9916
2.4	0.9918	0.9920	0.9922	0.9925	0.9927	0.9929	0.9931	0.9932	0.9934	0.9936
2.5	0.9938	0.9940	0.9941	0.9943	0.9945	0.9946	0.9948	0.9949	0.9951	0.9952
2.6	0.9953	0.9955	0.9956	0.9957	0.9959	0.9960	0.9961	0.9962	0.9963	0.9964
2.7	0.9965	0.9966	0.9967	0.9968	0.9969	0.9970	0.9971	0.9972	0.9973	0.9974
2.8	0.9974	0.9975	0.9976	0.9977	0.9977	0.9978	0.9979	0.9979	0.9980	0.9981
2.9	0.9981	0.9982	0.9982	0.9983	0.9984	0.9984	0.9985	0.9985	0.9986	0.9986
3.0	0.9987	0.9987	0.9987	0.9988	0.9988	0.9989	0.9989	0.9989	0.9990	0.9990
3.1	0.9990	0.9991	0.9991	0.9991	0.9992	0.9992	0.9992	0.9992	0.9993	0.9993
3.2	0.9993	0.9993	0.9994	0.9994	0.9994	0.9994	0.9994	0.9995	0.9995	0.9995
3.3	0.9995	0.9995	0.9995	0.9996	0.9996	0.9996	0.9996	0.9996	0.9996	0.9997
3.4	0.9997	0.9997	0.9997	0.9997	0.9997	0.9997	0.9997	0.9997	0.9997	0.9998

Trigonometric Identities and Relationships

Definitions

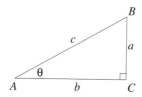

$$\sin\theta = \frac{\text{length of opposite side}}{\text{length of hypotenuse}} = \frac{a}{c}$$

$$\cos\theta = \frac{\text{length of adjacent side}}{\text{length of hypotenuse}} = \frac{b}{c}$$

$$\tan\theta = \frac{\text{length of opposite side}}{\text{length of adjacent side}} = \frac{a}{b} = \frac{\sin\theta}{\cos\theta}$$

$$\cot\theta = \frac{\text{length of adjacent side}}{\text{length of opposite side}} = \frac{b}{a} = \frac{1}{\tan\theta} = \frac{\cos\theta}{\sin\theta}$$

$$\sec\theta = \frac{\text{length of hypotenuse}}{\text{length of adjacent side}} = \frac{c}{b} = \frac{1}{\cos\theta}$$

$$\csc\theta = \frac{\text{length of hypotenuse}}{\text{length of opposite side}} = \frac{c}{a} = \frac{1}{\sin\theta}$$

Basic Transformations

$$\sin(-\theta) = -\sin\theta \qquad\qquad \cos(-\theta) = \cos\theta \qquad\qquad \tan(-\theta) = -\tan\theta$$

$$\sin\!\left(\theta \pm \tfrac{\pi}{2}\right) = \pm\cos\theta \qquad\qquad \cos\!\left(\theta \pm \tfrac{\pi}{2}\right) = \mp\sin\theta$$

Pythagorean Identities

$$\cos^2\theta + \sin^2\theta = 1 \qquad\qquad 1 + \tan^2\theta = \sec^2\theta \qquad\qquad 1 + \cot^2\theta = \csc^2\theta$$

Sum and Difference Identities

$$\cos(\alpha \pm \beta) = \cos\alpha \cdot \cos\beta \mp \sin\alpha \cdot \sin\beta \qquad\qquad \sin(\alpha \pm \beta) = \sin\alpha \cdot \cos\beta \pm \cos\alpha \cdot \sin\beta$$

$$\tan(\alpha \pm \beta) = \frac{\tan\alpha \pm \tan\beta}{1 - \tan\alpha \cdot \tan\beta}$$

Double- and Half-Angle Identities

$$\sin(2\theta) = 2\sin\theta \cdot \cos\theta \qquad\qquad \cos(2\theta) = \cos^2\theta - \sin^2\theta \qquad\qquad \tan(2\theta) = \frac{2\tan\theta}{1 - \tan^2\theta}$$

$$= 2\cos^2\theta - 1$$

$$= 1 - 2\sin^2\theta$$

$$\sin^2\theta = \frac{1}{2}\left[1 - \cos(2\theta)\right] \qquad\qquad \cos^2\theta = \frac{1}{2}\left[1 + \cos(2\theta)\right]$$

$$\sin\!\left(\frac{\theta}{2}\right) = \pm\sqrt{\frac{1 - \cos\theta}{2}} \qquad\qquad \cos\!\left(\frac{\theta}{2}\right) = \pm\sqrt{\frac{1 + \cos\theta}{2}}$$

$$\tan\!\left(\frac{\theta}{2}\right) = \pm\sqrt{\frac{1 - \cos\theta}{1 + \cos\theta}} = \frac{1 - \cos\theta}{\sin\theta} = \frac{\sin\theta}{1 + \cos\theta}$$

Function Sum Relations

$$\sin\alpha + \sin\beta = 2\sin\left(\frac{\alpha+\beta}{2}\right)\cos\left(\frac{\alpha-\beta}{2}\right)$$

$$\cos\alpha + \cos\beta = 2\cos\left(\frac{\alpha+\beta}{2}\right)\cos\left(\frac{\alpha-\beta}{2}\right)$$

$$a\sin\theta + b\cos\theta = \sqrt{a^2+b^2}\cdot\sin(\theta+\alpha),\ \text{where}\ \alpha = \tan^{-1}\left(\frac{b}{a}\right),$$

$$\text{if}\ a \geq 0,\ \text{or}\ \tan^{-1}\left(\frac{b}{a}\right) + \pi,\ \text{if}\ a < 0$$

Triangle Formulas

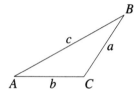

Law of Sines $\dfrac{a}{\sin A} = \dfrac{b}{\sin B} = \dfrac{c}{\sin C}$

Law of Cosines $c^2 = a^2 + b^2 - 2ab\cos C$

Sound Samples

Data Set for a Piano (Synthesized on a Casio Keyboard) Playing a C
Time is in milliseconds.

Time	Amplitude	Time	Amplitude	Time	Amplitude	Time	Amplitude
0.000	− 0.720	2.491	0.735	4.983	0.501	7.474	− 0.974
0.104	− 0.457	2.595	1.038	5.087	0.149	7.578	− 0.764
0.208	− 0.213	2.699	0.940	5.190	− 0.359	7.682	− 0.559
0.311	0.027	2.803	0.852	5.294	− 0.760	7.785	− 0.364
0.415	0.369	2.907	0.686	5.398	− 0.877	7.889	− 0.085
0.519	0.706	3.010	0.427	5.502	− 0.960	7.993	0.232
0.623	0.921	3.114	0.154	5.606	− 0.906	8.097	0.603
0.727	1.048	3.218	− 0.154	5.709	− 0.642	8.201	0.867
0.830	0.970	3.322	− 0.569	5.813	− 0.398	8.304	1.023
0.934	0.847	3.426	− 0.916	5.917	− 0.256	8.408	1.023
1.038	0.725	3.529	− 1.121	6.021	− 0.017	8.512	0.965
1.142	0.413	3.633	− 1.121	6.125	0.408	8.616	0.813
1.246	0.071	3.737	− 0.916	6.228	0.735	8.720	0.603
1.349	− 0.261	3.841	− 0.652	6.332	0.896	8.824	0.300
1.453	− 0.706	3.945	− 0.457	6.436	0.979	8.927	− 0.046
1.557	− 1.018	4.048	− 0.164	6.540	0.970	9.031	− 0.550
1.661	− 1.121	4.152	0.203	6.644	0.867	9.135	− 0.926
1.765	− 0.994	4.256	0.520	6.747	0.608	9.239	− 1.082
1.869	− 0.847	4.360	0.755	6.851	0.423	9.343	− 1.057
1.972	− 0.554	4.464	1.004	6.955	0.090	9.446	− 0.911
2.076	− 0.383	4.567	1.067	7.059	− 0.320	9.550	− 0.730
2.180	− 0.129	4.671	1.082	7.163	− 0.745	9.654	− 0.525
2.284	0.149	4.775	0.965	7.266	− 1.033	9.758	− 0.281
2.388	0.506	4.879	0.735	7.370	− 1.092	9.862	0.032

Data Set for a Piano (Synthesized on a Casio Keyboard) Playing a C#
Time is in milliseconds.

Time	Amplitude	Time	Amplitude	Time	Amplitude	Time	Amplitude
0.000	1.346	2.491	−1.258	4.983	0.296	7.474	0.901
0.104	1.175	2.595	−1.297	5.087	0.833	7.578	0.296
0.208	1.038	2.699	−1.253	5.190	1.175	7.682	−0.374
0.311	0.706	2.803	−1.165	5.294	1.350	7.785	−0.896
0.415	0.076	2.907	−0.970	5.398	1.399	7.889	−1.136
0.519	−0.564	3.010	−0.569	5.502	1.311	7.993	−1.170
0.623	−1.013	3.114	−0.027	5.606	1.140	8.097	−1.160
0.727	−1.194	3.218	0.579	5.709	0.716	8.201	−1.131
0.830	−1.223	3.322	1.057	5.813	0.066	8.304	−0.970
0.934	−1.219	3.426	1.321	5.917	−0.589	8.408	−0.628
1.038	−1.145	3.529	1.429	6.021	−1.043	8.512	−0.085
1.142	−0.886	3.633	1.360	6.125	−1.199	8.616	0.510
1.246	−0.462	3.737	1.223	6.228	−1.199	8.720	1.004
1.349	0.105	3.841	1.018	6.332	−1.140	8.824	1.263
1.453	0.637	3.945	0.506	6.436	−1.062	8.927	1.370
1.557	1.013	4.048	−0.183	6.540	−0.818	9.031	1.346
1.661	1.209	4.152	−0.808	6.644	−0.379	9.135	1.204
1.765	1.321	4.256	−1.150	6.747	0.188	9.239	1.013
1.869	1.238	4.360	−1.263	6.851	0.740	9.343	0.506
1.972	1.096	4.464	−1.243	6.955	1.145	9.446	−0.164
2.076	0.774	4.567	−1.228	7.059	1.355	9.550	−0.779
2.180	0.173	4.671	−1.121	7.163	1.404	9.654	−1.111
2.284	−0.486	4.775	−0.818	7.266	1.316	9.758	−1.223
2.388	−1.009	4.879	−0.315	7.370	1.179	9.862	−1.209

Data Set for a Piano (Synthesized on a Casio Keyboard) Playing a D
Time is in milliseconds.

Time	Amplitude	Time	Amplitude	Time	Amplitude	Time	Amplitude
0.000	1.082	2.491	− 0.862	4.983	1.013	7.474	− 0.974
0.104	0.779	2.595	− 0.720	5.087	0.935	7.578	− 0.911
0.208	0.427	2.699	− 0.481	5.190	0.745	7.682	− 0.784
0.311	− 0.017	2.803	− 0.139	5.294	0.423	7.785	− 0.569
0.415	− 0.486	2.907	0.300	5.398	− 0.022	7.889	− 0.247
0.519	− 0.828	3.010	0.740	5.502	− 0.506	7.993	0.169
0.623	− 0.974	3.114	1.023	5.606	− 0.877	8.097	0.613
0.727	− 0.965	3.218	1.136	5.709	− 1.043	8.201	0.916
0.830	− 0.862	3.322	1.106	5.813	− 1.013	8.304	1.053
0.934	− 0.681	3.426	0.984	5.917	− 0.891	8.408	1.053
1.038	− 0.413	3.529	0.745	6.021	− 0.720	8.512	0.935
1.142	− 0.027	3.633	0.335	6.125	− 0.466	8.616	0.706
1.246	0.432	3.737	− 0.169	6.228	− 0.090	8.720	0.325
1.349	0.808	3.841	− 0.642	6.332	0.354	8.824	− 0.154
1.453	1.028	3.945	− 0.940	6.436	0.764	8.927	− 0.618
1.557	1.116	4.048	− 1.023	6.540	1.018	9.031	− 0.921
1.661	1.077	4.152	− 0.979	6.644	1.111	9.135	− 1.028
1.765	0.930	4.256	− 0.882	6.747	1.077	9.239	− 0.965
1.869	0.613	4.360	− 0.691	6.851	0.945	9.343	− 0.852
1.972	0.164	4.464	− 0.423	6.955	0.657	9.446	− 0.662
2.076	− 0.330	4.567	− 0.017	7.059	0.252	9.550	− 0.383
2.180	− 0.730	4.671	0.432	7.163	− 0.252	9.654	0.017
2.284	− 0.921	4.775	0.794	7.266	− 0.691	9.758	0.466
2.388	− 0.965	4.879	0.989	7.370	− 0.921	9.862	0.852

Data Set for a Piano (Synthesized on a Casio Keyboard) Playing a D#
Time is in milliseconds.

Time	Amplitude	Time	Amplitude	Time	Amplitude	Time	Amplitude
0.000	0.901	2.491	− 0.672	4.983	0.647	7.474	− 0.330
0.104	0.745	2.595	− 0.486	5.087	0.315	7.578	0.002
0.208	0.466	2.699	− 0.237	5.190	− 0.046	7.682	0.374
0.311	0.183	2.803	0.154	5.294	− 0.388	7.785	0.691
0.415	− 0.203	2.907	0.540	5.398	− 0.589	7.889	0.867
0.519	− 0.506	3.010	0.794	5.502	− 0.681	7.993	0.926
0.623	− 0.598	3.114	0.945	5.606	− 0.701	8.097	0.847
0.727	− 0.750	3.218	0.984	5.709	− 0.647	8.201	0.618
0.830	− 0.662	3.322	0.843	5.813	− 0.496	8.304	0.305
0.934	− 0.574	3.426	0.559	5.917	− 0.198	8.408	− 0.056
1.038	− 0.354	3.529	0.208	6.021	0.173	8.512	− 0.364
1.142	− 0.037	3.633	− 0.134	6.125	0.525	8.616	− 0.574
1.246	0.291	3.737	− 0.427	6.228	0.764	8.720	− 0.667
1.349	0.657	3.841	− 0.593	6.332	0.867	8.824	− 0.647
1.453	0.808	3.945	− 0.623	6.436	0.852	8.927	− 0.574
1.557	0.886	4.048	− 0.593	6.540	0.716	9.031	− 0.369
1.661	0.823	4.152	− 0.496	6.644	0.437	9.135	− 0.066
1.765	0.647	4.256	− 0.266	6.747	0.081	9.239	0.296
1.869	0.349	4.360	0.120	6.851	− 0.261	9.343	0.657
1.972	− 0.061	4.464	0.457	6.955	− 0.515	9.446	0.877
2.076	− 0.344	4.567	0.745	7.059	− 0.667	9.550	0.970
2.180	− 0.618	4.671	0.930	7.163	− 0.711	9.654	0.916
2.284	− 0.730	4.775	0.960	7.266	− 0.657	9.758	0.750
2.388	− 0.740	4.879	0.862	7.370	− 0.540	9.862	0.481

Data Set for a Flute (Synthesized on a Casio Keyboard) Playing an A
Time is in milliseconds.

Time	Amplitude	Time	Amplitude	Time	Amplitude	Time	Amplitude
0.000	0.501	2.491	− 0.916	4.983	0.330	7.474	2.596
0.104	0.437	2.595	0.291	5.087	0.300	7.578	2.347
0.208	0.393	2.699	1.790	5.19	0.271	7.682	2.112
0.311	0.354	2.803	2.821	5.294	0.242	7.785	1.912
0.415	0.310	2.907	2.562	5.398	0.208	7.889	1.722
0.519	0.286	3.01	2.317	5.502	0.183	7.993	1.551
0.623	0.256	3.114	2.088	5.606	0.149	8.097	1.394
0.727	0.232	3.218	1.883	5.709	0.017	8.201	1.253
0.830	0.213	3.322	1.692	5.813	− 1.209	8.304	1.131
0.934	0.183	3.426	1.521	5.917	− 2.576	8.408	1.023
1.038	0.149	3.529	1.370	6.021	− 3.534	8.512	0.911
1.142	0.090	3.633	1.233	6.125	− 3.299	8.616	0.823
1.246	− 0.891	3.737	1.111	6.228	− 2.967	8.72	0.740
1.349	− 2.274	3.841	1.004	6.332	− 2.659	8.824	0.676
1.453	− 3.299	3.945	0.901	6.436	− 2.391	8.927	0.608
1.557	− 3.392	4.048	0.813	6.540	− 2.132	9.031	0.550
1.661	− 3.055	4.152	0.735	6.644	− 1.912	9.135	0.501
1.765	− 2.742	4.256	0.667	6.747	− 1.707	9.239	0.452
1.869	− 2.454	4.36	0.598	6.851	− 1.521	9.343	0.403
1.972	− 2.200	4.464	0.545	6.955	− 1.350	9.446	0.364
2.076	− 1.966	4.567	0.486	7.059	− 0.916	9.550	0.335
2.180	− 1.756	4.671	0.437	7.163	0.310	9.654	0.296
2.284	− 1.570	4.775	0.408	7.266	1.814	9.758	0.271
2.388	− 1.389	4.879	0.359	7.37	2.860	9.862	0.242

Data Set for a Flute (Synthesized on a Casio Keyboard) Playing a C
Time is in milliseconds.

Time	Amplitude	Time	Amplitude	Time	Amplitude	Time	Amplitude
0.000	0.232	2.491	0.930	4.983	− 0.965	7.474	0.286
0.104	− 1.419	2.595	0.833	5.087	0.466	7.578	0.247
0.208	− 2.957	2.699	0.755	5.190	2.107	7.682	0.169
0.311	− 3.524	2.803	0.676	5.294	2.576	7.785	− 1.082
0.415	− 3.167	2.907	0.618	5.398	2.337	7.889	− 2.723
0.519	− 2.840	3.010	0.559	5.502	2.112	7.993	− 3.602
0.623	− 2.547	3.114	0.510	5.606	1.902	8.097	− 3.231
0.727	− 2.283	3.218	0.457	5.709	1.717	8.201	− 2.904
0.830	− 2.044	3.322	0.403	5.813	1.541	8.304	− 2.601
0.934	− 1.824	3.426	0.369	5.917	1.394	8.408	− 2.332
1.038	− 1.619	3.529	0.335	6.021	1.253	8.512	− 2.078
1.142	− 1.111	3.633	0.286	6.125	1.131	8.616	− 1.863
1.246	0.364	3.737	0.247	6.228	1.013	8.720	− 1.639
1.349	1.990	3.841	0.134	6.332	0.906	8.824	− 0.984
1.453	2.625	3.945	− 1.311	6.436	0.828	8.927	0.496
1.557	2.376	4.048	− 2.825	6.540	0.740	9.031	2.083
1.661	2.151	4.152	− 3.553	6.644	0.672	9.135	2.571
1.765	1.937	4.256	− 3.192	6.747	0.603	9.239	2.317
1.869	1.751	4.360	− 2.860	6.851	0.550	9.343	2.088
1.972	1.575	4.464	− 2.562	6.955	0.496	9.446	1.883
2.076	1.419	4.567	− 2.298	7.059	0.442	9.550	1.692
2.180	1.272	4.671	− 2.059	7.163	0.403	9.654	1.531
2.284	1.150	4.775	− 1.834	7.266	0.354	9.758	1.375
2.388	1.033	4.879	− 1.619	7.370	0.320	9.862	1.243

Data Set for a Flute (Synthesized on a Casio Keyboard) Playing an F
Time is in milliseconds.

Time	Amplitude	Time	Amplitude	Time	Amplitude	Time	Amplitude
0.000	− 2.703	2.491	− 3.563	4.983	0.442	7.474	0.672
0.104	− 2.195	2.595	− 3.201	5.087	0.325	7.578	0.598
0.208	− 1.873	2.699	− 2.874	5.190	− 1.243	7.682	0.535
0.311	− 0.413	2.803	− 2.581	5.294	− 3.001	7.785	0.471
0.415	1.717	2.907	− 2.293	5.398	− 3.397	7.889	0.403
0.519	2.313	3.010	− 2.049	5.502	− 3.055	7.993	− 0.213
0.623	2.098	3.114	− 1.331	5.606	− 2.737	8.097	− 2.186
0.727	1.883	3.218	0.730	5.709	− 2.449	8.201	− 3.578
0.830	1.697	3.322	2.425	5.813	− 2.190	8.304	− 3.231
0.934	1.531	3.426	2.205	5.917	− 1.883	8.408	− 2.894
1.038	1.380	3.529	1.995	6.021	− 0.427	8.512	− 2.596
1.142	1.243	3.633	1.800	6.125	1.741	8.616	− 2.322
1.246	1.116	3.737	1.624	6.228	2.332	8.720	− 2.059
1.349	1.004	3.841	1.463	6.332	2.107	8.824	− 1.365
1.453	0.906	3.945	1.321	6.436	1.902	8.927	0.569
1.557	0.813	4.048	1.184	6.540	1.712	9.031	2.435
1.661	0.735	4.152	1.062	6.644	1.536	9.135	2.220
1.765	0.662	4.256	0.965	6.747	1.389	9.239	2.010
1.869	0.593	4.360	0.862	6.851	1.248	9.343	1.814
1.972	0.530	4.464	0.774	6.955	1.126	9.446	1.634
2.076	0.466	4.567	0.706	7.059	1.018	9.550	1.468
2.180	0.393	4.671	0.623	7.163	0.916	9.654	1.321
2.284	− 0.339	4.775	0.559	7.266	0.818	9.758	1.194
2.388	− 2.322	4.879	0.506	7.370	0.740	9.862	1.077

Data Set for a Organ (Synthesized on a Casio Keyboard) Playing an A
Time is in milliseconds.

Time	Amplitude	Time	Amplitude	Time	Amplitude	Time	Amplitude
0.000	− 0.921	4.983	− 0.203	9.965	0.760	14.95	0.760
0.208	− 0.891	5.190	2.024	10.17	1.121	15.16	1.199
0.415	0.081	5.398	0.550	10.38	0.955	15.36	1.189
0.623	1.751	5.606	0.066	10.59	0.789	15.57	0.125
0.830	1.258	5.813	0.891	10.80	0.667	15.78	− 1.980
1.038	0.808	6.021	0.550	11.00	0.510	15.99	− 1.922
1.246	1.028	6.228	1.287	11.21	− 1.140	16.19	− 2.195
1.453	0.828	6.436	0.974	11.42	− 1.297	16.40	0.364
1.661	0.711	6.644	− 1.922	11.63	− 2.088	16.61	1.204
1.869	0.569	6.851	− 1.770	11.83	− 0.173	16.82	− 0.916
2.076	− 0.628	7.059	− 2.264	12.04	0.730	17.02	− 0.081
2.284	− 1.282	7.266	− 0.579	12.25	− 1.414	17.23	0.242
2.491	− 1.663	7.474	1.492	12.46	− 0.828	17.44	1.209
2.699	− 0.779	7.682	− 0.447	12.66	0.002	17.65	1.824
2.907	1.092	7.889	− 0.598	12.87	0.481	17.86	0.628
3.114	− 0.672	8.097	0.300	13.08	1.565	18.06	− 0.921
3.322	− 1.111	8.304	0.633	13.29	0.164	18.27	− 0.598
3.529	0.017	8.512	1.941	13.50	− 1.736	18.48	− 0.720
3.737	− 0.061	8.720	1.448	13.70	− 1.473	18.69	1.814
3.945	1.453	8.927	− 0.955	13.91	− 1.433	18.89	1.473
4.152	1.072	9.135	− 0.589	14.12	0.930	19.10	0.564
4.360	− 1.770	9.343	− 0.916	14.33	1.932	19.31	1.057
4.567	− 1.810	9.550	1.048	14.53	0.042	19.52	0.847
4.775	− 1.673	9.758	1.629	14.74	0.569	19.72	0.725

Data Set for a Organ (Synthesized on a Casio Keyboard) Playing a C
Time is in milliseconds.

Time	Amplitude	Time	Amplitude	Time	Amplitude	Time	Amplitude
0.000	0.276	4.983	2.122	9.965	− 3.485	14.95	0.266
0.208	0.169	5.190	1.731	10.17	− 2.806	15.16	0.247
0.415	− 2.396	5.398	1.424	10.38	− 2.190	15.36	0.193
0.623	− 3.250	5.606	1.165	10.59	− 1.648	15.57	− 0.379
0.830	0.686	5.813	0.945	10.80	− 1.531	15.78	− 2.894
1.038	1.111	6.021	0.745	11.00	− 1.214	15.99	− 1.292
1.246	− 0.315	6.228	0.559	11.21	1.482	16.19	1.048
1.453	0.818	6.436	0.481	11.42	2.752	16.40	0.725
1.661	0.862	6.644	0.432	11.63	1.321	16.61	0.125
1.869	0.716	6.851	0.315	11.83	− 1.980	16.82	0.711
2.076	0.545	7.059	− 0.213	12.04	− 1.590	17.02	0.598
2.284	− 3.602	7.266	0.300	12.25	1.473	17.23	0.486
2.491	− 2.908	7.474	0.276	12.46	2.484	17.44	− 0.896
2.699	− 2.317	7.682	0.237	12.66	1.976	17.65	− 3.382
2.907	− 1.477	7.889	0.120	12.87	1.639	17.86	− 2.718
3.114	− 1.551	8.097	− 2.772	13.08	1.346	18.06	− 1.658
3.322	− 1.238	8.304	− 2.562	13.29	1.092	18.27	− 1.878
3.529	− 0.076	8.512	1.223	13.50	0.886	18.48	− 1.516
3.737	2.977	8.720	0.960	13.70	0.691	18.69	− 1.175
3.945	2.347	8.927	− 0.286	13.91	0.510	18.89	3.099
4.152	− 2.059	9.135	0.872	14.12	0.462	19.10	2.542
4.360	− 1.653	9.343	0.750	14.33	0.393	19.31	0.134
4.567	− 0.095	9.550	0.613	14.53	0.266	19.52	− 1.971
4.775	2.659	9.758	0.315	14.74	0.193	19.72	− 1.546

Data Set for a Organ (Synthesized on a Casio Keyboard) Playing an F
Time is in milliseconds.

Time	Amplitude	Time	Amplitude	Time	Amplitude	Time	Amplitude
0.000	− 1.907	4.983	2.244	9.965	− 1.770	14.95	1.028
0.208	− 0.457	5.190	1.761	10.17	− 1.375	15.16	0.828
0.415	0.134	5.398	− 0.589	10.38	− 1.199	15.36	0.046
0.623	2.034	5.606	− 1.858	10.59	1.018	15.57	− 2.225
0.830	1.658	5.813	− 0.320	10.80	1.761	15.78	− 1.306
1.038	1.004	6.021	− 0.388	11.00	1.004	15.99	− 1.433
1.246	− 1.643	6.228	1.101	11.21	− 1.722	16.19	− 0.637
1.453	− 0.720	6.436	1.932	11.42	− 0.911	16.40	1.790
1.661	− 0.852	6.644	1.536	11.63	− 0.320	16.61	1.433
1.869	− 0.217	6.851	− 0.408	11.83	0.217	16.82	− 0.515
2.076	1.868	7.059	− 1.346	12.04	1.814	17.02	− 1.726
2.284	1.507	7.266	− 0.535	12.25	1.487	17.23	− 0.437
2.491	0.647	7.474	− 0.559	12.46	1.116	17.44	− 0.300
2.699	− 1.453	7.682	0.940	12.66	− 1.443	17.65	0.935
2.907	− 0.418	7.889	1.643	12.87	− 0.593	17.86	1.600
3.114	− 0.354	8.097	1.311	13.08	− 0.623	18.06	1.277
3.322	0.354	8.304	− 0.447	13.29	− 0.237	18.27	− 0.183
3.529	1.453	8.512	− 0.755	13.50	1.570	18.48	− 1.302
3.737	1.170	8.720	− 0.071	13.70	1.272	18.69	− 0.388
3.945	− 0.545	8.927	− 0.027	13.91	0.916	18.89	− 0.535
4.152	− 2.532	9.135	1.346	14.12	− 1.189	19.10	0.872
4.360	− 1.399	9.343	1.126	14.33	− 0.056	19.31	1.365
4.567	− 1.658	9.550	0.862	14.53	− 0.061	19.52	1.087
4.775	− 0.564	9.758	− 1.839	14.74	0.564	19.72	− 0.281

Data Set for a Trumpet (Synthesized on a Casio Keyboard) Playing an A
Time is in milliseconds.

Time	Amplitude	Time	Amplitude	Time	Amplitude	Time	Amplitude
0.000	1.389	2.491	− 0.222	4.983	− 1.297	7.474	0.364
0.104	1.101	2.595	0.090	5.087	− 1.463	7.578	0.085
0.208	0.706	2.699	0.408	5.190	− 2.029	7.682	0.300
0.311	0.002	2.803	0.589	5.294	− 2.239	7.785	0.247
0.415	− 0.813	2.907	0.491	5.398	− 1.790	7.889	0.188
0.519	− 1.477	3.010	0.388	5.502	− 0.808	7.993	0.550
0.623	− 1.995	3.114	0.232	5.606	0.740	8.097	0.330
0.727	− 2.132	3.218	0.242	5.709	1.810	8.201	− 0.882
0.830	− 2.054	3.322	0.335	5.813	1.834	8.304	− 2.083
0.934	− 1.394	3.426	0.608	5.917	1.634	8.408	− 2.620
1.038	− 0.129	3.529	0.598	6.021	1.448	8.512	− 2.293
1.142	1.336	3.633	− 0.173	6.125	1.072	8.616	− 0.393
1.246	1.878	3.737	− 1.267	6.228	0.227	8.720	2.117
1.349	1.692	3.841	− 2.488	6.332	− 0.291	8.824	1.932
1.453	1.507	3.945	− 2.562	6.436	− 0.623	8.927	1.731
1.557	1.248	4.048	− 1.140	6.540	− 0.432	9.031	1.531
1.661	0.476	4.152	1.692	6.644	− 0.374	9.135	1.263
1.765	− 0.115	4.256	1.844	6.747	− 0.510	9.239	0.672
1.869	− 0.418	4.360	1.658	6.851	− 0.418	9.343	− 0.076
1.972	− 0.530	4.464	1.487	6.955	− 0.339	9.446	− 0.921
2.076	− 0.515	4.567	1.297	7.059	− 0.178	9.550	− 1.487
2.180	− 0.457	4.671	1.013	7.163	0.227	9.654	− 1.927
2.284	− 0.486	4.775	0.540	7.266	0.447	9.758	− 2.195
2.388	− 0.393	4.879	− 0.305	7.370	0.623	9.862	− 2.034

Data Set for a Trumpet (Synthesized on a Casio Keyboard) Playing a C
Time is in milliseconds.

Time	Amplitude	Time	Amplitude	Time	Amplitude	Time	Amplitude
0.000	0.481	2.491	1.516	4.983	0.120	7.474	− 0.847
0.104	1.810	2.595	1.360	5.087	− 0.149	7.578	0.061
0.208	1.751	2.699	1.194	5.190	− 0.388	7.682	1.150
0.311	1.560	2.803	0.725	5.294	− 0.330	7.785	1.873
0.415	1.389	2.907	− 0.066	5.398	0.183	7.889	1.712
0.519	0.945	3.010	− 0.496	5.502	0.886	7.993	1.526
0.623	− 0.037	3.114	− 0.906	5.606	1.243	8.097	1.297
0.727	− 0.672	3.218	− 1.385	5.709	0.857	8.201	0.510
0.830	− 0.843	3.322	− 1.673	5.813	− 0.159	8.304	− 0.276
0.934	− 0.432	3.426	− 1.678	5.917	− 1.502	8.408	− 0.706
1.038	0.105	3.529	− 1.482	6.021	− 1.883	8.512	− 0.745
1.142	0.247	3.633	− 1.072	6.125	− 0.647	8.616	− 0.466
1.246	− 0.042	3.737	− 0.237	6.228	1.629	8.720	− 0.071
1.349	− 0.364	3.841	0.945	6.332	1.556	8.824	0.066
1.453	− 0.242	3.945	1.819	6.436	1.399	8.927	− 0.051
1.557	0.252	4.048	1.707	6.540	1.223	9.031	− 0.256
1.661	0.818	4.152	1.516	6.644	0.633	9.135	− 0.173
1.765	1.121	4.256	1.302	6.747	− 0.213	9.239	0.193
1.869	0.794	4.360	0.676	6.851	− 0.584	9.343	0.755
1.972	− 0.154	4.464	− 0.129	6.955	− 0.974	9.446	1.057
2.076	− 1.463	4.567	− 0.672	7.059	− 1.492	9.550	0.716
2.180	− 2.010	4.671	− 0.974	7.163	− 1.741	9.654	− 0.330
2.284	− 0.872	4.775	− 0.691	7.266	− 1.722	9.758	− 1.702
2.388	1.473	4.879	− 0.056	7.370	− 1.448	9.862	− 1.966

Data Set for a Trumpet (Synthesized on a Casio Keyboard) Playing an F
Time is in milliseconds.

Time	Amplitude	Time	Amplitude	Time	Amplitude	Time	Amplitude
0.000	− 0.838	2.491	− 1.629	4.983	0.120	7.474	1.424
0.104	1.131	2.595	− 1.790	5.087	− 0.510	7.578	1.209
0.208	1.707	2.699	− 1.546	5.190	− 1.082	7.682	0.706
0.311	1.546	2.803	− 0.711	5.294	− 1.468	7.785	0.413
0.415	1.385	2.907	0.559	5.398	− 1.751	7.889	− 0.169
0.519	1.228	3.010	1.653	5.502	− 1.678	7.993	− 0.769
0.623	0.955	3.114	1.639	5.606	− 1.106	8.097	− 1.272
0.727	0.183	3.218	1.477	5.709	0.042	8.201	− 1.643
0.830	− 0.364	3.322	1.316	5.813	1.277	8.304	− 1.766
0.934	− 0.320	3.426	1.116	5.917	1.717	8.408	− 1.477
1.038	− 0.330	3.529	0.442	6.021	1.565	8.512	− 0.535
1.142	− 1.209	3.633	− 0.247	6.125	1.389	8.616	0.711
1.246	− 2.493	3.737	− 0.403	6.228	1.223	8.720	1.736
1.349	− 2.254	3.841	− 0.213	6.332	0.760	8.824	1.614
1.453	0.081	3.945	− 0.647	6.436	− 0.017	8.927	1.448
1.557	1.761	4.048	− 1.985	6.540	− 0.447	9.031	1.282
1.661	1.590	4.152	− 2.611	6.644	− 0.300	9.135	1.048
1.765	1.414	4.256	− 1.160	6.747	− 0.339	9.239	0.315
1.869	1.214	4.360	1.766	6.851	− 1.316	9.343	− 0.330
1.972	0.720	4.464	1.692	6.955	− 2.518	9.446	− 0.423
2.076	0.413	4.567	1.512	7.059	− 2.039	9.550	− 0.242
2.180	− 0.144	4.671	1.326	7.163	0.471	9.654	− 0.774
2.284	− 0.755	4.775	0.950	7.266	1.770	9.758	− 2.147
2.388	− 1.277	4.879	0.481	7.370	1.600	9.862	− 2.557

Answers to Selected Problems

Exercise Set 1.2, *page 12*

3. $R_n = R_{n-1} - \frac{1}{3}$, where R is in seconds. This model implies that after a long time the mile could be run in 0 or even negative seconds. Clearly, this does not make sense. There must be some lower limit to how fast the mile can be run.

6. $A_0 = 100$, $A_n = A_{n-1} + 0.05 \cdot A_{n-1}$

n	A_n	n	A_n
0	100	8	147.746
1	105	9	155.133
2	110.25	10	162.889
3	115.763	11	171.034
4	121.551	12	179.586
5	127.628	13	188.565
6	134.01	14	197.993
7	140.71	15	207.893

It takes about 15 years to double. An equation that describes this data is: $A_n = 100(1.05)^n$, where n is a non-negative integer representing number of years.

9. $B_0 = 1000$
$B_n = 1.10 \cdot B_{n-1} + 500$
$B_{10} = 10,562.45$
$B_{20} = 35,365.00$
$B_{30} = 99,696.41$

12. a. $B_0 = 500$, $B_n = 0.8 \cdot B_{n-1} + 50$, population stabilizes at 250.

b. Population will increase to 250 and then stabilize there.

c. The initial population does not affect the long-term trend. No matter what the value of B_0, the population stabilizes at 250 butterflies.

Exercise Set 1.3, *page 24*

3. a. 4.08%

b. $B(t) = 200 \cdot e^{0.04t}$, t in years, B in dollars

6. a. 12.68%   b. 5 years and 10 months
   c. $330.04   d. 11.94%

8. 100 years

10. a. $B_n = B_{n-1} + (-0.1B_{n-1} + 100) \cdot \Delta t$
    b. $B_0 = 200$

For $\Delta t = 1$:

t_n	B_n
0	200
1	280
2	352
3	416.8
4	475.12
5	527.608
6	574.847
7	617.362
8	655.626
9	690.064
10	721.057

For $\Delta t = 0.1$:

t_n	B_n
0	200
1	276.494
2	345.674
3	408.240
4	464.823
5	515.995
6	562.275
7	604.129
8	641.981
9	676.214
10	707.174

For $\Delta t = 0.01$:

t_n	B_n
0	200
1	276.166
2	345.081
3	407.434
4	463.851
5	514.897
6	561.082
7	602.871
8	640.681
9	674.891
10	705.844

c.

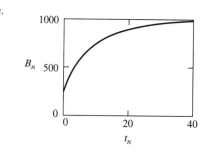

Exercise Set 2.1, *page 34*

2.

x_i	y_i	Δx	Δy	$\dfrac{\Delta y}{\Delta x}$
-3	9			
		1	-5	-5
-2	4			
		1	-3	-3
-1	1			
		1	-1	-1
0	0			
		1	1	1
1	1			
		1	3	3
2	4			
		1	5	5
3	9			

x_i	y_i	Δx	Δy	$\dfrac{\Delta y}{\Delta x}$
-6	36			
		2	-20	-10
-4	16			
		2	-12	-6
-2	4			
		2	-4	-2
0	0			
		2	4	2
2	4			
		2	12	6
4	16			
		2	20	10
6	36			

The larger Δx values are associated with larger average rates of change.

5. a. 6   b. 7
   c. 7.5   d. 7.95
   e. 7.999

Exercise Set 2.2.A, *page 39*

1. a.

e.

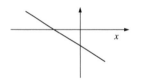

2. a.

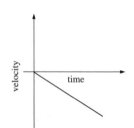

g.

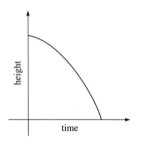
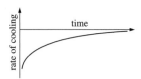

3. b.

c.

e.

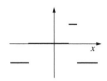

6. b.

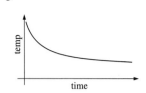

c.

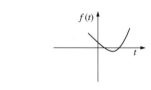

Exercise Set 2.2.B, *page 43*

2. b.

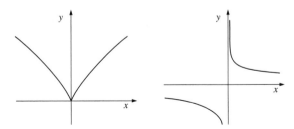

d.

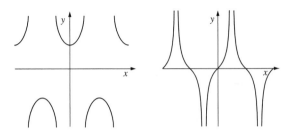

4. Each graph may vary somewhat, but should be of the following form.
   a. linear   b. parabolic
   c. increasing   d. decreasing
   e. increasing but flattening out

Exercise Set 2.3, *page 50*

2. a. For values close to $x = 1$, $\ln x \approx x - 1$. The closer x is to 1, the better the approximation is.
   b. 0.05. Near $x = 1$ use $y = x - 1$ as an approximation to $y = \ln x$.

4. a. The slopes approach 3.
   b. $y = 3x + 5$
   c. The line tangent to $y = \sqrt{x}$ matches the square root graph over a larger interval than the line tangent to $f(x) = x^2 + 3x + 5$ matches the parabola; this is because the square root function is more linear around $x = 9$ than the parabola is around $x = 0$.

6. a. ∞ (gets infinitely steep with positive slope)
   b. $-\infty$ (gets infinitely steep with negative slope)
   c. A tangent line does not exist at $x = 0$ because the slope does not approach one value around $x = 0$.

8. a. $\dfrac{12 - 6}{2} = 3$
   b. It does not exist because the slopes are not approaching the same value from both sides. From the left of $x = 2$, the function has a slope of 6; from the right the function has a slope of 0. So at $x = 2$, we cannot determine a value for the instantaneous rate of change.
   c. It does not exist.

Exercise Set 2.4, *page 58*

1. e.

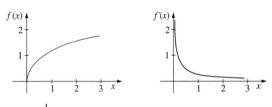

$$f'(x) = \frac{1}{2\sqrt{x}}$$

$$\lim_{\Delta x \to 0} \frac{\sqrt{x + \Delta x} - \sqrt{x}}{\Delta x} = \lim_{\Delta x \to 0} \left(\frac{\sqrt{x + \Delta x} - \sqrt{x}}{\Delta x} \right) \left(\frac{\sqrt{x + \Delta x} + \sqrt{x}}{\sqrt{x + \Delta x} + \sqrt{x}} \right)$$

$$= \lim_{\Delta x \to 0} \frac{x + \Delta x - x}{\Delta x(\sqrt{x + \Delta x} + \sqrt{x})}$$

$$= \lim_{\Delta x \to 0} \frac{1}{\sqrt{x + \Delta x} + \sqrt{x}} = \frac{1}{2\sqrt{x}}$$

4. $\dfrac{d}{dx}(x^n) = nx^{n-1}$, true for all integers n.

6. $\dfrac{d}{dx}(4x^2) = \lim_{\Delta x \to 0} \dfrac{4(x + \Delta x)^2 - 4x^2}{\Delta x}$

$$= \lim_{\Delta x \to 0} \frac{4x^2 + 8x\Delta x + 4(\Delta x)^2 - 4x^2}{\Delta x}$$

$$= \lim_{\Delta x \to 0} 8x + 4\Delta x = 8x$$

This result is 4 times the derivative of $f(x) = x^2$. The 4 causes all y values to be four times as great so the slopes of the tangent lines change accordingly.

9. The function will flatten out at $x = a$. This could be a turning point.

Exercise Set 2.5.A, *page 67*

1. a. $f'(x) = e^x$ c. $f'(x) = 2x$
   e. $f'(x) = -2x$ g. $f'(x) = \dfrac{3}{2\sqrt{x}}$

3. a. $\dfrac{dy}{dx} = \cos\left(x - \dfrac{\pi}{4}\right)$ c. $\dfrac{dy}{dx} = 5(4x)^4 \cdot 4$

   e. $\dfrac{dy}{dx} = \dfrac{3}{\sqrt{6x}}$ g. $\dfrac{dy}{dx} = \cos(x + 3)$

   i. $\dfrac{dy}{dx} = 3(-6x + 7)^2 \cdot (-6)$ k. $\dfrac{dy}{dx} = -0.5\sin(0.5x - 1)$

   m. $\dfrac{dy}{dx} = \dfrac{5}{2\sqrt{5x + 3}}$ o. $\dfrac{dy}{dx} = \dfrac{2}{2x - 5}$

7. b. $\dfrac{df(x + k)}{dx} = \lim\limits_{\Delta x \to 0} \dfrac{f(x + \Delta x + k) - f(x + k)}{\Delta x}$.

   Let $u = x + k$, so $\Delta u = \Delta x$. Then

   $$\lim_{\Delta x \to 0} \frac{f(x + k + \Delta x) - f(x + k)}{\Delta x} = \lim_{\Delta u \to 0} \frac{f(u + \Delta u) - f(u)}{\Delta u}$$
   $$= f'(u) = f'(x + k)$$

Exercise Set 2.5.B, *page 72*

1. a. $\dfrac{dy}{dx} = 3 \cdot \cos x + 2 \cdot \sin x$ c. $\dfrac{dy}{dx} = -\dfrac{1}{x^2} + \dfrac{1}{2\sqrt{x}}$

   e. $\dfrac{dy}{dx} = 1 - e^{-x}$

2. a. $\dfrac{dy}{dx} = x \cdot \cos x + \sin x$

   c. $\dfrac{dy}{dx} = \dfrac{1}{x} \cdot e^x + e^x \cdot \left(-\dfrac{1}{x^2}\right)$

   e. $\dfrac{dy}{dx} = (\cos x)(10x + 3) + (5x^2 + 3x - 1)(-\sin x)$

   g. $\dfrac{dy}{dx} = \sqrt{x} \cdot \ln 2 \cdot 2^x + 2^x \cdot \dfrac{1}{2\sqrt{x}}$

   h. $\dfrac{dy}{dx} = -2 \cdot \dfrac{1}{x} \cdot \sin(2x) - \cos(2x) \cdot \dfrac{1}{x^2}$

   i. $\dfrac{dy}{dx} = \dfrac{4}{x}\left(\dfrac{1}{2\sqrt{x}} + e^x\right) + (1 + \sqrt{x} + e^x)\left(-\dfrac{4}{x^2}\right)$

   k. $\dfrac{dy}{dx} = 7e^{-2x} - 2(7x - 4)e^{-2x}$

5. $y = 12.5x + 12.5$

7. a. $P'(t) = \left(\dfrac{25}{t}\right)(\cos t) + (\sin t)\left(-\dfrac{25}{t^2}\right)$
   $P'(1) \approx -7.53$, $P'(2) \approx -10.88$, $P'(3) \approx -8.64$,
   $P'(4) \approx -2.90$, $P'(5) \approx 2.38$

   b. It decreases rapidly in the first 3 hours (indicated by negative derivative) but then starts decreasing less rapidly and
   somewhere between 4 and 5 hours after opening the number starts increasing (indicated by positive derivative).

Exercise Set 2.5.C, *page 78*

1. a. $5(x^2 + 5x - 1)^4(2x + 5)$ c. $e^{x^2 + 7} \cdot 2x$

   e. $-\dfrac{2x - 1}{(x + 1)^2} + \dfrac{2}{x + 1}$ g. $\dfrac{7}{x} + \dfrac{1}{7x} \cdot 7 \left(\text{or } \dfrac{8}{x}\right)$

   i. $-x^3 \sin(x^2) \cdot (2x) + \cos(x^2) \cdot (3x^2)$

   k. $\dfrac{-2}{(1 + \ln x)^3} \cdot \dfrac{1}{x}$

   m. $\dfrac{1}{2(e^x \cdot \cos x)^{1/2}}\left(-e^x \cdot \sin x + \cos x \cdot e^x\right)$

3. a. $f'(x) = n \cdot x^{n-1}$, $f''(x) = n \cdot (n - 1) \cdot x^{n-2}$
   b. $f'(x) = \cos x$, $f''(x) = -\sin x$
   c. $f' = e^{-x^2} \cdot (-2x)$

   $f'' = e^{-x^2} \cdot (-2) + (-2x) \cdot e^{-x^2}(-2x)$
   $= -2e^{-x^2} + 4x^2 e^{-x^2}$

6. a. The oxygen content is lowest at $t = 10$.
   b. Over $0 \le t \le 15$, $C'(t)$ is greatest at $t = 15$.

Exercise Set 2.6, *page 87*

2. We know $\cos x = \sin\left(x + \dfrac{\pi}{2}\right)$. Since we showed (by using the
   limit definition of the derivative) in problem 1 that
   $\dfrac{d}{dx}(\sin x) = \cos x$ then

   $$\frac{d}{dx}\left[\sin\left(x + \frac{\pi}{2}\right)\right] = \cos\left(x + \frac{\pi}{2}\right).$$

   Then, using a trig identity, we see that

   $$\cos\left(x + \frac{\pi}{2}\right) = \cos x \cdot \cos \frac{\pi}{2} - \sin x \cdot \sin \frac{\pi}{2} = -\sin x.$$

5. a. does not exist

    c. does not exist if $m < n$, 0 if $m > n$, 1 if $m = n$

    e. 0

    g. does not exist

    i. 1

    k. 0

6. a. 0

    c. ∞ (See part e.)

Exercise Set 2.7.A, *page 92*

2. a. $\lim_{x \to 1} 2x + 3 = 5$, so $L = 5$.

    b. $f(x)$ is between 4.8 and 5.2, so $f(x)$ is within 0.2 of L.

    c. x is between 0.95 and 1.05, so x is within 0.05 of 1.

4. a.

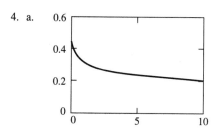

    b. 0.25

    c. $q(x)$ is between 0.251 and 0.249, so $q(x)$ is within 0.001 of the limit (0.25) found in part b.

    d. $\lim_{x \to 4} \dfrac{\sqrt{x} - 2}{x - 4} = \lim_{x \to 4} \dfrac{\sqrt{x} - 2}{\left(\sqrt{x} - 2\right) \cdot \left(\sqrt{x} + 2\right)}$

         $= \lim_{x \to 4} \dfrac{1}{\sqrt{x} + 2} = \dfrac{1}{4}$

6. If x is between 2.9833 and 3.0166, $f(x)$ is between 5.9 and 6.1.

8. a. $\lim_{x \to 2} \dfrac{x^3 - 8}{x - 2}$ represents the limiting value of the slopes of secants between $(2, 8)$ and (x, x^3); this limiting value gives the slope of the tangent at $(2, 8)$.

    b. $\lim_{x \to 2} \dfrac{x^3 - 8}{x - 2}$ is equal to the derivative of $g(x)$ evaluated at $x = 2$; since $g'(x) = 3x^2$, $g'(2) = 12$.

Exercise Set 2.7.B, *page 97*

2. 6

3. a. $\lim_{t \to 0^+} g(t) = \lim_{t \to 0^+} t^2 = 0$

      $\lim_{t \to 0^-} g(t) = \lim_{t \to 0^-} t + 1 = 1$

    b. $\lim_{t \to 0} g(t)$ does not exist because $\lim_{t \to 0^+}$ and $\lim_{t \to 0^-}$ are different; the function is discontinuous at $t = 0$.

5. Yes. $\lim_{x \to 0} f(x) = \lim_{x \to 0} \dfrac{\sin x}{x} = 1$. Assigning the function the value of 1 at $x = 0$ makes the limit equal the value of the function and provides continuity.

8. No. If a function is continuous at $x = 1$, then the limit will exist there and, in fact, the limit will equal $f(1)$.

10. $a = -\dfrac{1}{4}$, $b = 4$

END OF CHAPTER EXERCISES
Review and Extensions, *page 100*

2. $y = x - 0.285$

3. a. $x(\ln 2)(2^x) + 2^x$

    b. $(\tan x)(-e^{-x}) + e^{-x}(\sec^2 x)$

    c. $\dfrac{-x^2 + 8x - 8}{(x^2 - 8)^2}$

    d. $\dfrac{1}{3}(x^2 + 4x + 7)^{-2/3}(2x + 4)$

    e. $-(\sin x^2 + x)^{-2}\left[(\cos x^2)2x + 1\right]$

5. a. does not exist

    b. 0

    c. $\dfrac{2}{3}$

6. a.

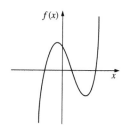

    b.

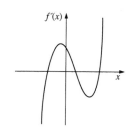

8. 0

9. $\dfrac{d}{dx}(x^n) = nx^{n-1}$ $\dfrac{d}{dx}(e^x) = e^x$

$\dfrac{d}{dx}(\sin x) = \cos x$ $\dfrac{d}{dx}(\ln x) = \dfrac{1}{x}$

$\dfrac{d}{dx}(\cos x) = -\sin x$ $\dfrac{d}{dx}(a^x) = a^x(\ln a)$

$\dfrac{d}{dx}(\tan x) = \sec^2 x$

10. 0.5

Exercise Set 3.1, *page 108*

1. a. $\dfrac{dy}{dx} = 2xe^{x^2}$; $\dfrac{d^2y}{dx^2} = 2e^{x^2} + 4x^2 e^{x^2}$

    c. $\dfrac{dy}{dx} = -\dfrac{1}{1-x}$; $\dfrac{d^2y}{dx^2} = -\dfrac{1}{(1-x)^2}$

2. a. $\dfrac{d^n f}{dx^n} = n!$ c. $\dfrac{d^n f}{dx^n} = (-1)^{(n+1)} \cdot \dfrac{(n-1)!}{x^n}$

    e. $\dfrac{d^n f}{dx^n} = \sin\left(x + n \cdot \dfrac{\pi}{2}\right)$

Exercise Set 3.2, *page 119*

1. a. Maximum is $(-1, -2)$; minimum is $(1, -6)$; inflection point is $(0, -4)$.

    c. Maximum is $\left(-\sqrt{\dfrac{1}{3}},\ 1 + \dfrac{2}{3}\sqrt{\dfrac{1}{3}}\right)$; minimum is $\left(\sqrt{\dfrac{1}{3}},\ 1 - \dfrac{2}{3}\sqrt{\dfrac{1}{3}}\right)$; inflection point is $(0, 1)$.

    e. Maximum is $(0, 1)$; inflection points are $\left(\sqrt{\dfrac{1}{2}},\ \sqrt{\dfrac{1}{e}}\right)$ and $\left(-\sqrt{\dfrac{1}{2}},\ \sqrt{\dfrac{1}{e}}\right)$.

    g. There are no maximum or minimum points; inflection points are $(k\pi,\ k\pi)$ where k is an integer.

    i. Maximum is $(0, 4)$; inflection points are $\left(\pm\sqrt{\dfrac{1}{3}},\ 3\right)$.

    k. Minimum is $\left(-1.5,\ \ln\left(\dfrac{7}{4}\right)\right)$; inflection points are $\left(\dfrac{-3 \pm \sqrt{7}}{2},\ \ln\left(\dfrac{7}{2}\right)\right)$.

    m. Maxima is $\left(\dfrac{\pi}{2} + k\pi,\ 1\right)$, where k is an integer; minima is $(k\pi,\ 0)$, where k is an integer; there are no inflection points.

---

4. $x = \pm\sqrt{4.5}$

6. a. Turning points at $x = \pm 0.8603,\ \pm 3.4256,\ \pm 6.4373,$ *etc.* These points cannot be found analytically. Zoom and trace or a numerical method is needed to approximate these values. The envelope curves touch just to the left of each turning point for $x > 0$ and just to the right for $x < 0$. The envelope curves do not touch at the turning points.

7. $r = \left(\dfrac{V}{4\pi}\right)^{1/3}$, $h = \left(\dfrac{V}{\pi}\right)^{1/3} \cdot 4^{2/3}$

9. a. $v = \dfrac{dx}{dt} = 2\pi t\cos(\pi t^2)$. The particle moves the fastest when v is a maximum or a minimum; graph v to find that $v = \mp 2\pi$ at $t = \pm 1$.

    b. $a = 2\pi\cos(\pi t^2) - (2\pi t)^2\sin(\pi t^2)$. Acceleration is greatest when a is maximum; graph a to find that $a = 2\pi$ at $t = 0$.

    c. Particle changes direction when $x'(t) = v$ changes sign. $t = 0,\ \dfrac{\sqrt{2}}{2},\ -\dfrac{\sqrt{2}}{2}$

    d. Particle moves to the left when v is negative: $-\dfrac{\sqrt{2}}{2} < t < 0$ and $\dfrac{\sqrt{2}}{2} < t \le 1$

11. height $= \dfrac{2R}{\sqrt{3}}$, radius $= \sqrt{\dfrac{2}{3}}R$

13. a.

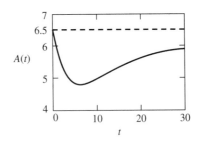

    b. Normal pH level is 6.5. It will take 204 minutes for the pH level to return to 6.4.

    c. Mercury should suggest brushing as soon as possible before the acid level is at its greatest, since it takes so long to return to normal.

    d. Acid level increases (pH gets lower) until $t = 6$; it decreases for $t > 6$.

    e. $f'(t)$ is negative for $t < 6$, positive for $t > 6$; therefore the mouth begins recovery at $t = 6$.

Exercise Set 3.3, *page 127*

2. 7500 items

4. 3000 units

7. Note that $P'(x) = pQ'(x) - c$, so $P''(x) = pQ''(x)$. If $Q'(x)$ has a maximum at $x = a$, then $Q''(a) = 0$ and $Q''(x) > 0$ for $x < a$ and $Q''(x) < 0$ for $x > a$. Thus $P''(a) = pQ''(a) = p \cdot 0 = 0$, $P''(x) > 0$ for $x < a$, and $P''(x) < 0$ for $x > a$. Therefore $P'(a)$ is a maximum. Since $P'(a)$ is a maximum at the point of diminishing returns, profits are increasing the fastest at this point.

Exercise Set 3.4, *page 131*

2. $a = 6.5$, $b = 9$, and $c = 9$

3. $E(v) = cD\dfrac{v^5}{v - r}$

Exercise Set 3.5.A, *page 136*

1. b. $t = 1.644$
   c. $x = 20.086$

5. This scheme generates a sequence of x values that approach a zero of $f'(x)$.

7. $x_n = \dfrac{1}{3}\left(2x_{n-1} + \dfrac{k}{x_{n-1}^2}\right)$

Exercise Set 3.5.B, *page 143*

1. The zeros of this function are −3 and 3. The only value that will cause Newton's method to fail on the first iteration is $x = 0$. All starting values greater than zero converge to 3, and all starting values less than zero converge to −3.

Exercise Set 3.6, *page 148*

1. $\dfrac{dr}{dt} = \dfrac{1}{4\pi}$ m/min and $\dfrac{dC}{dt} = \dfrac{1}{2}$ m/min

3. $\dfrac{dP}{dt} = 8.75 \ \dfrac{\text{lb/in}^2}{\text{sec}}$

5. Height is decreasing at a rate of $\dfrac{5}{9\pi}$ cm/sec.

8. a. $\dfrac{d\theta}{dt} \approx 0.075$ rad/sec, approximately 4.3 degrees/sec

   b. $\dfrac{d\theta}{dt} \approx 0.0946$ rad/sec

11. $\dfrac{dA}{dt} \approx 78540 \ \text{ft}^2/\text{hr}$

13. $\dfrac{ds}{dt} = k(-h)$

15. a. $\dfrac{dd}{dt} = -2.99$, so demand is decreasing by almost 3000 items per year.

    b. $\dfrac{dp}{dt} \approx 1.91$, so price is increasing at a rate of $1.91 per year.

Exercise Set 3.7, *page 157*

3. a. $\dfrac{dy}{dx} = \dfrac{-(2x + y)}{x - \dfrac{1}{2\sqrt{y}}}$

   c. $\dfrac{dy}{dx} = \dfrac{1}{x}$

   e. $\dfrac{dy}{dx} = \dfrac{1}{\sqrt{1 - x^2}}$

   g. $\dfrac{dy}{dx} = -\dfrac{1}{x\sqrt{1 - (\ln x)^2}}$

   i. $\dfrac{dy}{dx} = (\sin x)^x[x \cot x + \ln(\sin x)]$

6. $\sqrt{ab + b^2}$ cm

END OF CHAPTER EXERCISES
Review and Extensions, *page 177*

1.

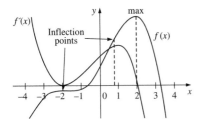

2. a. $f(x) = x^4$
   b. $f''(x)$ must change signs at $x = a$.

3. relative maximum at $\left(e, \dfrac{1}{e} \right)$

   point of inflection at $\left(e^{3/2}, \dfrac{3}{2e^{3/2}} \right)$.

4. $h = \sqrt[3]{\dfrac{6V}{\pi}}$ $r = \sqrt{\dfrac{3V}{\pi \left(\dfrac{6V}{\pi} \right)^{1/3}}}$

5. $a = 3,\ b = -2$

6. a. $x_0 = 1$
      $x_3 \approx 1.1640$

7. Approximately -16 ft/sec

8. a. $\dfrac{dy}{dx} = \dfrac{-4}{\sqrt{1 - 16x^2}}$ b. $\dfrac{dy}{dx} = x^{\sqrt{3x}} \left(\dfrac{\sqrt{3}}{\sqrt{x}} + \dfrac{3 \ln x}{2\sqrt{3x}} \right)$

9. $y = -x$

10. $y = x - 4$

11. a.

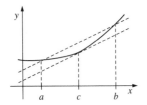

b. This theorem tells us that at some point on the interval (a, b) the slope of the tangent line is the same as the slope of the secant line.

c. $c = \sqrt{3}$

d. $c = \dfrac{1}{2}(x_2 + x_1)$

e. The Mean Value Theorem does not apply because the function f is not differentiable for all values of x on $(-2, 2)$. Specifically $f'(0)$ does not exist.

Exercise Set 4.1, *page 183*

1. $y = x^2 + 3x + c$
   $y = x^2 + 3x + 4$

3. $y = 2\sqrt{x} + c$
   $y = 2\sqrt{x} - 2$

5. $y = \dfrac{x^3}{3} + \dfrac{3x^2}{2} + c$
   $y = \dfrac{x^3}{3} + \dfrac{3x^2}{2}$

Exercise Set 4.2, *page 189*

1. d. $\dfrac{dy}{dx} = \cos x$

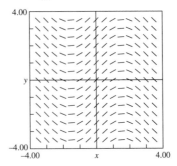

4. d. $\dfrac{dy}{dx} = \cos x$
   $y = \sin x + C$

The slope field appears to have waves throughout with maxima at $x = \frac{\pi}{2} + 2k\pi$ and minima at $x = \frac{3\pi}{2} + 2k\pi$. This is consistent with the maxima and minima of the sine curve regardless of the value of C.

5. At this point in the course, students might not be able to solve any of the differential equations beginning with part f.

   k. The solutions of this differential equation appear to be functions with an oblique asymptote around $y = -x - 1$. For initial conditions above this line, the solutions are concave up. This is easy to see since the second derivative is

   $$\frac{d^2 y}{dx^2} = 1 + (x + y),$$

   so when $x + y > -1$ we are above the asymptote and the second derivative is positive. For initial conditions below the line, the solutions are decreasing functions and concave down as both derivatives will be negative.

8. If the initial temperature is less than 25 degrees, the graph of the particular solution will be an increasing function with a horizontal asymptote at $T = 25$. This behavior results from the fact that for $T < 25$ the factor $(T - 25)$ will be negative, thus producing positive values of $\frac{dT}{dt}$. As T values approach 25, these negative rates of change approach zero, thus producing the leveling off at $T = 25$.

10. a. If the initial value of the population is greater that 80, the graph of the particular solution will decrease and level off at $P = 80$. This results from the fact that the factor $\left(1 - \frac{P}{80}\right)$ is negative while the other two factors in the differential equation are positive.

    b. If the initial population is between 10 and 80, the graph of the particular solution will be an increasing function with two horizontal asymptotes, one at $P = 80$ and one at $P = 10$. If P values are small (close to 10), the population increases slowly because the factor $\left(\frac{P}{10} - 1\right)$ is close to zero. The population increases at a faster rate for population values midway between $P = 10$ and $P = 80$ as no factor is close to zero. As P values approach 80, the graph levels off as the factor $\left(1 - \frac{P}{80}\right)$ approaches zero. This causes the rate of growth to approach zero.

    c. If the initial value of the population is less than 10 but greater than 0, the particular solution will decrease and level off at $P = 0$. This results from the fact that the factor $\left(\frac{P}{10} - 1\right)$ is

negative while the other two factors are positive. Initial values of P less than 0 would not make sense in the context of this problem even though the differential equation would work. The slopes for P values less than 0 are positive since two of the factors are negative and one positive.

Exercise Set 4.3, *page 199*

1.

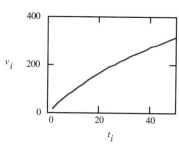

$$t_{25} = 50 \text{ and } v_{25} = 313.406$$

3.

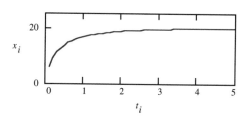

5. a. If $\Delta x = 0.1$, $x_{90} = 10$, $y_{90} = 4.18$

   b. $y = \sqrt{x} + 1$, $y(10) = 4.162$

   c. Approximations from Euler's method will be too large due to the fact that this curve is concave down. Every tangent line will overshoot the actual function, making every approximation a little larger than the actual value.

7. a. $\frac{dP}{dt} = 0.03P + 100000$

   b. With $\Delta t = 0.5$, in 146 years $P > 1,000,000,000$.

Exercise Set 4.5, *page 219*

1. a. $q(x) = 0.5x^2 + x + 1$
   c. $q(x) = x^2 - x + 1$

3. $v_i = v_{i-1} + \dfrac{dv}{dt} \cdot \Delta t + \dfrac{1}{2}\dfrac{d^2v}{dt^2} \cdot (\Delta t)^2$

   $x_i = x_{i-1} + \dfrac{dx}{dt} \cdot \Delta t + \dfrac{1}{2}\dfrac{d^2x}{dt^2} \cdot (\Delta t)^2$

   $v_i = v_{i-1} + \left(-\dfrac{k}{m}x_{i-1}\right) \cdot \Delta t + \dfrac{1}{2}\left(-\dfrac{k}{m}v_{i-1}\right) \cdot (\Delta t)^2$

   $x_i = x_{i-1} + v_{i-1} \cdot \Delta t + \dfrac{1}{2}\left(-\dfrac{k}{m}x_{i-1}\right) \cdot (\Delta t)^2$

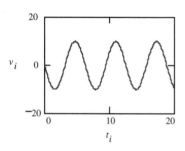

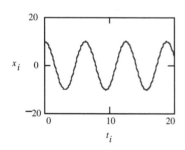

The graphs using the quadratic approximations are not expanding as quickly as those in the text using the linear approximations. Because the graphs for harmonic motion have so many turning points, the "tangent parabolas" provide much better approximations than tangent lines.

5.
$$p(x) = \frac{1}{6}f'''(a)(x-a)^3 + \frac{1}{2}f''(a)(x-a)^2 + f'(a)(x-a) + f(a)$$

Exercise Set 4.6, *page 228*

3. When $t = 1$ the path of the projectile makes an angle of $0.6°$ with the horizontal.

   When $t = 2$ the measure of the acute angle formed by the horizontal and the tangent line, measuring in the clockwise direction, is $8.8°$.

   When $t = 5$ the path of the projectile makes an angle of $33.1°$ with the horizontal, measuring in the clockwise direction.

6. a. $\dfrac{dy}{dt} = 0$ if $t = \dfrac{\pi}{4} + k\pi$, k an integer. The points are $(-2\sqrt{2}, -1) \approx (-2.828, -1)$; $(-\sqrt{2}, 1) \approx (-1.414, 1)$; $(\sqrt{2}, 1) \approx (1.414, 1)$;   and $(2\sqrt{2}, -1) \approx (2.828, -1)$. Since the slope of a tangent line is given by $\dfrac{dy}{dx} = \dfrac{dy/dt}{dx/dt}$, slopes will equal zero and tangent lines will be horizontal at these points.

   b. $\dfrac{dx}{dt} = 0$ if $t \approx -1.249 + k\pi$, k an integer. The points are approximately $(3.162, -0.6)$ and $(-3.162, -0.6)$. Slopes of tangent lines will be undefined at these points, so these tangent lines will be vertical and equations will be of the form $x = k$.

8. $y = -0.094(x - 0.50) + 2.98$ and $y = 0.094(x - 0.50) - 2.98$.

Exercise Set 4.8, *page 242*

1. a. $p(x) = 1 + \dfrac{x}{2} - \dfrac{x^2}{2^2 \cdot 2!} + \dfrac{3 \cdot x^3}{2^3 \cdot 3!} - \dfrac{15 \cdot x^4}{2^4 \cdot 4!}$

3. a. $p(x) = 1 + x + \dfrac{x^2}{2} + \dfrac{x^3}{3!} + \dfrac{x^4}{4!} + \dfrac{x^5}{5!} + \cdots + \dfrac{x^n}{n!}$
   c. As the degree of the Taylor polynomial increases, there is a corresponding increase in the set of x values over which the polynomial is a good approximation for $f(x) = e^x$. It appears that we could get a good approximation for any x value if we use enough terms in the polynomial.

5. a. $p(x) = 1 + x - \dfrac{x^2}{2} + \dfrac{x^3}{3} - \dfrac{x^4}{4} + \dfrac{x^5}{5}$
   b. $p(0.75) \approx 1.5777$
   c. $y_3 \approx 1.6167$
   d. $f(0.75) = \ln(1.75) + 1 \approx 1.5596$
   e. To improve the accuracy using Taylor polynomials you would need to add more terms and thus increase the

degree. To improve the accuracy using the linear Euler's method you would need to increase the number of iterations, that is, decrease the size of Δx.

END OF CHAPTER EXERCISES

Review and Extensions, *page 243*

1. a. $y = 2x + \frac{1}{3}\sin(3x) + 1$ b. $P = 10e^{0.085t}$

2. a. $f_4 = 109.00$

    d. $\frac{df}{dt} = 0.05f\left(1 - \frac{f}{800}\right) - 0.10f$

4. $y_1 = 1.875$
    $y_2 = 3.922$

5. $p(x) = 1 + 2(x-1) + \frac{1}{2}(x-1)^2$

6. a.   slope ≈ -3.58
    b.   Any point using $t \approx \pm 1.23 + 2k\pi$, where k is an integer.

7. a. $p(x) = 1 - \frac{1}{2}x + \frac{1}{8}x^2 - \frac{1}{48}x^3$
    b. $x \le 2.6451$

8. a. $p(x) = 3 + \frac{1}{6}(x-9) - \frac{1}{216}(x-9)^2 +$
             $\frac{1}{3888}(x-9)^3 - \frac{5}{279936}(x-9)^4$

    b. $\sqrt{8} \approx 2.8284286$
         relative error $\approx -5.36 \cdot 10^{-7}$

Exercise Set 5.1, *page 251*

1. a.   1   c.   369.333
    e.   3   g.   2.12

Exercises Set 5.2, *page 256*

2. 10.667   3.   4.5

5. 0.414   7.   8

9. 8.75

Exercise Set 5.3, *page 261*

1. a.   Displacement $= -1.5$ Total distance $= 5.167$
    c.   Displacement $= 0$ Total distance $= 4$

3. Between the points of intersection at $x = -1$ and $x = 1$, the graph of $f(x)$ is always above (or equal to) the graph of $g(x)$. Because the graphs are symmetric, the area to the left of the y-axis is equal to that on the right. Therefore, the total area is twice the area on the right: $2\int_0^1 (x^2 - x^4)dx$.

5. 3300

Exercise Set 5.4, *page 267*

1. Lowest fifth   3.82
    Second fifth   9.52
    Third fifth   15.82
    Fourth fifth   24.02
    Highest fifth   46.82

3. As n approaches 1, the Gini index approaches 0.
    As n approaches ∞, the Gini index approaches 1.

5. $1 - \frac{2}{n+1}$

7. For 1980, the Lorenz curve is $y = x^{0.756}$; area $= 0.0695$ and Gini index $= 0.1390$. For 1990, the Lorenz curve is $y = x^{0.613}$; area $= 0.1200$ and Gini index $= 0.2400$. The population distribution is more uneven in 1990 than it was in 1980.

Exercise Set 5.5, *page 276*

1. $\frac{1}{2}e^{3x^2+4x} + C$ 3. $\frac{2}{3}(1 + f(x))^{3/2} + C$

5. $\frac{1}{2}e^{x^2+2x} + C$ 7. $\frac{1}{2}e^{x^2} + C$

9. 2.899   11. $-\frac{1}{3}(4-x^2)^{3/2} + C$

13. $-\frac{1}{3}e^{-x^3} + C$ 15. $\tan t - t + C$

17. $\frac{1}{2}[\ln(\sin x)]^2 + C$

19. $-\frac{2}{3}\cos^3 t + C$

21. $\frac{x^{n+1}}{n+1} + C$

23. 0.3209

Exercise Set 5.6.A, *page 280*

1. a. $y = -\dfrac{1}{\frac{1}{2}x^2 + C}$

    c. $y = ae^{-\cos x}$

    e. $y = e^{\pm\sqrt{x^2+C}}$

    g. $y = ae^{\frac{1}{2}x^2}$

3. a. $T = ae^{kt} + A$

    c. $T = 70e^{-1.6946t} + 70$

5. a. $B_0 = 0$ B_n = balance after n months

    $B_n = B_{n-1}(1.01) + 100$

   b. $\dfrac{dB}{dn} = 0.01B + 100.$

   c. $B(n) = 10{,}000e^{0.01n} - 10{,}000$

   d. Using recursion: $B_{240} = 98{,}925.54$. Using part c results, $B(240) = 100{,}231.76$. The recursive equation generates new values only at one month intervals. Since the balance only changes once per month, this is probably more realistic. The model $B(n) = 10{,}000e^{0.01n} - 10{,}000$ allows us to generate $B(n)$ without knowing $B(n-1)$. This may be more efficient than the recursive equation if technology is not available to evaluate the recursive equations.

Exercise Set 5.6.B, *page 287*

1. a. $2\ln|x-3| - \ln|x+2| + C$ c. $3\ln|x| + 7\ln|x+1| + C$

3. $\dfrac{ds}{dt} = ks\left(1 - \dfrac{s}{500}\right)$ $s(0) = 5$ $s = \dfrac{2525.\overline{25}e^{kt}}{500 + 5.\overline{05}e^{kt}}$

    To find k, we would need to know another point on the function.

6. S = proportion of computers in use at time t. $\dfrac{dS}{dt} = kS(1-S)$, so $S(t) = \dfrac{ae^{kt}}{1 + ae^{kt}}$.

    To find a, you would need to know the proportion of schools using computers at $t = 0$. The value of k depends on how fast computers are being incorporated into schools. If another data point was known (for instance, how many computers are in use at a school after several years from the starting time), we could solve for k.

Exercise Set 5.7, *page 297*

1. $\frac{1}{3}\tan x^3 + C$

3. 0.255

5. $\frac{1}{3}\sec^{-1}\left|\frac{2}{3}t\right| + C$

7. $\frac{1}{24}\ln\left|\frac{4x+3}{4x-3}\right| + C$

9. $\frac{1}{64}\left[\frac{x}{2}\left(32x^2 + 25\right)\sqrt{16x^2 + 25} - \frac{625}{8}\ln\left|4x + \sqrt{16x^2 + 25}\right|\right] + C$

11. $-\frac{1}{2}\sin x \cos x + \frac{1}{2}x + C$

13. $-\frac{1}{4}\sin^3 x \cos x - \frac{3}{8}\sin x \cos x + \frac{3}{8}x + C$

15. $\frac{\sin x}{2} - \frac{\sin(7x)}{14} + C$

17. $\frac{1}{6}\left(\frac{1}{16}\sec^{-1}\left|\frac{3x^2}{16}\right|\right) + C$

19. $\frac{1}{64}\ln|x| - \frac{1}{128}\ln|8 - x| - \frac{1}{128}\ln|8 + x| + C$

21. $\dfrac{-\cos(x^2)}{2} - \dfrac{\cos(7x^2)}{14} + C$

23. $\frac{1}{3}\ln|\sec(3x)| + C$

25. 14.825

Exercise Set 5.8, *page 304*

3. The analytic solution is 156. Using the left endpoints and 100 rectangles gives us a value of 153.53. Using the right endpoints, we need to change our summation to $\sum_i f(x_{i+1}) \cdot \Delta x$. The solution using 100 rectangles is 158.49. If we use midpoints, the solution using 100 rectangles is 155.995. As the number of rectangles increases, the estimates get closer to the analytic solution.

5. a. With $n = 50$, the value is 1.63.

   c. With $n = 200$, the value is -12.00.

   e. With $n = 5000$, the value is 0.33.

   g. With $n = 100$, the value is 0.34.

Exercise Set 5.9, *page 313*

1. $h(t) = \frac{-9.789}{2}t^2 + 12.008t + 93.8.$

3. $CO_2(t) = 51.04e^{0.02t} + 268.827.$ The ambient level of CO_2 is 268.827.

END OF CHAPTER EXERCISES
Review and Extensions, *page 315*

1. a. 42

2. a. Area $= 113\frac{1}{3}$ miles.
   Confidence in this result is moderate since the speeds are so variable and the Δt's are quite large.

3. $A = \int_0^{\frac{3\pi}{2}} (\sin x + 1 - \cos x)\,dx + \int_{\frac{3\pi}{2}}^{2\pi} [\cos x - (\sin x + 1)]\,dx$

4. a. $\frac{dP}{dt} = 0.1P - 20$
   b. $P = a \cdot e^{0.1t} + 200$ where $a = \pm e^C.$

5. a. $\frac{1}{7}\sin^7 x + C$ b. $\sec x + C$
   c. $\frac{x^4}{4} - \frac{3}{2}x^2 + 8\ln|x| + C$ d. $\frac{4}{3}\ln|x - 5| - \frac{1}{3}\ln|x - 2| + C$
   e. $\frac{1}{2}[\ln(2x)]^2 + C$
   f. $-\frac{1}{6}\sin^2(x^2 + 7)\cos(x^2 + 7) - \frac{1}{3}\cos(x^2 + 7) + C$
   g. $\frac{1}{8}\left[\frac{x}{4}(8x^2 - 9)\sqrt{9 - 4x^2} + \frac{81}{8}\sin^{-1}\left(\frac{2x}{3}\right)\right] + C$

6. $y = -a \cdot e^{-\frac{1}{10}(\ln t)^2} + 3$

7. displacement : 0   distance: 8

8. a. area ≈ 1.27 b. area ≈ 1.76

Exercise Set 6.1, *page 323*

1. 7.32 mm

3. $P(\text{sum} < 5) = 0.125,$ $P(\text{sum} < 10) = 0.5$

5. 0.804

6. a. 0.1916   c. 0.0065
   e. $\frac{1}{120}$ h. $0.\overline{148} = \frac{4}{27}$

7. 0.637

Exercise Set 6.2, *page 330*

1. a. 0.003   b. 0.152

Exercise Set 6.3, *page 335*

1. a. 0.68, 0.95, 0.997, 0.9999
   d. The probability of getting a value within one standard deviation of the mean is 0.68, within two standard deviations of the mean is 0.95, within three standard deviations of the mean is 0.997, and within four standard deviations of the mean is 0.9999.

3. a. between 85 and 115 (within one standard deviation of the mean)
   c. Both are very unlikely. 140 is 2.67 standard deviations above the mean, so only about 0.4% score above 140. 150 is 3.33 standard deviations above the mean, and only about 0.04% score above 150.

5. The probability of a box weighing less than 15 oz. is 0.048. The probability of a box weighing between 15.9 and 16.1 oz. is about 0.13. Companies set the target weight higher than the weight indicated since it is likely that a box will be lower than the target weight. To avoid complaints and to comply with regulations, the company will set its target higher than 16 oz.

7. 27.38 ounces

9. Answers will vary for different data sets, but the mean should be approximately 50 and the standard deviation should be approximately 5. Thus a density function similar to
$$f(x) = \frac{1}{5\sqrt{2\pi}}e^{-\frac{1}{2}\left(\frac{x-50}{5}\right)^2}$$
should fit the data fairly well.

Exercise Set 6.4, *page 345*

1. a. $S = \dfrac{70}{9}$

    c. $S = \dfrac{31}{99}$

    e. $S = \dfrac{4}{3}$

2. a. 1

    c. 10, 100, 1000

    e. $\dfrac{1}{p-1}$

    g. $\dfrac{1}{2}$

    i. 0

    k. diverges

    m. diverges

3. a. 6.93

    c. 0

5. a. $-\dfrac{1}{9}$

    c. 0.120

    d. 1.04

Exercise Set 6.5, *page 350*

1. diverges

3. converges

5. converges

Exercise Set 6.6, *page 356*

1. Most of the data lie to the left, that is, "low" values are more likely than "high". In the normal distribution, the values cluster around the mean and are symmetric about it. High and low values are equally likely.

5. a. 0.393

    c. 0.081

7. 0.16

9. 1.54

11. a. $1 - e^{-\lambda}$

     b. $1 - e^{-\lambda}$

Exercise Set 6.7, *page 363*

1. c. 150 hours

    e. The mean and the median are unequal because the PDF is not symmetric.

3. a. 2.1 minutes

    b. 3 minutes

Exercise Set 6.8, *page 367*

1. a. $\dfrac{x^2}{2}\ln x - \dfrac{x^2}{4} + C$

    c. $-x^2\cos x + 2x\sin x + 2\cos x + C$

    e. 1

    g. 0.439

    i. $\dfrac{-\sin x \cos x + x}{2} + C$

    k. $x\ln(x^2 - 1) - 2x - \ln|x - 1| + \ln|x + 1| + C$

    m. $-\dfrac{2}{9}$

3. b. 0.2354

    c. The 20% of the shortest times between calls should be within the first 0.95 minutes. The 40% of the shortest times should be within the first 2.17 minutes. For 60%, time is 3.89 minutes, for 80%, time is 6.84 minutes. There is no time k for which $P(0 \le t \le k) = 1$.

    e. Since the probabilities based on the exponential PDF are very close in value to the data provided, we can assume that the assumption is valid.

5. Using the PDF provided, we could calculate theoretical probabilities associated with maximum wind velocities and compare these probabilities with what has been observed historically.

Exercise Set 6.10, *page 377*

1. Consumption from 1980 to 1989 at pre-1974 rate: 406.477. Consumption from 1980 to 1989 at post-1974 rate: 293.338. Oil saved: 113.139 billion barrels.

3. 90

5. a. $F(0) = 0$ c. $F(4) = 0$

8. $\dfrac{dF}{dx} = e^{\cos(x^2)} \cdot 2x$

Exercise Set 6.11, *page 383*

2. a. $\dfrac{16}{3}$ c. $\dfrac{4\sqrt{3}}{3}$

4. $\dfrac{128\pi}{5}$

5. $\dfrac{3\pi}{10}$

Exercise Set 6.12, *page 387*

1. The average temperature for the first six hours after sunrise is about 39.82°. The average for the last six hours before sunset is about 55.95°.

3. The average distance for the first 2 seconds of flight is about 10.47 m. The average for the remainder of the flight is about 7.36 m. The average for the first part of the flight takes into account a longer time of flight so it would be "under-represented" if you just averaged the two averages. (That is, averaging treats each number equally. Averaging the averages only works if the averages cover equal intervals.)

5. a. i. 0.632   ii. 0.100   iii. $\dfrac{1}{k}\left(1 - \dfrac{1}{e^k}\right)$
   b. As $k \to \infty$, the average value approaches 0.

7. b. 1.266
   c. The areas under each are equal.

9. a. −48.8 m/sec
   b. −48.8 m/sec
   c. The value in part a is how fast the object would need to fall to reach the same point in 12 seconds if its velocity were constant. It is the average of the velocity values between $t = 0$ and $t = 12$. The value in part b is the slope of the secant line connecting the initial and final positions and represents the average rate-of-change of height over 12 seconds.

Exercises Set 6.13, *page 397*

1. a. 9.073   c. $\dfrac{4}{3}$

3. $\displaystyle\int_0^{\frac{2v_0 \sin\theta}{g}} \sqrt{(v_0\cos\theta)^2 + (-gt + v_0\sin\theta)^2}\, dt$

5. 1219.7 meters

7. 22.103. This is an ellipse centered at the origin.

END OF CHAPTER EXERCISES
Review and Extensions, *page 399*

1. a. $\dfrac{1}{2}e^{x^2} + c$
   b. e^2
   c. $\dfrac{1}{2}$
   d. $-\dfrac{1}{2}x\cos(2x) + \dfrac{1}{4}\sin(2x) + c$
   e. 0
   f. $x\sin^{-1}x + \sqrt{1 - x^2} + c$

3. $k = 6.93$

4. $\pi\displaystyle\int_0^{\pi/4}(\cos^2 x - \sin^2 x)\, dx$

5. a. Approximately 0.048   b. $m \approx 12.99$

7. a. 3   b. $c = \sqrt{3}$

Exercise Set 7.1, *page 412*

2. $x - \dfrac{x^2}{2} + \dfrac{x^3}{3} - \dfrac{x^4}{4} + \cdots + \dfrac{(-1)^{n+1}x^n}{n} + \cdots$; interval of convergence: $-1 < x \le 1$

3. a. $\displaystyle\sum_{n=0}^{\infty}(-1)^n\dfrac{x^{2n}}{n!}$; interval of convergence: all reals
   c. $\displaystyle\sum_{n=0}^{\infty}(-1)^n\dfrac{x^{2(2n+1)}}{(2n+1)!}$; interval of convergence: all reals
   e. $\displaystyle\sum_{n=0}^{\infty}(-1)^n t^{2n}$; with interval of convergence: $-1 < t < 1$
   g. Interval of convergence: $-\sqrt{\dfrac{1}{\pi}} + 1 < x < \sqrt{\dfrac{1}{\pi}} + 1$
   i. Interval of convergence: $-2 < x \le 0$

Exercise Set 7.2, *page 415*

2. a. $\sin(-2x) = \sum_{n=0}^{\infty}(-1)^{n+1}\frac{(2x)^{2n+1}}{(2n+1)!}$; interval of convergence: all reals

c. $\frac{1}{1+2x^2} = \sum_{n=0}^{\infty}(-1)^n(2x^2)^n$; interval of convergence: $-\sqrt{\frac{1}{2}} < x < \sqrt{\frac{1}{2}}$

4. a. $\int \sin x^2\, dx = C + \sum_{n=0}^{\infty}(-1)^n \frac{x^{4n+3}}{(4n+3)(2n+1)!}$

c. $\int \frac{dx}{1+x^4} = C + \sum_{n=0}^{\infty}(-1)^n \frac{x^{4n+1}}{4n+1}$

5. a. 0.3103
   c. 0.4940

6. a. $p(x) = \sum_{n=1}^{\infty}(-1)^{n-1}\frac{x^n}{n}$; interval of convergence: $-1 < x \le 1$.
   $p'(x) = \sum_{n=1}^{\infty}(-1)^{n-1}x^{n-1}$; interval of convergence: $-1 < x < 1$.

c. $p(x) = \sum_{n=0}^{\infty}x^{2n}$; interval of convergence: $-1 < x < 1$.
   $p'(x) = \sum_{n=0}^{\infty}2nx^{2n-1}$; interval of convergence: $-1 < x < 1$.

Differentiation may affect the endpoints of the interval of convergence, making an endpoint which was included in the interval no longer included for the derivative series.

Exercise Set 7.3, *page 425*

2. a. $y = 4e^x - x - 2$
   c. $y = \frac{1}{2}e^{2x} - \frac{1}{2} - x - \frac{x^2}{2}$

5. $A(s) = a_0 \cos s + b_0 \sin s$ and $B(s) = b_0 \cos s - a_0 \sin s$

Exercise Set 7.4, *page 434*

1. a. $f(x) = -0.57 + 1.27\cos(x) + 0.14\cos(3x)$
   c. $f(x) = 1.57 - \sin(2x)$

3. a.
$f_{10}(x) = 51.95 + 48\cos(x) - 3\cos(2x) + 0.59\cos(3x) - 0.19\cos(4x) + 0.08\cos(5x) - 0.04\cos(6x) + 0.02\cos(7x) - 0.01\cos(8x) + 0.01\cos(9x) - 0.00\cos(10x)$

Exercise Set 7.5, *page 437*

1. a. $F_3(x) = 0.50 - 0.32\sin(2\pi x) - 0.16\sin(4\pi x) - 0.11\sin(6\pi x)$
   c. $F_3(x) = 0.25 + 0.32\cos(2\pi x) - 0.32\sin(2\pi x) - 0.32\sin(4\pi x) - 0.11\cos(6\pi x) - 0.11\sin(6\pi x)$

3.
$F_{13}(x) = 0.50 + 0.64\cos(x) - 0.21\cos(3x) + 0.13\cos(5x) - 0.09\cos(7x) + 0.07\cos(9x) - 0.06\cos(11x) + 0.05\cos(13x)$

Exercise Set 7.6, *page 441*

2.
$F_4(x) = 60.38 + 5.15\cos\left(\frac{\pi}{12}x\right) + 2.33\sin\left(\frac{\pi}{12}x\right) + 0.35\cos\left(\frac{2\pi}{12}x\right) - 0.18\sin\left(\frac{2\pi}{12}x\right) - 0.08\cos\left(\frac{3\pi}{12}x\right) - 0.40\sin\left(\frac{3\pi}{12}x\right) - 0.04\cos\left(\frac{4\pi}{12}x\right) - 0.22\sin\left(\frac{4\pi}{12}x\right)$

Index